AF578220

211 Current Topics in Microbiology and Immunology

Springer
Berlin
Heidelberg
New York
Barcelona
Budapest
Hong Kong
London
Milan
Paris
Santa Clara
Singapore
Tokyo

Molecular Aspects of Myeloid Stem Cell Development

Edited by L. Wolff and A. S. Perkins

With 98 Figures and 16 Tables

Linda Wolff, Ph.D.

Senior Investigator
Laboratory of Genetics
Bldg. 37, Rm. 1B16
National Cancer Institute
National Institutes of Health
37 Convent Drive MSC 4255
Bethesda, MD 20892-4255
USA

Archibald S. Perkins, M.D., Ph.D.

Assistant Professor
Department of Pathology
Yale University School of Medicine
P.O. Box 208023
New Haven, CT 06520-8023
USA

Cover illustration: The cover depicts the morphological changes that accompany differentiation of the myeloblast cell line, 32Dcl3, after treatment with granulocyte colony stimulating factor (G-CSF). Cells were stained using Diff Quick (Scientific Products, Inc.). Top panel, no treatment; bottom panel treatment with G-CSF for 14 days. Photographs were kindly provided by Juraj Bies.

Cover design: Künkel+Lopka, Ilvesheim

ISSN 0070-217X
ISBN 3-540-60414-6 Springer-Verlag Berlin Heidelberg New York

Library of Congress Catalog Card Number 15-12910
Printed in Germany

Typesetting: Camera-ready by authors
SPIN: 10498140 27/3020 – 5 4 3 2 1 0 – Printed on acid-free paper

Preface

A workshop on "MOLECULAR ASPECTS OF MYELOID STEM CELL DEVELOPMENT" was held at the Historic Inns in Annapolis, MD during April 30–May 3, with approximately 70 persons attending. The enormous success of this meeting was attributable to the recent outburst of information generated from two areas of study: transcriptional regulation of developmentally important myeloid genes and molecular dissection of chromosomal abnormalities in human myeloid leukemias. It was reassuring that studies of normal myelopoiesis and abnormal myelopoiesis, as observed in dysplasias and neoplasia, are revealing several interrelated mechanisms of gene regulation. However, great challenges await us as we try to determine how these and other mechanisms in this intricate system orchestrate the ultimate fate of cells e.g. proliferation, differentiation or apoptosis. This volume of Current Topics in Microbiology encapsulates many of the key aspects of the workshop summarizing research over the past few years on the ontogeny and pathology of myeloid cells.

Investigations into the complex process of normal myeloid cell maturation were first made possible through development of in vitro clonigenic assays. These established the relationships of colony stimulating factors, interleukins and their combinations to the specific fates of myeloid progenitors derived from the bone marrow and spleen. Molecular studies became feasible, however, through the later establishment of several cell culture systems involving immortalized cells which, under the proper induction, could fairly accurately recapitulate the myelopoietic process. These have without a doubt opened up the field to investigations of receptors, signal transducers and nuclear regulators and their wide use was evident at the workshop.

Recent studies of gene expression in vitro highlight the fact that programatic gene transcription is the result of combinatorial effects of various transcription factors. Many factors that are important in the process are not necessarily myeloid specific, and may be utilized in other organ systems or for other processes. It is believed that specific decisions at important junctures of the cell's life are carried out by a precise blend of such factors in context of the existing cellular environment. Commitment at each level of the differentiation hierarchy presumably limits choices available at the next level.

Some fairly precise ideas concerning the role of specific transcription factors can be gained through the trendy approach of targeted gene disruption and has already led to some notable obversations. The transcription factors, c-Myb and PU.1(Spi1), have been shown to be absolutely required for development of specific subtypes of myeloid cells, thereby demonstrating that they cannot be functionally replaced by other genes through redundancy. It appears that this represents just the beginning of the use of "knockouts" for determining key elements in the process of ontogeny.

Rapid advances in the molecular basis for myeloid proliferative disorders such as AML and CML are due to the availability of enhanced techniques for localizing and cloning genetic regions encompassing translocation breakpoints. The unraveling of the genetic alterations in these diseases, as well as in animal models, has led to the determination of numerous oncogenes and suppressor genes that can contribute to the neoplastic processes. It is clear that leukemogenesis represents deregulation of the mechanisms of cellular developmental control, and involves many of the same key regulatory molecules as the normal pathway. Thus, it has already been shown that c-Myc, c-Myb, AML1(CBFα), CBFβ, EVI1, and RARα play roles in both normal and abnormal processes.

Grand themes highlighted here are rapidly blurred as one starts to inspect the details. Complexities abound, as well as unanswered questions. Many of the prime movers of myelopoiesis are still at large; a clearer role of stromal cells is needed; dissection of signaling pathways is in its infancy; the composition and dynamics of the key regulatory transcription factor complexes are unresolved. As these and other details emerge, the hopes for specific and curative pharmacological intervention in myeloid cell disease brighten.

Acknowledgements

This workshop was made possible through the sponsorship of the Division of Cell Biology, Diagnostics and Centers of the National Cancer Institute and we are extremely grateful to its director Dr. Alan Rabson for his support and enthusiasm. Additional support for this meeting was provided by AMGEN, Inc., Bristol-Myers Squibb Pharmaceutical Institute, Merck, Sandoz Research Institute, Bayer

Corporation, and The Upjohn Company. Special thanks are due to Pairin Lancaster of the Cygnus Corporation whose careful and considerate planning helped to make the meeting a success. We are very grateful to Mary Millison for her assistance in correspondence and in preparing this book.

ARCHIBALD S. PERKINS
LINDA WOLFF

Table of Contents

Regulation of Myeloid Differentiation, Growth Arrest and Apoptosis

Receptors and Signal Transduction

Myb: Proliferation and Maintenance of Stage Specific Genes

Myeloid-Specific Transcription

Molecular Aspects of Myeloid Leukemia. I. Murine Studies

Molecular Aspects of Myeloid Leukemia. II. Human Studies

List of Contributors

(Their addresses can be found at the beginning of their respective chapters.)

Regulation of Myeloid Differentiation, Growth Arrest and Apoptosis

Molecular Control of Development in Normal and Leukemic Myeloid Cells by Cytokines, Tumor Suppressor and Oncogenes

L. Sachs

Department of Molecular Genetics and Virology, Weizmann Institute of Science, Rehovot 76100, Israel

The establishment of a cell culture system for the clonal development of hematopoietic cells made it possible to discover the proteins that regulate cell viability, multiplication and differentiation of different hematopoietic cell lineages and the molecular basis of normal and abnormal development in blood-forming tissues (reviewed in [1-5]). These regulators include cytokines now called colony stimulating factors (CSFs) and interleukins (ILs) and hematopoiesis is controlled by a network of cytokine interactions [1-7]. This multigene network includes positive regulators such as CSFs and ILs and negative regulators such as transforming growth factor β and tumor necrosis factor. The cytokine network which has arisen during evolution allows considerable flexibility depending on which part of the network is activated and the ready amplification of response to a particular stimulus. Hematopoietic cytokines can induce the expression of transcription factors [8]. Cytokine signalling through transcription factors can thus ensure the autoregulation and transregulation of cytokine genes that occur in the network.

Cytokines that regulate normal hematopoiesis can control the abnormal growth of certain types of leukemic cells and suppress malignancy by inducing differentiation [1-5]. Genetic abnormalities that give rise to malignancy in these leukemic cells can be bypassed and their effects nullified by inducing differentiation and programmed cell death (apoptosis).

The hematopoietic cytokines discovered in culture are active *in vivo* and are being used clinically to correct defects in hematopoiesis in patients with various abnormalities including cancer. The results provide new approaches to therapy (reviewed in [1-5]). The existence of a network and the ability of cytokines to suppress apoptosis in normal and leukemic cells, including apoptosis induced by irradiation and cancer chemotherapeutic compounds [9-11], have to be taken into account in the clinical use of cytokines for therapy.

Studies on the cytokines that control hematopoiesis including CSFs and ILs, have shown that development of hematopoietic cells requires cytokines to suppress apoptosis and to induce cell multiplication and differentiation [1-5]. Apoptosis, multiplication and differentiation are separately regulated [11-13]. The same cytokines suppress apoptosis in normal and leukemic myeloid cells [9-13].

The tumor suppressor gene wild-type p53 induces apoptosis [14]. The oncogene mutant p53, like *bcl*-2, suppresses the enhancing effect on apoptosis of deregulated c-*myc* [15], and this allows induction of cell multiplication and inhibition of differentiation which are other functions of deregulated c-*myc*. The use of p53 knock-out mice and mutant leukemic cells have shown different apoptotic pathways [16,17]. Cytokine downregulation of *bcl*-2 increased cell susceptibility to apoptosis. This was suppressed by dexamethasone [17] by upregulating another apoptosis suppressing gene *bcl*-X_L [18]. Interactions between cytokines, steroids and genes that control apoptosis are thus regulators of development in myeloid cells.

References

1. Sachs L (1987) The molecular control of blood cell development. Science 238:1374-1379
2. Sachs L (1990) The control of growth and differentiation in normal and leukemic blood cells. The 1989 Alfred P Sloan Prize of the General Motors Cancer Research Foundation. Cancer 65:2196-2206
3. Sachs L (1992) The molecular control of hematopoiesis: From clonal development in culture to therapy in the clinic. Int J Cell Cloning 10:196-204
4. Sachs L (1993) The molecular control of hemopoiesis and leukemia. CR Acad Sci Paris, Sciences de la vie 316:882-891
5. Sachs L (1995) The adventures of a biologist: Prenatal diagnosis, hematopoiesis, leukemia, carcinogenesis and tumor suppression. Foundations in Cancer Research. Advances Cancer Research 66:1-40
6. Lotem J, Shabo Y, Sachs L (1991) The network of hematopoietic regulatory proteins in myeloid cell differentiation. Cell Growth Differ 2:421-427
7. Blatt C, Lotem J, Sachs, L (1992) Inhibition of specific pathways of myeloid cell differentiation by an activated Hox-2.4 homeobox gene. Cell Growth Differ 3:671-676
8. Shabo Y, Lotem J, Sachs L (1990) Induction of genes for transcription factors by normal hematopoietic regulatory proteins in the differentiation of myeloid leukemic cells. Leukemia 4:797-801
9. Lotem J, Cragoe EJ, Sachs L (1991) Rescue from programmed cell death in leukemic and normal myeloid cells. Blood 78:953-960
10. Lotem J, Sachs L (1992) Hematopoietic cytokines inhibit apoptosis induced by transforming growth factor β1 and cancer chemotherapy compounds in myeloid leukemic cells. Blood 80:1750-1757
11. Sachs L, Lotem J (1993) Control of programmed cell death in normal and leukemic cells: New implications for therapy. Blood 82:15-21
12. Sachs L, Lotem, J (1995) Apoptosis in the hemopoietic system. In: Gregory CD (ed) Apoptosis and the Immune Response. Wiley, New York, pp 371-403
13. Lotem J, Sachs L (1995) Interferon-γ inhibits apoptosis induced by wild-type p53, cytotoxic anti-cancer agents and viability factor deprivation in myeloid cells. Leukemia 9:685-692

14. Yonish-Rouach E, Resnitzky D, Lotem J, Sachs L, Kimchi A, Oren M (1991) Wild type p53 induces apoptosis of myeloid leukaemic cells that is inhibited by interleukin 6. Nature 352:345-347
15. Lotem J, Sachs L (1993) Regulation by *bcl*-2, c-*myc* and p53 of susceptibility to induction of apoptosis by heat shock and cancer chemotherapy compounds in differentiation competent and defective myeloid leukemic cells. Cell Growth Differ 4:41-47
16. Lotem J, Sachs L (1993) Hematopoietic cells from mice deficient in wild-type p53 are more resistant to induction of apoptosis by some agents. Blood 82:1092-1096
17. Lotem J, Sachs L (1994) Control of sensitivity to induction of apoptosis in myeloid leukemic cells by differentiation and *bcl*-2 dependent and independent pathways. Cell Growth Differ 5:321-327
18. Lotem J, Sachs L (1995) Regulation of *bcl*-2, *bcl*-X_L and *bax* in the control of apoptosis by hematopoietic cytokines and dexamethasone. Cell Growth Differ 6:647-653

Retinoic Acid Receptors in Hematopoiesis

Steven J. Collins and Schickwann Tsai, Clinical Research Division, Fred Hutchinson Cancer Research Center: Irwin Bernstein, Pediatric Oncology Program, Fred Hutchinson Cancer Research Center, Seattle Washington

Utilizing dominant negative RA receptors to explore the role of RA in hematopoiesis

Retinoic acid receptors (RARs) are critical transcriptional regulators that are involved in the development and differentiation of a wide variety of different cells (Evans 1988). Several lines of evidence suggest that RARs may be involved in the regulation of hematopoiesis. RARs (particularly RARα) are expressed in virtually all hematopoietic cell types (de The et al., 1989). Moreover, in acute promyelocytic leukemia (APL) patients that harbor the 15/17 chromosome translocation generating the aberrant RARα fusion transcript (designated PML-RARα), RA induces terminal granulocytic differentiation of the leukemia cells (Huang et al. 1988; Castaigne et al., 1990; Warrell et al., 1991). Our laboratory has been studying the role of retinoic acid (RA) and the RA receptors (RARs) in regulating various aspects of hematopoiesis. Our approach has involved introducing a dominant negative RA receptor construct into certain cell lines representing different hematopoietic lineages as well as into normal mouse bone marrow and then assessing any phenotypic changes in these transduced cells. We have constructed a retroviral vector (designated LRARα403SN), which harbors RARα with a COOH terminal truncation (Figure 1). We have observed that this construct exhibits dominant negative activity against the normal RARα in both mouse NIH3T3 cells and the HL-60 myeloid leukemia cell line (Tsai et al., 1992).

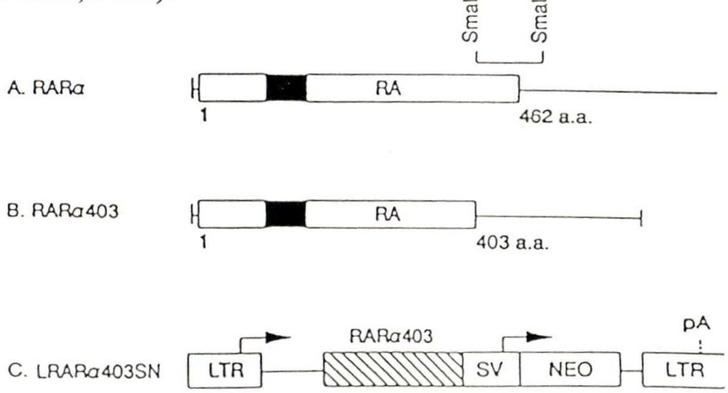

Figure 1. Structure of the retroviral vector LRARα403SN. (A) The open bar represents the reading frame encoding the normal human RARα. (B) The Sma I fragment truncated RAR (designated RARα 403) lacks the 59 COOH terminal amino acids of RARα. (C) Inserting this truncated RARα403 cDNA into the LXSN vector results in the dominant negative retroviral vector designated LRARα403SN.

Inhibition of neutrophil differentiation by a dominant negative RA receptor

We initially introduced the dominant negative LRARα403SN retroviral vector into the multipotent IL-3 dependent FDCP mix A4 murine hematopoietic cell line. This cell line ordinarily commits spontaneously to neutrophilic and monocytic differentiation at moderate frequencies, but upon introduction of the dominant negative RARα construct they exhibit a rapid switch to the development of terminally differentiated basophils/mast cells (Tsai et al., 1992). This observation suggested that in this particular cell line the normal RARα transcription factor appears to promote the differentiation of neutrophils and monocytes but suppress the development of basophils/mast cells. Suppression of the normal RA receptor by introducing the dominant negative construct resulted in a block to neutrophil/monocyte differentiation and the promotion of basophil/mast cell development.

We next used this retroviral vector to introduce the dominant negative RA receptor construct into normal mouse bone marrow and cultured these infected cells in GM-CSF. Uninfected and control (LXSN) infected bone marrow when cultured in GM-CSF proliferated and terminally differentiated into neutrophils, macrophages and eosinophils with all cells dying after 10-14 days. In contrast numerous cells in the LRARα403SN (the dominant negative) -infected marrow culture exhibited polyclonal expansion and developed into a continuously dividing GM-CSF dependent cell line (designated MPRO) with the morphological characteristics of promyelocytes (Tsai and Collins, 1993).

Interestingly these GM-CSF dependent, continuously proliferating MPRO (for Mouse Promyelocytes) cells can be induced to terminally differentiate to neutrophils with relatively high "pharmacological" concentrations of RA (10^{-6} to 10^{-5}M). These observations indicate that the introduction of the dominant negative RA receptor construct into normal GM-CSF dependent mouse progenitors blocks the GM-CSF induced neutrophilic differentiation of these cells. This suggests that the normal RA receptor in synergy with GM-CSF is essential for normal neutrophil differentiation. Blocking this RA receptor activity, which is likely triggered by the physiological levels of RA present in serum (10^{-9} - 10^{-8}M), appears to encourage self-renewal and inhibit differentiation of the GM-CSF dependent promyelocytes.

Consistent with these observations are further observations we have made in transactivation assays utilizing a CAT reporter construct driven by a promoter harboring a tandem array of four RA receptor response elements (RRE_4 - CAT). We have observed that the "physiological " levels of RA (10^{-9} - 10^{-8}M) will transactivate this CAT reporter construct in the murine FDCP mix A4 cell line (Figure 2A). In contrast these same concentrations of RA have little effect on the same reporter construct introduced into the mouse promyelocyte (MPRO) cell line that has been transduced with the dominant negative RA receptor construct (Figure 2B). However, higher "pharmacological" levels of RA (10^{-6} to 10^{-5}M) will transactivate this reporter construct in the MPRO cells, which is virtually the same concentration of RA that induces terminal neutrophilic differentiation of these cells (Figure 2B). These observations indicate that the dominant negative construct effectively inhibits the RA activity contributed by the endogenous "physiological" serum RA concentrations (10^{-9} - 10^{-8}M). Nevertheless the activity of the dominant negative construct is not absolute- that is at higher concentrations of RA (10^{-6} to 10^{-5}M) the effect of the dominant negative construct can be suppressed or neutralized.

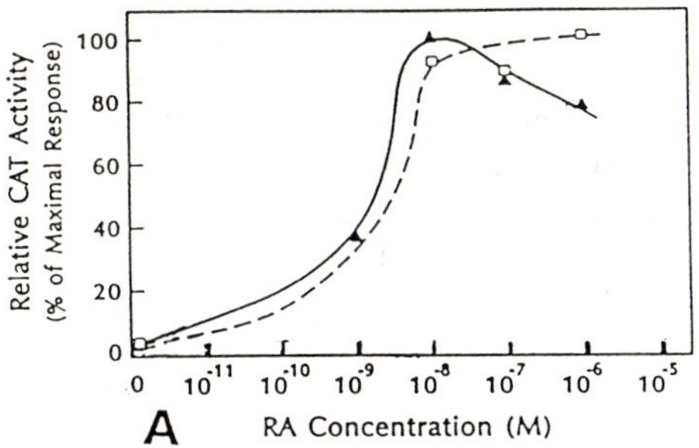

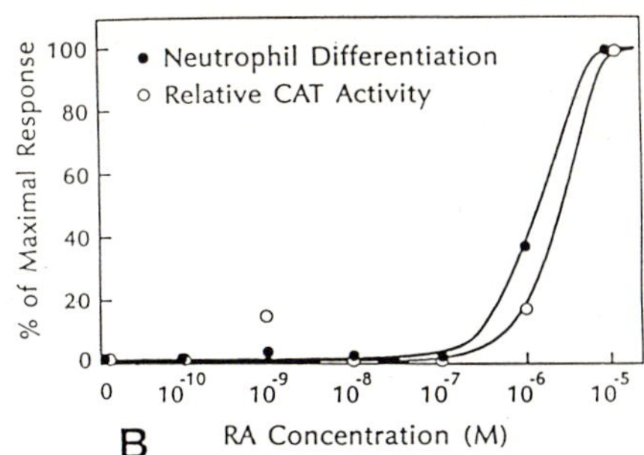

Figure 2 (A) Dose response curve of RA-induced transactivation of the pRRE4-CAT reporter construct electroporated into FDCP mix A4 cells, which harbor presumably functionally normal RA receptors. (B) Dose-response curve of RA-induced neutrophil differentiation and transactivation of pRRE4 - CAT in MPRO cells, which harbor the dominant negative RARα403 construct.

The dominant negative RA receptor construct promotes SCF-dependent lymphohematopoietic stem cell proliferation

We then infected normal mouse bone marrow with the retroviral vector harboring the dominant negative RA receptor and cultured the infected cells in stem cell factor (SCF). Utilizing this approach we have reproducibly obtained SCF-dependent cell lines (designated EML) that harbor the dominant negative proviral genome and exhibit the characteristics of lymphohematopoietic stem cells (Tsai et al., 1994). Flow cytometry indicates that approximately 50% of the EML cells express the B cell B220 antigen while approximately 50% express the "ter119" antigen which is associated with erythroid precursors, suggesting that cells of both B-lymphoid and erythroid lineages are spontaneously generated in the EML cell line.

The lymphoid potential of the EML cells is further demonstrated by culturing these cells with a bone marrow stromal cell line, W20, together with IL-7. Under these conditions the EML cells exhibit D-J rearrangement together with enhanced recombination activating gene-1 (RAG-1) mRNA expression (Tsai et al., 1994).

The erythroid potential of the ter119 ⊕ EML cells is further demonstrated by culturing the cells in liquid suspension in Epo (plus SCF). Under these conditions many erythroblasts at all stages of hemoglobinization appear as clusters of 4-32 cells constituting about 20% of all cells. Moreover when EML cells are cultured in methylcellulose medium containing SCF and Epo, typical burst forming unit-erythroid (BFU-E)-derived colonies are observed. These observations indicate that the SCF-

In contrast with its spontaneous generation of large numbers of B-lymphoid and erythroid progenitors, the SCF-dependent EML cells produce very few progenitors for the neutrophil and macrophage lineages (CFU-GMs) (Table 1). Interestingly, this deficiency of CFU-GMs in EML cells can be overcome with a combination of IL-3 and high "pharmacological" concentrations of RA (10^{-6} - 10^{-5}M) leading to a dramatic increase in CFU-GMs (Table 1). This latter observation together with our previous observations that the dominant negative effect of the RARα403 construct could be overcome or bypassed by pharmacological concentrations (10^{-6} - 10^{-5}M) of RA in both the transactivation and differentiation assays (Figure 2 above) indicates that the inhibition of CFU-GM formation in EML is attributable to the activity of the dominant negative RARα403, and this inhibition can be overcome or bypassed by pharmacological concentrations of RA.

SCF	+	+	+	+
IL-3	–	–	+	+
RA (10^{-5} M)	–	+	–	+
Total Cells on Day 3 ($\times 10^6$)	2.0	0.6	3.9	3.1
Total CFU-GM	2.5	0	1,115	32,040
Total SCF-supported clonogenic cells	5,200	1,000	7,400	2,000

Table 1. Effects of retinoic acid on the production of CFU-GM in EML cells. EML cells were cultured for three days in liquid media containing the indicated growth factors ± RA. Cells were then harvested and aliquots were cultured in methylcellulose supplemented with GM-CSF or SCF to assay for CFU-GM and SCF-supported clonogenic cells respectively.

Taken together our studies utilizing the dominant negative RA receptor construct indicate a dramatic and unexpected role for normal RA receptors in regulating hematopoiesis. When the dominant negative construct is introduced into normal GM-CSF dependent mouse hematopoietic progenitors, the terminal differentiation of these cells is blocked and their self renewal is enhanced (MPRO cells). When this same RA receptor dominant negative construct is introduced into more primitive SCF-dependent cells, their self-renewal is also enhanced and their differentiation, particularly toward the neutrophil/monocyte/macrophage lineage is blocked (EML cells). Thus the normal RA receptors in synergy with specific hematopoietic growth factors appear to promote the differentiation vs. self renewal of immature hematopoietic precursors. Moreover, our results summarized to date (Figure 3) indicate a particular role for RA receptors in regulating various stages of neutrophil differentiation.

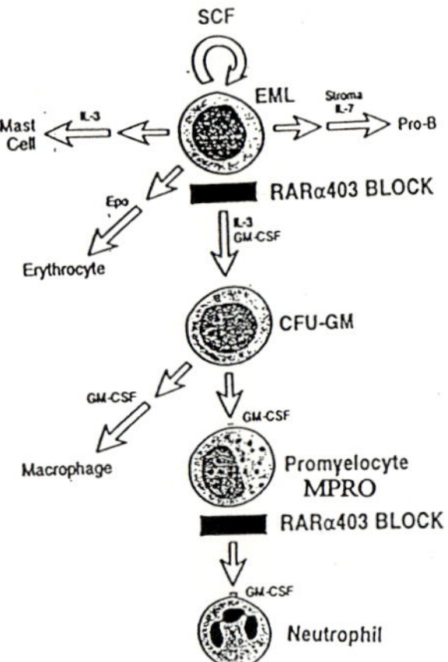

Figure 3. The dominant negative RARα403 construct imposes a developmental block to neutrophil differentiation at both the pre-CFU-GM and the neutrophilic promyelocyte stages. Both blocks can be overcome by "pharmacological" concentrations (10^{-6} - 10^{-5}M) of RA leading to the production of mature neutrophils (in MPRO cells) or the production of CFU-GM (in EML cells).

Synthetic retinoids exhibiting RA receptor antagonistic activity

The above observations indicate that inhibiting RA receptor activity in hematopoietic stem cells will inhibit their differentiation and enhance their self-renewal. Since enhancing the self renewal of hematopoietic stem cells could have considerable clinical utility, we have initiated studies to determine whether certain retinoids displaying specific RA receptor antagonistic activity might enhance the self renewal of hematopoietic stem cells in vitro.

The synthetic retinoid R0-41-5253 has been reported to act as a specific antagonist to the RA receptor α by binding to its ligand binding domain without inducing functional activation of the receptor (Apfel et al., 1992). To confirm these observations we obtained this synthetic compound from Hoffman-LaRoche and studied its ability to suppress RA receptor activity in transactivation assays. We transfected CV-1 cells with a CAT reporter construct harboring the TK promoter and a synthetic RA response element. Enhanced CAT activity was noted in the transfected CV-1 cells when stimulated with 10^{-8}M to 10^{-6}M RA (Figure 4). This RA-inducible CAT activity was most likely mediated through endogenous RA receptors in CV-1 cells. This RA-induced reporter gene activity was suppressed when the transfected target cells were cultured in the synthetic retinoid RO-41-5253 (5 X 10^{-6}M). Our experiments confirm the previously observed ability of this synthetic retinoid to suppress the transactivation function of normal RA receptors.

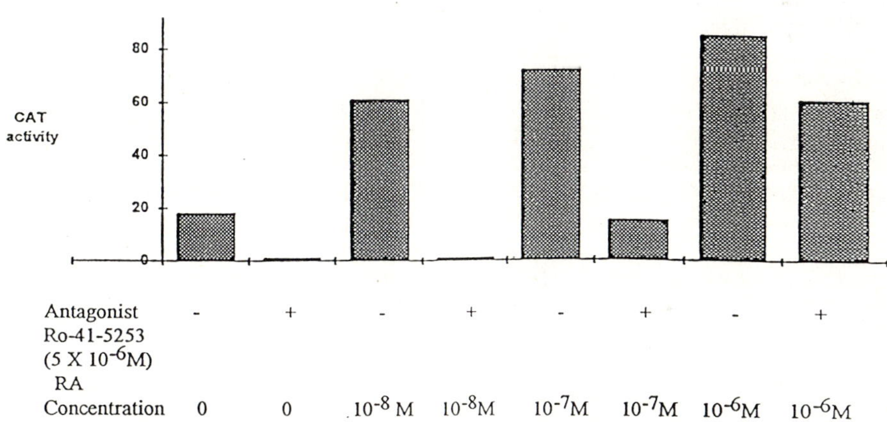

Figure 4. Inhibition of RA-mediated transactivation by the synthetic retinoid RO-41-5253. CV-1 cells were transfected with the RRE_4-CAT reporter construct and then treated with the indicated concentrations of RA with and without the addition of 5 X 10^{-6}M RO-41-5253. After two days of culture CAT activity was measured in cell lysates.

We also tested the ability of this synthetic retinoid to inhibit RA-induced differentiation of HL-60 cells (Collins et al., 1977; Breitman et al., 1980), an activity that we have previously demonstrated is directly mediated through the RA receptor α (Collins et al., 1990). We noted that RO-41-5253 inhibited the RA - induced granulocytic differentiation of HL-60 (Figure 5).

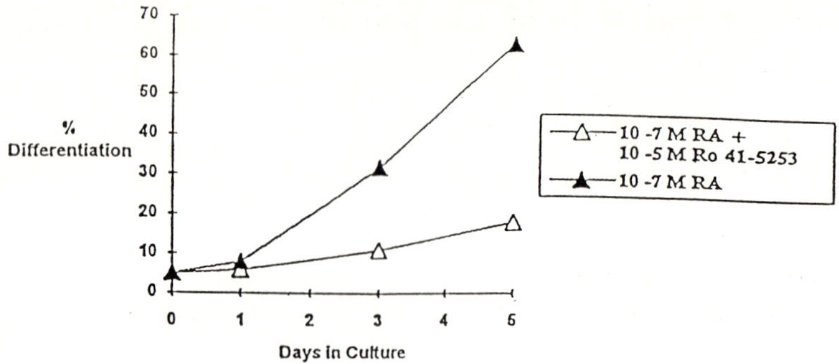

Figure 5. HL-60 cells were incubated in 10-7M RA with and without the addition of 10-5M RO-41-5253. At the indicated times the per cent of morphological granulocytic differentiation was assessed on Wright stained cytospin preparations of the cultured cells.

Thus our preliminary studies with this synthetic retinoid indicate that it indeed inhibits RA receptors both in transcriptional transactivation assays (Figure 4) and in functional activities related to RA receptor mediated cell differentiation (Figure 5).

Effect of RA antagonist on hematopoietic stem cell proliferation and differentiation.

The studies outlined above indicate that retroviral vector mediated transduction of a dominant negative retinoic acid receptor construct encourages the self renewal and inhibits the differentiation of certain hematopoietic precursor cells. We hypothesize that certain synthetic retinoids that antagonize retinoic acid activity might similarly encourage hematopoietic precursor cell self renewal in short term cultures of hematopoietic stem cells. Below we describe studies utilizing the synthetic retinoid antagonist RO-41-5253 that we have performed which indeed suggest that RA antagonists might exert significant biological effects on hematopoietic stem cell differentiation and self renewal.

The CD34 antigen is expressed on immature hematopoietic precursor cells including hematopoietic stem cells (Andrews et al 1990). In initial studies we isolated CD34 positive cells from normal human bone marrow and cultured these cells in liquid suspension (at an initial cell concentration of 50 cells/well) in the presence of multiple hematopoietic growth factors including kit ligand (KL), IL-3 and G-CSF. We wished to determine how the addition of the RA antagonist RO-41-5253 would alter the magnitude and kinetics of CFC production in these cultures. In our initial experiment we noted enhanced CFC production, including BFU-E, CFU-E and possibly CFU-GM in cultures treated with different concentrations of the RA antagonist with the most prominent effect observed at the higher concentration (10^{-6}M) of this antagonist. (Figure 6).

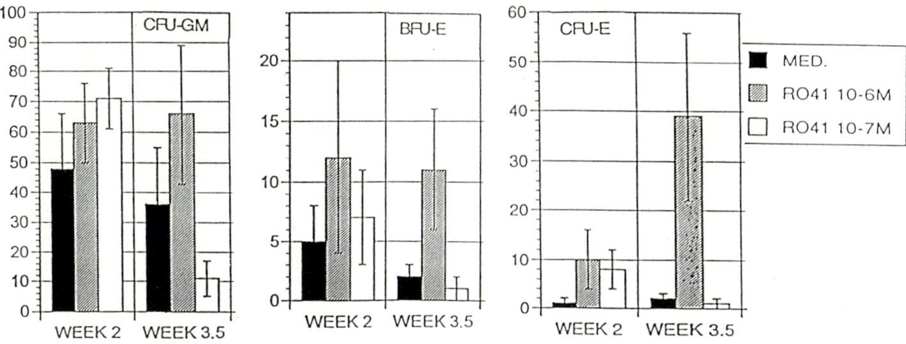

Figure 6. CD34 positive cells were isolated and cultured in microtiter wells (50 cells/well) in the presence of IL-3, G-CSF and kit ligand (KL). At the indicated time intervals aliquots of these cells were harvested and plated into a single dish of methylcellulose containing IL-3, IL-6, GM-CSF, SCF, and Epo. Shown are the number of CFC per 100 cells initially cultured. The RA antagonist was added twice weekly at the indicated concentration.

These initial observations indicated that the RA antagonist prolonged CFC production with significant increases in erythroid precursor production at the later time point and a trend towards increased CFU-GM production. Since the RA antagonist might have been inhibiting the differentiation of primitive hematopoietic precursors present in the cultured CD34+ population enabling the continued production of CFC, we next tested the effect of the RA antagonist on enriched populations of more primitive hematopoietic precursors. We isolated the known primitive subset of CD34+ cells based on light scatter properties of small-medium lymphocyte-sized cells, and lacking expression of maturation linked surface antigens (lin$^-$) as well as CD38 (Bernstein et al., 1991; Terstappen et al., 1991). We cultured these FACS sorted CD34+,lin-,CD38- , small-medium lymphocyte sized cells in liquid suspension in multiple HGF's including SCF, IL-1,IL-3,IL-6, G-and GM-CSF. As described in our original proposal, under these conditions these cells proliferate and differentiate over a 1-2 week period to produce colony forming progenitors including CFU-GM and BFU-E . We wished to determine how the addition of the RA antagonist RO-41-5253 would alter CFC production in these cultures (Figure 7, A and B).

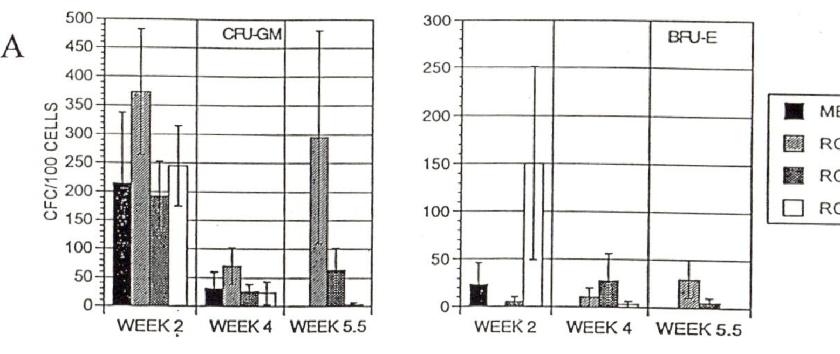

B

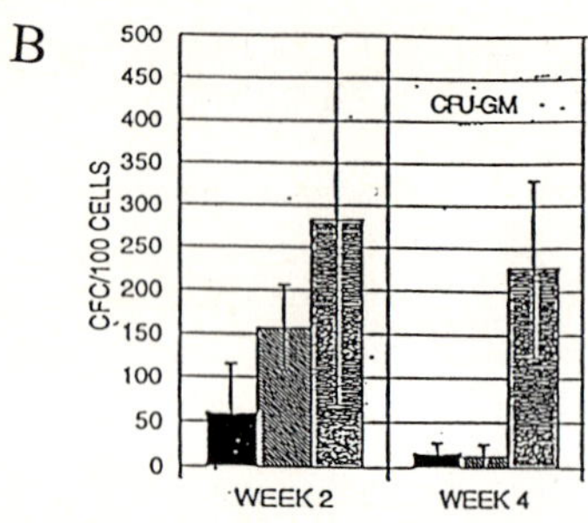

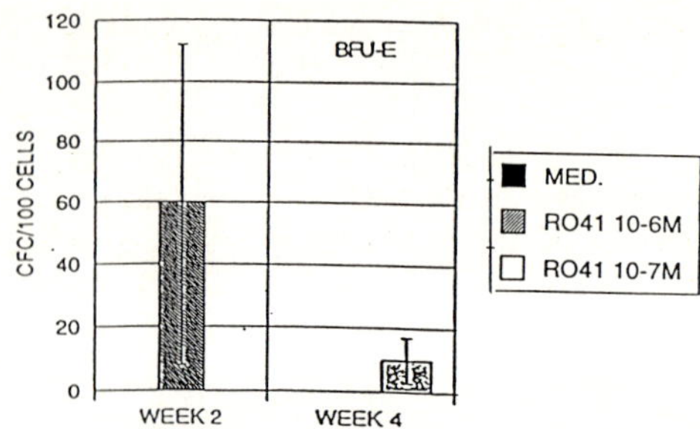

Figure 7. Human CD34+, Lin⁻,CD38⁻ small-medium lymphocyte sized cells were plated at 5 cells/well in IMDM-20% FBS with IL-1, IL-3,IL-6, G-CSF,GM-CSF and SCF all at 100 ng/ml with the indicated concentration of RA antagonist added twice weekly to the cultures. At the indicated weekly intervals, cells from each well were harvested and plated into a single dish of methylcellulose containing IL-3, IL-6, GM-CSF, SCF and Epo. Shown are the number of CFC per 100 cells initially cultured. Experiments A and B refer to separate experiments performed on two different donor marrow cells. .

The Figure 2 experiments indicate that highly enriched primitive hematopoietic precursor cells cultured in the indicated cytokine "cocktail" exhibit prolonged production of CFC with the addition of the RA antagonist RO-41-5253. (Note that in both experiments both CFU-GM and BFU-E production are present after 3-5.5 weeks of culture primarily in those cultures containing the RA antagonist). These observations indicate that the RA antagonist may prevent the terminal differentiation of CFU-GM or CFU-GM precursors, thereby enhancing the continued production of CFU-GM. Whether the associated enhanced generation of erythroid presursors is due to an increased probability of generating erythroid precursors from the same multipotent precursor that give rise to CFU-GM is being evaluated in single cells assays using methods we have previously described (Bernstein et al., 1991).

Although relatively straightforward in conception, these experiments are complex in their execution largely because of the relatively high number of potential experimental variables. These include: the type and concentration of cultured target cells, the type and concentration of hematopoietic growth factors, and the concentration and frequency of addition of the R0-41-5253. Nevertheless we are quite encouraged by these initial observations suggesting that the RA antagonists may exhibit a significant biological effect in encouraging the self renewal and proliferation and inhibiting the differentiation of primitive hematopoietic precursors in liquid suspension culture.

REFERENCES

Andrews R, Singer J, Bernstein I: Human hematopoietic precursors in long-term culture: single CD34+ cells that lack detectable T cell, B cell, and myeloid cell antigens produce multiple colony-forming cells when cultured with marrow stromal cells. J Exp Med 172: 355, 1990.

Apfel C., Bauer F., Crettaz M., et al. A retinoic acid receptor α antagonist selectively counteracts retinoic acid effects. Proc. Natl. Acad. Sci. USA 89: 7129-7133, 1992.

Bernstein I, Andrews R, Zsebo K: Recombinant human stem cell factor enhances the formation of colonies by CD34+lin⁻cells, and the generation of colony-forming cell progeny from CD34+lin⁻ cells cultured with IL-3, G-CSF or GM-CSF. Blood 77: 2316, 1991.

Breitman T, Selonick S, Collins S: Induction of differentiation of the human promyelocytic leukemia cell line (HL-60) by retinoic acid. Proc Natl Acad Sci USA 77: 2936, 1980.

Castaigne S, Chomienne C, Daniel M et al. All-trans retinoic acid as a differentiation therapy for acute promyelocytic leukemia. I. Clinical Results Blood 76: 1704, 1990.

Collins SJ, Robertson K, and Mueller L: Retinoic acid-induced granulocytic differentiation of HL-60 myeloid leukemia cells is mediated directly through the retinoic acid receptor (RARα). Mol Cell Biol 10: 2154, 1990.

Collins S, Gallo R, and Gallagher R: Continuous growth and differentiation of human myeloid leukemia cells in suspension culture. Nature 270: 347, 1977.

de The H, Marchio A, Tiollais P, and Dejean A. Differential expression and ligand regulation of the retinoic acid receptor alpha and beta genes. EMBO J. 8: 429., 1989.

Evans R: The steroid and thyroid hormone receptor superfamily. Science 240: 889, 1988.

Huang M, Ye YC, Chen SR et al. Use of all trans retinoic acid in the treatment of acute promyelocytic leukemia. Blood 72: 567, 1988.

Terstappen LWMM, Huang S, Safford M et al. Sequential generations of hematopoietic colonies derived from single nonlineage-committed CD34+CD38- progenitor cells. Blood 77: 1218, 1991.

Tsai S, Bartelmez S, Heyman R, Damm K, Evans R, Collins S: A mutated retinoic acid receptor alpha exhibiting dominant negative activity alters the lineage development of a multipotent hematopoietic cell line. Genes and Development 6: 2258, 1992.

Tsai S, and Collins S: A dominant negative retinoic acid receptor blocks neutrophil differentiation at the promyelocyte stage. Proc Natl Acad Sci USA 90: 7153, 1993.

Tsai S., Bartelmez, S, Sitnicka E., and Collins S. Lymphohematopoietic progenitors immortalized by a retroviral vector harboring a dominant negative retinoic acid receptor can recapitulate lymphoid, myeloid and erythroid development. Genes and Development 8: 2831, 1994.

Warrell R, Frankel S, Miller W et al. Differentiation therapy of acute promyelocytic leukemia with treretinoin (all-trans retinoic acid). New Engl. J. Med 324: 1385, 1991

Role of c-myc in Myeloid Differentiation, Growth Arrest and Apoptosis

B. Hoffman[1], D.A. Liebermann[1], M. Selvakumaran[1] and H.Q. Nguyen[1,2]

[1]Fels Institute for Cancer Research and Molecular Biology, Temple University School of Medicine, 3307 N. Broad St., Phil., PA 19140

[2]current address: Amgen, Inc., Thousand Oaks, CA 91320

Introduction

Hematopoiesis is a profound example of cell homeostasis which is regulated throughout life, whereby a hierarchy of hematopoietic progenitor cells in the bone marrow (BM) proliferate and terminally differentiate along multiple, distinct cell lineages. This includes the proliferation and differentiation of myeloid precursor cells into granulocytes and macrophages (33, 40, 42). Clearly, a variety of control mechanisms are needed to maintain steady state levels of mature blood cells, as well as to stimulate the rapid production of specific cell types as needed. To achieve this requires the participation of many factors, including positive and negative regulators of growth and differentiation, which determine survival, growth stimulation, differentiation, functional activation, and programmed cell death (apoptosis) (41, 44).

The proto-oncogene c-*myc* has been shown to play a pivotal role in growth control, differentiation and apoptosis, and its abnormal expression has been associated with many naturally occurring neoplasms (4, 14, 15, 45). C-*myc* is expressed in almost all proliferating normal cell types, and is down-regulated in many cell types when they are induced to terminally differentiate (14, 20, 25). The c-myc protein contains three structural domains that are homologous to domains found in characterized transcription factors, including a leucine zipper, a helix-loop-helix motif, and an adjacent domain rich in basic amino acids (24, 36). It is localized to the nucleus, and its binding to DNA is sequence-specific (9). Recent experimental evidence suggests that c-myc functions as a transcriptional regulator as part of a network of interacting factors (7, 14).

Given the central role c-myc plays in growth control, differentiation and apoptosis, understanding how c-myc functions will increase our understanding about normal cell development, and how alterations in these processes can lead to malignancy. In this paper, we describe recent work carried out in our laboratory on the role of c-myc in myeloid cell development, including its role as a regulator of differentiation and apoptosis.

Terminal Differentiation of Myeloid Leukemia Cells is Blocked at an Intermediate Stage by Constitutive c-myc

C-*myc* is expressed in proliferating M1 myeloid leukemic cells, and is down-regulated following induction of terminal differentiation by Interleukin-6 (IL-6) or Leukemia Inhibitory Factor (LIF), when there is no detectable c-*myc* mRNA or protein by 18 hours following stimulation for differentiation (20). Unlike the parental M1 cells, M1myc cell lines, established following transfection with an expression vector containing the c-*myc* gene under control of the B-*actin* promoter, constitutively synthesized c-*myc* . Deregulated and continued expression of c-*myc* blocked terminal differentiation induced by either IL-6 or LIF at an intermediate stage in the progression from immature blasts to mature macrophages (20). The proto-oncogene c-*myb*, whose transcripts are found primarily in tissues of hematopoietic origin (17), is down-regulated more rapidly than c-*myc* following induction of differentiation of M1 cells. Consistent with this observation, IL-6 or LIF treatment of M1myc cells resulted in down-regulation of c-*myb*. Differentiation primary response genes (MyD), which are induced within 30 min and in the absence of de novo protein synthesis following stimulation of M1 cells with IL-6 or LIF, have been isolated in this laboratory (1, 2, 26, 27, 29-32). Many of these genes continue to be expressed for at least 3 days. All MyD genes tested, including *junB*, c-*jun*, *junD*, *MyD88*, *IRF*-1, *Egr*-1 and *Waf1* (20, 43, 46 and unpublished data) were induced in M1myc cells following stimulation with IL-6, and those which continued to be expressed in M1 cells also continued to be expressed in M1myc cells. Fc receptors were induced to normal levels; however, C3 receptor induction was markedly inhibited in M1myc cells. Induction of *lysozyme* and increased levels of *ferritin* light-chain transcripts start to be detected by 2 days following stimulation for terminal differentiation of M1 cells and continue to increase. No such induction was detected for M1myc cells following treatment with differentiation inducers. Taken together, these data demonstrate that the block in myeloid differentiation caused by deregulated expression of c-*myc* occurs at the time that c-*myc* is normally suppressed in IL-6- or LIF-treated M1 cells (Fig. 1).

Withdrawal from the cell cycle is believed to be a prerequisite for terminal differentiation of most cell types (16). M1myc cells treated with IL-6 or LIF failed to exit the cell cycle, consistent with the notion that enforced expression of c-*myc* precludes cells from arresting in G0/G1 (20).

M1myc cells continued to proliferate following stimulation with either IL-6 or LIF, albeit with an increased doubling time relative to untreated M1 and M1myc cells, while maintaining the intermediate stage morphology. Usually c-*myb* expression is associated with proliferating cells and inhibition of its expression is

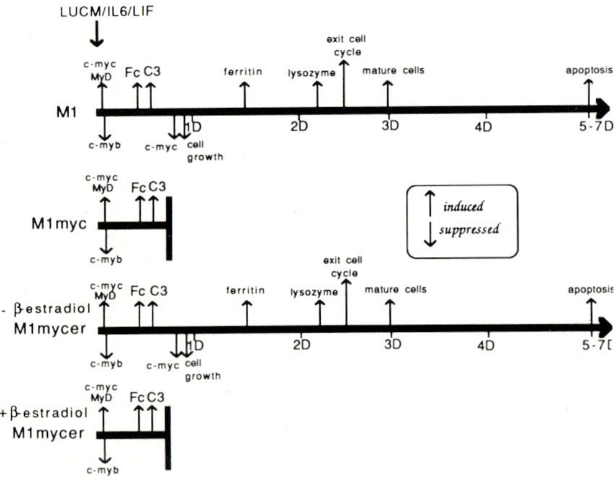

Fig. 1. Schematic representation of the myeloid developmental program induced by IL-6/LIF/LUCM, comparing M1 and M1myc cells with M1mycer cells without or with β-estradiol. (LUCM, conditioned medium of mouse lungs, is a potent physiological source of hematopoietic growth and differentiation inducers that has been shown to contain IL-6 and LIF)

associated with the terminally differentiated and growth arrested state (23, 25). Therefore, the finding that IL-6 or LIF treated M1myc cells continued to proliferate in the absence of c-*myb* expression was quite surprising. However, lack of c-*myb* expression may account for the observed reduced growth rate of the cells, consistent with the observation that anti-sense c-*myb* oligomers which diminish the level of c-myb protein, also inhibit proliferation, to varying degrees, of human myeloid leukemic cell lines (3). Alternatively, the reduced growth rate may be due to the synthesis of IFNβ in M1myc cells, partially accounting for the growth inhibition associated with terminal myeloid differentiation (2). Each mechanism is not necessarily mutually exclusive, and other mechanisms may also be involved.

Dissection of the Pleiotropic Effects of Deregulated c-myc During Myeloid Leukemia Cell Differentiation

The results of the experiments discussed thus far have shown that the c-myc transcription factor, which is involved in the control of hematopoietic cell growth (14), also serves as a negative regulator of terminal differentiation and its associated growth arrest, in which failure to suppress c-*myc* expression blocks IL-6/LIF -induced terminal differentiation and growth arrest (20). By the time c-*myc* is normally suppressed, both M1 and M1myc cells have undergone a multitude of changes, including induction of many transcription factors (2, 20, 25-7, 29, 31). In spite of this, failure to suppress c-*myc* blocks differentiation at an intermediate stage.

Analysis of the Effects of Deregulated c-myc Using High-Resolution 2D Gel Electrophoresis

To extend our understanding of the pleiotropic effects exerted by deregulated c-*myc* on differentiation, we employed high resolution 2D protein gel electrophoresis, which provided many more differentiation associated markers for analysis. Using this approach, 124 differentiation associated protein changes were analyzed (21), in which many of these changes are a manifestation, either direct or indirect, of developmentally regulated changes in gene expression.

The use of high resolution 2D gel analysis has revealed that failure to suppress c-*myc* in M1myc cells had massive pleiotropic consequences, affecting in various ways (delaying, partially or completely blocking) most of the protein changes that take place during IL-6 induced differentiation of M1 myeloid leukemia cells. Many of the early protein changes, detected 1 day following induction for M1 differentiation, also were observed in M1myc cells, yet with altered (delayed, partially blocked) expression kinetics. Only a few of the later protein changes, detected in M1 3 or 5 days following induction for differentiation, were observed in M1myc cells. The extent of the pleiotropic effects on differentiation associated protein changes resulting from the failure to suppress c-*myc* was unexpectedly high, given that c-*myc* is normally suppressed at a point in time when many changes in gene expression have already taken place, including the induction of a multitude of transcription factors. High-resolution 2D-protein gel analysis also has revealed that although most of the developmental-associated protein changes which take place during myeloid differentiation following c-*myc* suppression are c-*myc* suppression dependent, some are c-*myc* suppression independent. It is interesting to point out that previous studies, using 2D-gel electrophoresis to analyze protein changes during the developmental program of differentiation inducible (M1D+) and naturally occurring differentiation defective M1 clones, had led to the conclusion

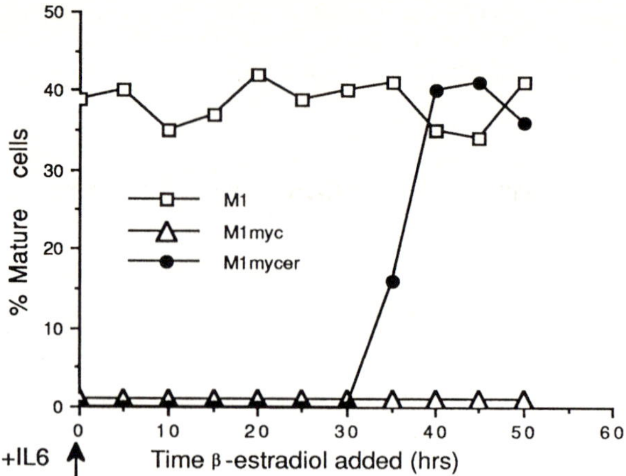

Fig. 2. The effect on differentiation of M1mycer cells by adding β-estradiol to the culture medium at various times after the addition of IL-6. Four days after the addition of IL-6, the percentage of mature cells was determined.

that differentiation is composed of separately programmed pathways of gene expression (28). The existence of c-*myc* suppression dependent and independent protein changes provides a definitive example supporting this notion.

Conditional c-myc Expression as a Tool to Dissect the Role of c-myc in Differentiation

Obviously, identifying c-myc target genes is an essential step towards better understanding the molecular mechanisms by which c-myc regulates normal hematopoiesis, as well as how deregulation of c-*myc* contributes to the development of leukemias. The strategy pursued by us to clone and identify c-myc target genes involves the use of conditional mutants of the myeloblastic leukemia M1 cells, expressing a chimeric *mycer* transgene, established in our laboratory. The chimeric *mycer* transgene is composed of the hormone-binding domain of the β-estradiol receptor fused to the carboxy terminus of the c-*myc* coding region; its function is dependent on β-estradiol in the culture medium and not de novo protein synthesis, thereby allowing for precise regulation of c-myc activity (12, 13).

M1mycer cells behave as conditional c-*myc* mutants; they undergo terminal differentiation after treatment with IL-6, like parental M1, but proceed only to the intermediate stage after treatment with IL-6 when β-estradiol is included in the culture medium, like M1myc cells (43)(Fig. 1). The differentiation program can be blocked by activation of c-myc function as late as 30 hours after treatment with differentiation inducers, in spite of the fact that endogenous c-*myc* transcripts and protein are no longer detected by 18 hours. Thus, not only is the pleiotropic effect exerted by deregulated c-*myc* at the level of protein changes, assessed by 2D gel analysis, more extensive than anticipated, but this effect can be mediated as late as 30 hours after induction of differentiation (Fig. 2). Moreover, by manipulating the function of the *mycer* transgene product, we have shown that there is a 10-hour window during myeloid differentiation, from 30-40 hours after the addition of the differentiation inducer, when the terminal differentiation program switches from being dependent on c-*myc* suppression to becoming c-*myc* suppression independent. Therefore, activation of c-myc after 40 hours no longer has any apparent effect on mature macrophages (Fig. 2).

Thus, it can be seen that M1mycer cell lines provide a powerful tool to increase our understanding of the role of c-myc in normal myelopoiesis and leukemogenesis, while also providing a strategy to clone c-myc target genes (43).

Functional Analysis of Ornithine Decarboxylase (ODC), a c-myc Target Gene, in Myeloid Differentiation

Recently, evidence has accumulated via several strategies, including the use of the *mycer* transgene, to indicate that the ornithine decarboxylase (ODC) gene is a transcriptional target gene of c-myc (8, 38, 48). ODC catalyzes the conversion of ornithine to putrescine, the first and rate-limiting step in polyamine biosynthesis, and has been shown to be crucial to cell growth (47). Several lines of investigation have shown that ODC activity is critical for cell transformation (5, 19, 35). It has also been shown that ODC expression is sufficient to induce accelerated apoptosis following IL-3 withdrawal in IL-3-dependent 32Dcl3 myeloid cells, and is a mediator of c-myc -induced accelerated apoptosis of these cells (37). Several lines of experimentation were pursued to ascertain if deregulated ODC expression can negatively regulate myeloid differentiation, and if ODC is a mediator of the differentiation block due to deregulated c-*myc* expression.

We have shown that ODC activity is down-regulated following induction of M1 differentiation, and fails to be suppressed in M1myc cells following treatment with the differentiation inducer IL-6 (Table 1). We have also found that in M1mycer cells, when endogenous c-*myc* and ODC expression have been suppressed following IL-6 treatment for over 30 hours, treatment with β-estradiol and cyclohexamide resulted in induction of ODC transcripts, thereby demonstrating that ODC is a c-myc target gene in M1 cells (data not shown). M1ODC cell lines, which constitutively express ODC, have been established These cells can undergo terminal differentiation and growth arrest following IL-6 stimulation, exactly like parental M1 cells (Table 1). Therefore, ODC is not sufficent to block myeloid differentiation. Another question asked is if ODC expression is necessary for the c-*myc*-induced block in differentiation. The use of the ODC inhibitor α-difluoromethylornithine (DFMO), an irreversible inhibitor of ODC enzyme activity, strongly suggests that ODC is not necessary for the c-myc-induced differentiation block (data not shown). Our results demonstrate that different or overlapping, but

Table 1. ODC expression following treatment of M1, M1myc and M1ODC cell lines with IL-6[a]

Cell Line	IL-6[b]	ODC activity[c]	Cell type (%)[d]		
			Blast	Inter.	Mature
M1	--	190	>99	<1	0
M1	+	25	3	37	60
M1MYC	--	210	>99	<1	0
M1MYC	+	180	25	75	0
M1ODC	--	450	>99	<1	0
M1ODC	+	220	4	38	58

[a]All values represent the mean of at least three independent experiments.

[b]Cells were treated with IL-6 at 100 ng/ml for 3 days.

[c]ODC activity as pmole/30 min/mg protein.

[d]Morphology was determined by counting at least 300 cells on May-Grunwald-Giemsa-stained cytospin smears.

not identical, sets of c-myc target genes are responsible for the effect of c-myc on transformation, apoptosis and differentiation.

Blocking c-myc Expression Potentiates Differentiation of Normal and Leukemic Myeloid Cells

By deregulating expression of c-*myc*, it has been demonstrated that c-myc is a negative regulator of differentiation (20, 43). It can be asked what effect blocking c-*myc* expression will have on proliferation and differentiation. Previously, it had been reported that blocking c-*myc* expression in HL60 cells, in which the c-*myc* gene is amplified 8- to 30- fold and is highly expressed (10, 11), results in growth arrest, and, in some instances, induction of monocytic differentiation (22, 50). However, dysregulated expression of c-*myc* in HL60 cells may account for the transformed phenotype, and, therefore, the manifestations due to blocking its expression may be related to these potentially transforming lesions, rather than to the role of c-*myc* as a negative regulator of cell differentiation. Looking at the effect of blocking c-*myc* expression on both normal myeloid cells and the M1 cell line, in which the level of c-*myc* expression is comparable to proliferating normal myeloblasts (20, 25), will allow one to ascertain the effects of blocking c-*myc* expression on growth and differentiation, independent of its role in transformation.

C-*myc* transcripts are expressed in proliferating M1 cells and are no longer detectable by 18 hours after induction for terminal differentiation by IL-6 (Fig. 1) (20, 43). Down-regulation of transcripts of *max*, the molecular partner of c-myc, was observed; however, in contrast to c-*myc*, significant levels of *max* transcripts continue to be expressed several days following induction of terminal differentiation by IL-6 (43). Myeloblast enriched bone-marrow cells express c-*myc* and *max* transcripts, and both are down-regulated following treatment with GM-CSF. However, for both c-myc and its molecular partner max, transcripts are still detectable even after six days in the presence of GM-CSF, when there are still precursor cells in the population with proliferative capabilities (25, and data not shown).

Antisense methodologies were used to block expression of c-*myc* and *max*. Specifically, we added oligodeoxynucleotides complementary to the first AUG codon and 6 subsequent downstream codons of either c-*myc* or *max* to the culture medium. Indirect immunofluorescence demonstrated that antisense oligomers to both c-*myc* and *max* effectively blocked expression of their cognate proteins.

Blocking expression of either c-*myc* or *max* in M1 cells activated the differentiation program; the cells assumed an intermediate stage myeloid morphology, and there was induction of Fc and C3 receptors, markers of differentiation in M1 cells (Table 2). Simultaneously blocking protein expression of both c-*myc* and *max* in M1 cells resulted in induction of mature macrophages (Table 2). These data suggest, among other things, that c-myc and max may have other molecular partners besides each other. Blocking expression of either c-*myc* or *max* in myeloblast enriched bone-marrow concommitantly treated with GM-CSF, accelerated the differentiation program, as determined by morphological analysis. Thus, blocking c-*myc* or *max* expression in myeloid cells at specific stages of development activates or accelerates the terminal differentiation program. What effect blocking expression of c-*myc* and/or *max* has on the different c-myc target genes will enhance our understanding of the relationship between c-*myc* and control of differentiation. Understanding how blocking either c-*myc* or *max* expression activates the M1 differentiation program, circumventing events at the

Table 2. Effect of c-myc and max antisense oligonucleotides on M1 cells

Inducer	Oligomer	Cell no.[a] 10^6/ml	Fc[b] (%)	C3[b] (%)	Cell type (%)[a] Blast	Inter.	Mature
--	--	1.40	0.5	1.2	>99	<1	0
--	scrambled	1.30	0.7	1.1	>99	<1	0
--	AS-myc	0.87	9.0	14.0	75	25	0
--	AS-max	0.89	8.0	22.0	63	35	2
--	AS-myc+AS-max	0.53	32.0	51.0	45	10	45
IL-6	--	0.40	36.0	56.0	4	41	55

[a]Cell number and morphology were determined after three days.

[b]Fc and C3 receptors were determined after one day.

membrane, and how normal myeloid-enriched bone-marrow differentiation is accelerated, will increase our understanding of the regulation of differentiation, coupling of growth to hematopoietic differentiation, and how alterations in the controls can lead to leukemias.

Deregulated Expression of c-myc Blocks TGFβ-Induced Growth Arrest and Accelerates TGFβ-Induced Apoptosis

A role for c-myc in the regulation of apoptosis was found in growth arrested fibroblasts (15), factor deprived IL-3 dependent 32D myeloid cells (4), and in anti-TCR antibody stimulated T-cell hybridomas (45), where its deregulated expression has been shown to either accelerate or induce apoptosis. In addition, the proto-oncogene c-*myc* was suggested to play a pivotal role in TGFβ1-induced growth inhibition of various cell types (34, 39). That TGFβ1 induces both growth inhibition and apoptosis of M1 myeloid leukemia cells, and that the genetically engineered M1myc cell line is available, enabled us to assess a role for c-myc in TGFβ-induced apoptosis and growth arrest.

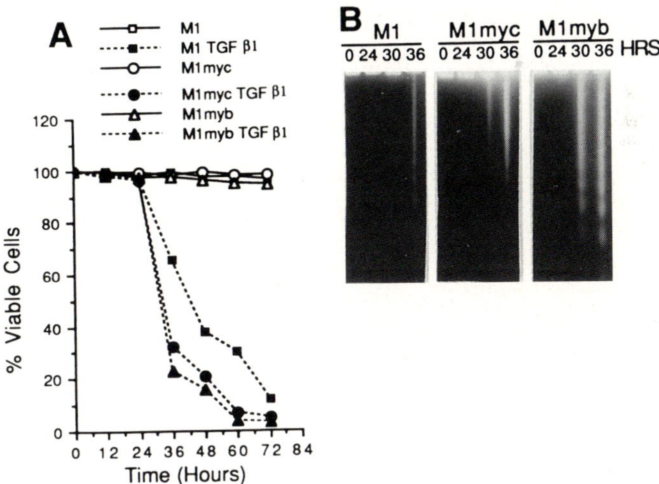

Fig. 3. Effect of deregulated expression of c-*myc* and c-*myb* on TGFβ1-induced loss of viability and apoptosis. (A) Percent viable cells at the indicated times following TGFβ1 treatmentof M1, M1myc and M1myb cells. (B) Induction of DNA fragmentation by TGFβ1.

TGFβ1 induces both growth inhibition and apoptosis of M1 myeloid leukemia cells; this is readily apparent when there is enforced overexpression of *bcl*-2, which allows M1 cells to survive but not to proliferate following treatment with TGFβ1 (44). TGFβ1 rapidly downregulates c-*myc* in M1 cells. Using M1myc cell lines, it was shown that deregulated expression of c-*myc* inhibited TGFβ1-induced growth arrest, as seen by thymidine incorporation and FACS analysis, and accelerated TGFβ1-induced apoptosis, determined by loss of viability and DNA fragmentation (44) (Fig. 3). The findings with TGFβ1 provide another example that continued expression of c-*myc* in the presence of growth arrest signals leads to accelerated apoptosis.

M1myb cell lines, expressing deregulated c-*myb*, was shown, like deregulated c-*myc*, to inhibit TGFβ1-induced growth arrest and to accelerate apoptosis, where the acceleration is even more rapid than with cells expressing deregulated c-*myc*. This is the first reported demonstration that c-*myb* can regulate the apoptotic response (44).

Recently, it has been reported that c-*myc*-induced apoptosis is mediated by p53 in fibroblasts (18, 49). This is clearly not the case for c-*myc*-mediated accceleration of TGFβ-induced apoptosis of myeloid cells, since M1 does not express p53.

Concluding Remarks

The proto-oncogene c-*myc* plays a pivotal role in growth control, differentiation, apoptosis and tumorigenesis. Towards understanding how these cellular processes are regulated, it is essential to understand the involvement of c-*myc*. Identifying the different c-myc target genes participating in each process, and how specific target genes are regulated by the c-myc family of transcription factors is crucial to achieve these goals. Determining if all the biological effects of c-*myc* are manifestations of expression of the same target genes, which result in different phenomenon due to the physiological state of the cell, or are caused by expression of either different or overlapping sets of target genes, will lead to a greater understanding of how c-*myc*, a central player in the regulation of growth, differentiation, and survival, exerts its pleiotropic effects. Our results on ODC are indicative of the involvement of different or overlapping, but not identical, sets of c-myc target genes in playing a role in the regulation of the different cellular processes. A network of interacting transcription factors appears to modulate the transcriptional activity of c-myc, and includes the basic helix-loop-helix leucine zipper protein Max (myn), which dimerizes to c-myc and is required for its transcriptional activity (7, 14); Mad and Mxi1 were identified as heterodimeric partners for Max which antagonize c-myc transcriptional activity (6, 51). Understanding the role of this network of factors in regulating c-myc target genes will also further our knowledge about how growth, differentiation and survival are regulated in both normal and neoplastic cells.

Acknowledgments

This work was supported by National Institutes of Health grants 1RO1CA51162 (BH), 1RO1CA43618 (DAL) and by National Cancer Institute Cancer Center

Support Grant No. 5 P30 CA12227. We also acknowledge Amgen's support of this work (BH and DAL).

References

1. Abdollahi A, Lord KA, Hoffman-Liebermann B and Liebermann DA (1991) Sequence and expression of a cDNA encoding MyD118: A novel myeloid differentiation primary response gene induced by multiple cytokines. Oncogene 6:165-167
2. Abdollahi A, Lord KA, Hoffman-Liebermann B and Liebermann DA (1991) Interferon regulatory factor 1 is a myeloid differentiation primary response gene induced by interleukin 6 and leukemia inhibitory factor: Role in growth inhibition. Cell Growth & Diff 2:401-407
3. Anfossi G, Gewirtz AM and Calabretta B (1989) An oligomer complementary to c-myb-encoded mRNA inhibits proliferation of human myeloid leukemia cell lines. Proc Natl Acad Sci USA 86:3379-3383
4. Askew DS, Ashmun RA, Simmons BC and Cleveland JL (1991) Constitutive c-myc expression in an IL-3 dependent myeloid cell line suppresses cell cycle arrest and accelerates apoptosis. Oncogene 6:1915-1922
5. Auvinen M, Paasinen A, Andersson LC and Holtta E (1992) Ornithine decarboxylase activity is critical for cell transformation. Nature 360:355-358
6. Ayer DE, Kretzner L and Eisenman RN (1993) Mad: A heterodimeric partner for max that antagonizes myc transcriptional activity. Cell 72:211-222
7. Ayer DE, Lawrence QA and Eisenman RN (1995) Mad-max transcriptional repression is mediated by ternary complex formation with mammalian homologs of yeast repressor Sin3. Cell 80:767-776
8. Bello-Fernandez C, Packham G and Cleveland JL. (1993) The ornithine decarboxylase gene is a transcriptional target of c-myc. Proc Nat Acad Sci USA 90:7804-7808
9. Blackwell TK, Kretzner L, Backwood EM, Eisenman RN and Weintraub H (1990) Sequence specific DNA binding by the c-myc protein. Science 250:1149-1151
10. Collins S and Groudine M (1982) Amplification of endogenous myc-related DNA sequences in a human myeloid leukemia cell line. Nature 298:679-681
11. Dalla-Favera R, Wong-Staal F and Gallo RC (1982) Oncogene amplification in promyelocytic leukemia cell line HL-60 and primary leukemic cells of the same patient. Nature 299:61-63
12. Eilers M, Picard D, Yamamoto KR and Bishop JM (1989) Chimaeras of myc oncoprotein and steroid receptors cause hormone-dependent transformation of cells. Nature 340:66-68
13. Eilers M, Schirm S and Bishop JM (1991) The myc protein activates transcription of the α-prothymosin gene. The EMBO J. 10:133-141
14. Evan GI and Littlewood TD (1993) The role of c-myc in cell growth. Curr Opinion Genet Dev 3:44-49
15. Evan GI, Wyllie AH, Gilbert CS, Littlewood TD, Land H, Brooks M, Waters CM, Penn LZ and Hancock DC (1992) Induction of apoptosis in fibroblasts by c-myc protein. Cell 69:119-128
16. Freytag SO (1988) Enforced expression of the c-myc oncogene inhibits cell differentiation by precluding entry into a distinct predifferentiation state in G0/G1. Mol Cell Biol 8:1614-1624
17. Gonda TJ, Sheiness DK and Bishop JM (1982) Transcripts from the cellular homologues of retroviral oncogenes: Distribution among tissues. Mol Cell Biol 2:617-624
18. Hermeking H and Eick D (1994) Mediation of c-myc-induced apoptosis by p53. Science 265:2091-2093
19. Hibshoosh H, Jonson M and Weinstein BI (1991) Effects of overexpression of ornithine decarboxylase (ODC) on growth control and oncogene-induced cell transformation. Oncogene 6:739-743
20. Hoffman-Liebermann B and Liebermann DA (1991) Interleukin-6 and leukemia inhibitory factor induced terminal differentiation of myeloid leukemia cells is blocked at an intermediate stage by constitutive c-myc. Mol Cell Biol 11:2375-2381
21. Hoffman-Liebermann B, Selvakumaran M, Buettger C, Matschinsky F and Liebermann DA. (1994) Dissection of the pleiotropic effects of deregulated c-myc during myeloid leukemia cell differentiation using high resolution 2D gel electrophoresis. Mol Cell Differ 2:185-200

22. Holt JT, Redner RL and Nienhuis AW (1988) An oligomer complementary to c-myc mRNA inhibits proliferation of HL-60 promyelocytic cells and induces differentiation. Mol Cell Biol 8:963-973
23. Kuehl WM, Bender TB, Stafford J, McClinton D, Segal S and Dmitrovsky E (1988) Expression and function of the c-myb oncogene during hematopoietic differentiation. Current topics in microbiology and immunology 141:318-323
24. Landshulz WH, Johnson PF and McKnight SL (1988) The leucine zipper: A hypothetical structure common to a new class of DNA binding proteins. Science 240:1759-1764
25. Liebermann DA and Hoffman-Liebermann B (1989) Proto-oncogene expression and dissection of the myeloid growth to differentiation developmental cascade. Oncogene 4:583-592
26. Liebermann DA and Hoffman B (1994) Genetic Programs in Myeloid Cell Differentiation. Current Opinion in Hematology 1:24-32
27. Liebermann DA and Hoffman B (1994) Differentiation primary response genes and proto-oncogenes as positive and negative regulators of terminal hematopoietic cell differentiation. Stem Cells 12:352-369
28. Liebermann D, Hoffman-Liebermann B and Sachs L (1980) Molecular dissection of differentiation in normal and leukemic myeloblasts: Separately programmed pathways of gene expression. Devel Biol 79:46-63
29. Lord KA, Hoffman-Liebermann B and Liebermann DA (1990) Complexity of the immediate early response of myeloid cells to terminal differentiation and growth arrest includes ICAM-1, Jun-B and histone variants. Oncogene 5:387-396
30. Lord KA, Hoffman-Liebermann B and Liebermann DA (1990) Sequence of MyDll6 cDNA: A novel myeloid differentiation primary response gene induced by IL6. Nucl Acids Res 18:2823
31. Lord KA, Abdollahi A, Hoffman-Liebermann B and Liebermann DA (1990) Dissection of the immediate early response of myeloid leukemia cells to terminal differentiation and growth inhibitory stimuli. Cell Growth & Diff 1:637-
32. Lord KA, Hoffman-Liebermann B and Liebermann D (1990) Nucleotide sequence and expression of a cDNA encoding MyD88: A novel myeloid differentiation primary response gene induced by IL-6. Oncogene 5:1095-1097
33. Metcalf D (1989) The molecular control of cell division, differentiation commitment and maturation in hematopoietic cells. Nature 339:27-30
34. Moses HL, Yang EY and Pietenpol JA (1990) TGFβ stimulation and inhibition of cell proliferation: new mechanistic insights. Cell 63:245-247
35. Moshier JA, Dosecu J, Skunca M and Luk GD (1993) Transformation of cells by ornithine decarboxylase overexpression. Cancer Res 53:2618-2622
36. Murre C, Schonleber-McCaw P and Baltimore D (1989) A new DNA binding and dimerization motif in immunoglobulin enhancer binding, dautherless, MyoD, and myc proteins. Cell 56:777-783
37. Packham G and Cleveland JL (1994) Ornithine decarboxylase is a mediator of c-myc-induced apoptosis. Mol Cell Biol 14:5741-5747
38. Pena A, Reddy CD, Wu S, Hickok NJ, Reddy PE, Yumet G, Soprano DR and Soprano KJ, Regulation of human ornithine decarboxylase expression by the c-myc.max protein complex. J Biol Chem 268:27277-27285
39. Pientenpol JA, Stein RW, Moran E, Yaciuk P, Schlegel R, Lyons RM, Pittelkow MR, Munger K, Howley PM and Moses HL (1990) TGFβ1 inhibition of c-myc transcription and growth in keratinocytes is abrogated by viral transforming proteins with pRB binding domains. Cell 61:777-785
40. Quesenberry PJ (1990) Hematopoietic stem cells, progenitor cells, and growth factors. in Williams WJ, Beutler E, Erslev AJ, Lichtman MA (eds): Hematology, McGraw-Hill, pp.129-147
41. Ruscetti FW, Dubois C, Falk LA, Jacobsen SE, Sing G, Longo DL, Wiltrout RH and Keller JR (1991) In vivo and in vitro effects of TGF-β1 on normal and neoplastic haemopoiesis. Clinical Applications of TGF-β, Ciba Foundation Symposium 157. Wiley, Chichester
42. Sachs L (1987) The molecular control of blood cell development. Science 238:1374,-1379
43. Selvakumaran M, Liebermann DA and Hoffman-Liebermann B (1993) Myeloblastic leukemia cells conditionally blocked by myc-estrogen receptor chimeric transgenes for terminal differentiation coupled to growth arrest and apoptosis. Blood 81:2257-2262

44. Selvakumaran M, Lin HK, TjinThamSjin R, Reed JC, Liebermann DA and Hoffman B (1994) The novel primary response gene MyD118 and the proto-oncogenes myb, myc, and bcl-2 modulate transforming growth factor β1-induced apoptosis of myeloid leukemia cells. Mol Cell Biol 14:2352-2360
45. Shi Y, Glynn JM, Guilbert LJ, Cotter TG, Bissonnette RP and Green DR (1992) Role for c-myc in activation-induced apoptotic cell death in T cell hybridomas. Science 257:212-214
46. Steinman RA, Hoffman B, Iro A, Guillouf C, Liebermann DA and El-Houseini ME (1994) Induction of p21 (WAF-1/CIP1) during differentiation. Oncogene 9:3389-3396
47. Tabor CW and Tabor H (1984) Polyamines. Annu Rev. Biochem. 53:749-790
48. Wagner AJ, Meyers C, Lamins LA and Hay N (1993) C-myc induces the expression and activity of ornithine decarboxylase. Cell Growth & Diff 4:879-883
49. Wagner AJ, Kokontis JM and Hay N (1994) Myc-mediated apoptosis requires wild-type p53 in a manner independent of cell cycle arrest and the ability of p53 to induce p21/waf1/cip1. Genes & Devel 8:2817-2830
50. Wickstrom EL, Bacon TA, Gonzalez A, Freeman DL, Lyman GH and Wickstrom E (1988) Human promyelocytic leukemia HL-60 cell proliferation and c-myc protein expression are inhibited by an antisense pentadecadeoxynucleotide targeted against c-myc mRNA. Proc Natl Acad Sci USA 85:1028-1032
51. Zervos AS, Gyuris J and Brent R (1993) Mxi1, a protein that specifically interacts with max to bind myc-max recognition sites. Cell 72:223-232

Self Renewal and Differentiation in Primary Avian Hematopoietic Cells: An Alternative to Mammalian *in Vitro* Models?

H. Beug[1], T. Metz[2], E.W. Müllner[3] and M.J. Hayman[4]

[1]Inst. of Molecular Pathology (IMP) and [3]Inst. of Molecular Biology, Vienna Biocenter; Dr. Bohr-Gasse, A-1030 Vienna, Austria; [2]Bender+Co Gmbh, Dr. Boehringer Gasse 5-11, A-1120 Vienna, Austria; [4]Department of Microbiology State University of New York Stony Brook New York, 11794 U.S.A.

Introduction

Hematopoietic cells are produced throughout the lifetime of an individual from a small set of stem cells found in the bone marrow. The progeny of these stem cells are required (i) to gradually lose multipotency, eventually developing into progenitors committed to a single lineage which then terminally differentiate and (ii) to make decisions regarding the balance between self-renewal (i.e. cell proliferation without detectably entering a differentiation pathway) and terminal differentiation to produce correct numbers of erythroid, myeloid and lymphoid cells. In leukemias and lymphomas, this delicate equilibrium obviously has been disturbed, resulting in an abnormal accumulation of immature cells that are apparently capable of self renewal, although they may be derived of and similar to either multipotent or committed hematopoietic progenitors (for reviews see [1, 2]).

Committed progenitors differ from the pluripotent stem cell in that the latter develops into all hematopoietic lineages while the former are restricted to one (or a few related) lineages. Secondly, pluripotent stem cells are capable of self renewal, while committed progenitors are thought to be unable to self renew, or do so only transiently (for review see [1, 3]. Commitment is thought to involve a finite number of cell divisions, during which the cells execute a predefined program of gene expression changes, eventually generating a terminally differentiated cell.

The idea of self renewal being restricted to the pluripotent stem cell does not apply to leukemias derived from committed progenitors (see [2] for review). This raises the question whether leukemic self renewal is an abnormal property caused by the genetic changes occurring in the leukemic cells and not shared by normal progenitors lacking those mutations [4, 5]. Alternatively, normal progenitors may possess a strictly regulated, normally silent ability to self renew and the mutations occurring in leukemia may just cause this normal progenitor self renewal to occur in a deregulated, constitutive fashion [6].

This short article focuses on *in vitro* cell systems currently employed as models to study these and related questions. Such models include colony assays using unpurified bone marrow cells as well as purified primary committed progenitors and stem cells of mouse and human origin. In the majority of studies, established, immortal hematopoietic cell lines are used derived from leukemias/ lymphomas or generated by introduction of foreign genes. The powerful *in vivo* models such as mice with gain- or loss-of-function mutations in genes controlling hematopoiesis will not be discussed here because of space limitations.

Here, we will concentrate on primary, non-immortalized hematopoietic cells of chick origin, that exhibit properties of either committed or multipotent progenitors. We will discuss whether this system may help to overcome some of the limitations and problems arising in using the above mammalian *in vitro* systems. We will focus on the prospects and limitations of the avian system rather than on specific results obtained. Another issue shortly discussed in this article will be, if and which changes are introduced in hematopoietic cells by the immortalization process itself and how such changes may limit the use of immortal lines as models for normal, nonimmortalized cells.

Primary Hematopoietic Progenitors.

Multipotent stem cells and committed progenitors are rare cells in hematopoietic organs, since they generate huge numbers of more differentiated offspring. Three approaches exist to study these cells *in vitro* despite their low numbers. Firstly, the cells can be purified and studied directly in colony assays. Secondly, some purified or nonpurified progenitors can be amplified *in vitro* with combinations of growth promoting agents. The most powerful method, however, is to stably introduce foreign proteins into the progenitors, that are capable of inducing self renewal in the desired progenitors. In the following, we will compare the possibilities to use the second and third approach in avian versus mammalian cells.

Normal Progenitors.

Recently, progress has been made in methods to purify and characterize pluripotent murine stem cells. [7]. However the cell numbers obtained are really low (1 stem cell per 106 bone marrow cells). Furthermore, efficient methods to significantly expand such cells do not exist. These limitations have so far severely hampered the molecular analysis of pluripotent stem cells.

Similar limitations exist for committed progenitor cells. A well-studied example are purified human erythroid progenitors, referred to as colony-forming unit erythrocyte (CFU-E) and burst-forming unit erythrocyte (BFU-E, [8, 9]). The former grow into compact colonies of 30-40 mature erythrocytes after 2-3 days, while the latter form large colonies consisting of thousands to ten-thousands of erythrocytes within 6-10 days [10, 11]. The cell numbers obtained in these purifications is again relatively low (BFU-E, 10^5-10^6, CFU-E < $5x10^7$). The same was true for cultures of early erythroid progenitors from anemia patients [12]. Using micromethods, some biological and molecular analyses were performed with such cells [13] but more extensive molecular studies were not possible

In the chicken, conditions for expansion of primary erythroid progenitors (SCF/TGFα progenitors) were recently described. Avian stem cell factor (SCF), transforming growth factor (TGF) α and estradiol induced the outgrowth of apparently normal erythroid progenitors from chick bone marrow. These cells, grow for the normal life span of chicken cells (the "Hayflick limit" [14]), corresponding to 30-40 generations *in vitro*. Thus, these cells can be theoretically expanded in suspension culture to 10^{10} - 10^{11} cells [15, 16]. When self-renewal factors (SCF, TGFα and estradiol) are removed and replaced by differentiation factors (anemic serum containing avian Epo and insulin), the normal progenitors rapidly downregulate the receptors for the self renewal factors (estrogen receptor, c-Kit, c-ErbB) and undergo terminal differentiation into erythrocytes within 4 to 5 days [16]. Recent results suggest, that these SCF/TGFα progenitors develop from normal CFU-E/BFU-E, if the latter cells are co-stimulated for 6-8 days with SCF, TGFα, estradiol and an unknown activity in chicken serum. [17]. This rather "unorthodox" combination of three, if not four growth factors may also be active to induce self renewal in normal human erythroid progenitors (M. von Lindern and H.B., manuscript in preparation).

Induction of Progenitor Outgrowth by Oncoproteins And Related Agents

Avian retroviruses cause erythroblastosis (AEV) and myeloid disorders (myelocytomatosis, MC29, MH2; myeloblastosis, AMV, E26) [18]. The AEV virus encodes the v-*erb*A and v-*erb*B oncogenes, a mutated receptor tyrosine kinase and a mutated nuclear thyroid hormone receptor. The oncoproteins v-Myc, v-Myb encoded by the MC29 and AMV/E26 represented mutated transcription factors of different types (for review see [19, 20]). The cell types obtained by *in vitro* transformation with v-ErbB plus v-ErbA were characterized as immature erythroid cells (apparently situated between BFU-E and CFU-E, the v-Myc transformed cells as monocyte-like cells (plus immature granuloblasts occurring in vivo) and those transformed by v-Myb as more immature monoblastic/myeloblastic cells [21]. All these cells could be readily propagated from colonies in semisolid medium and grown for 30-40 generations in culture before undergoing senescence. The isolation of temperature-sensitive versions of the v-ErbB, v-Sea, v-Myc and v-Myb oncoproteins allowed to induce apparently normal, mostly synchronous terminal differentiation in these cells after turning off oncoprotein function at the non permissive temperature (see the above reviews and [22, 23]).

In three cases, self renewal of multipotential progenitors could be induced by oncoprotein combinations. Firstly, the $p135^{gag\text{-}myb\text{-}ets}$ fusion protein of the E26 avian myeloblastosis virus induced the growth of multipotent cells from chicken blastoderm. Thc phenotype of the uninduced cells resembled that of very immature erythroid cells. When treated, however, with phorbol esters or superinfected with tyrosine kinase oncogenes, some of the cells differentiated into myeloblasts and eosinophils [24]. The system was improved by using a temperature sensitive mutant of $p135^{gag\text{-}myb\text{-}ets}$ in the v-Ets domain, ts1-1E26. Combinations of temperature shifts with drug treatment allowed to induce differentiation of these ts1-1E26 cells into erythrocytes, monocytes and eosinophils [25]. Multipotent progenitors capable of terminal differentiation into erythrocytes and macrophages could be induced by the ts mutant ts21-E26, (expressing a $p135^{gag\text{-}myb\text{-}ets}$ fusion protein with lesion in v-Myb) if cultivated in presence of stem cell factor (SCF), the ligand of the c-Kit receptor tyrosine kinase (M. von Lindern and H. Beug, unpublished).

A second type of multipotent progenitor was recently induced in chicken bone marrow by the nuclear oncoprotein v-Ski, again in cooperation with the ligand-activated receptor tyrosine kinase c-Kit. These cells are unique in that they have an *in vitro* life span of >100 generations, by far exceeding the normal *in vitro* life span of chicken cells (30-40 generations). In contrast to the E26 transformed cells, these multipotent progenitor cells were still dependent for self renewal on combinations of growth factors and hormones active on mammalian multipotent progenitors (SCF plus estradiol with cytokines like Il-6 or presumably Il-3, [26]). In addition these cells spontaneously differentiate along the erythroid, monocytic and granulocytic (mast cell) lineages. Furthermore clonal strains of Ski-induced multipotent progenitors can be "committed" to the erythroid or myeloid lineages by specific mixtures of growth factors and steroid hormones in presence of SCF, and then induced to terminal differentiation when treated with lineage-specific factors in absence of SCF. In presence of factors specific for the "wrong" lineage, the cells underwent apoptosis. Thus self-renewal, commitment and differentiation could be regulated in these multipotent cells by specific, physiological agents, similar as in the self-renewing erythroblasts [27].

The third type of self-renewing multipotential cells was induced by the v-Rel oncoprotein, a NFκB-related transcription factor oncoprotein [28]. Initially, these cells were characterized as preB-preT lymphoid cells [29]. Similar to the v-Ski/SCF induced cells they displayed a highly increased 80-100 generations or even unlimited *in vitro* life span (> 150 generations) without undergoing a Hayflick crisis at any time ([28, 29] and references therein). By employing a fusion protein of v-Rel with the hormone binding domain of the estrogen receptor (v-RelER) the function of which can be activated by estradiol, these cells were characterized as multipotent progenitors, that could give rise to immature B cells, dendritic cells and an unidentified granulocytic cell. [30]. Differentiation into the granulocytic and dendritic cell types occurred after inactivation of the v-RelER, while apoptosis rather than terminal B-lymphocytic differentiation was observed with respect to the B-cell lineage [30].

Primary, oncoprotein-transformed hematopoietic progenitors similar to the above chicken cells are rarely available in mice and not described in humans In the mouse, oncoproteins like v-Raf, v-Ras, v-Src, v-ErbB v-Abl and others induce colonies of proliferating, apparently immature erythroid progenitors in semisolid medium. Oncoprotein combinations (e.g. v-Raf plus v-Myc) lead to larger colonies and a more pronounced differentiation arrest [31]. In all these experiments, trials to expand the colonies into mass cultures of primary cells failed (see [32] for references). This was most likely due to the much shorter *in vitro* life span of mouse cells (7 to 15 generations; [33]). Since it requires 13 cell divisions to build up a typical macroscopic colony of 10^4 cells, the cells in such murine colonies may already have reached the end of their replicative life span. In some cases, erythroid cell lines could be grown from Myc-Raf- or Src-oncoprotein transformed erythroid cells [31, 34].

Primary colonies of myeloid murine bone marrow cells transformed by v-Myc [35] and a murine, E26 like virus could also not be expanded. In these cases, however, as well as with pre-B cells transformed by the v-Abl oncoprotein, the transformed cells readily grow into immortalized cell lines. The only cell system in which conditional oncoproteins have been used to allow differentiation induction in murine hematopoietic progenitors are preB cells transformed with a ts-mutant of the Abl oncoprotein. Efficient lymphoid differentiation of the preB cells after turning off Abl oncoprotein function required the presence of the Bcl-2 gene product to inhibit apoptosis (for review see [36]).

Immortalized cell lines of hematopoietic origin

As mentioned above, one of the major tools in studying hematopoiesis are established, immortalized cell lines of both murine and human origin. Human cell lines display characteristics of primitive erythroid (K562, KG-1) or myeloid cells (Hl-60, U937). Many lines with preB/preT or uncharacterized phenotypes have been also described. In addition, it is frequently but not always possible to grow human leukemia cells into immortalized cell lines (H. Messner, personal communication). Multipotent lines may exist (e.g. TF-1) but fail to show significant differentiation along specified lineages. Human lines generated by deliberate introduction of oncogenes do not exist to our knowledge.

In the mouse, a wealth of cell lines with erythroid, megakaryocytic, monocytic, granulocytic and B- or T-lymphocytic characteristics exist. They cannot be discussed further due to space limitations. In addition, several lines resemble multipotent progenitors capable of producing offspring differentiating along the erythroid plus several myeloid or even lymphoid lineages. In the following, a few aspects pertinent to the use of such immortalized lines as model systems will be discussed, in comparison to mortal, non immortalized cells or established cell lines of chicken origin.

What Oncoproteins Are Active in Hematopoietic Cell Lines ?
Many human and murine cell lines (e.g. FDCP-1 cells, 32D cells, BAF-B03 cells, etc.) arose from leukemias or related tumors or were spontaneously immortalized in culture. Consequently, the mutated genes causing self renewal and differentiation arrest are not known. In some cases, some of the contributing mutated genes could be identified. In K562 and HL60 cells, derived from chronic myelogenous leukemia, the Bcr-Abl fusion proteins generated by the CML-specific chromosomal translocations are expressed, but further contributing mutations are unknown. A similar situation holds for other human cell lines derived from leukemias with identified translocation-induced fusion gene products.

In the mouse, one of the best-understood examples are the Friend leukemia virus-(SFFV)-induced MEL cells. Initially, constitutive activation of the erythropoietin receptor (EpoR) by intracellular binding of the SFFV 55 kD envelope glycoprotein causes polyclonal expansion of erythroid progenitors. [37]. The later emergence of leukemic cells is associated with overexpression of a normal, non-mutated Spi-1/Pu1 protein, an Ets family transcription factor [38]. Furthermore, functional inactivation of the p53 tumor suppressor gene as the result of gene loss or mutation, occurring in the vast majority of Friend erythroleukemic cell lines [39]. Identification of other important oncoproteins able to induce progenitor self renewal has been largely assisted by cloning retroviral integration sites leading to oncoprotein activation (see this volume).

A very important role of p53 loss-of-function mutations has also emerged from trials to transform erythroid progenitors from foetal livers of wild-type or p53$^{-/-}$ mice, using the v-Myc and v-Raf oncoproteins. Transformed colonies arose with a *raf* and a *myc/raf* virus. The absence of p53 proved to be critical for expansion into mass cultures. Immortalized cell lines grew out following a typical Hayflick crisis in p53$^{-/-}$ cells, while cell lines formed rarely (v-Myc plus v-Raf) or not at all (v-Raf) from cells expressing wild-type p53. P53$^{-/-}$ cells transformed with v-Myc plus v-Raf instantaneously grew into lines with unlimited life span, that did not show any indication of a Hayflick crisis [32]. Interestingly, these lines grew with an unusually short cell cycle (11-13 hours) and, in contrast to the purely erythroid lines obtained with myc/raf in wild-type p53 expressing cells, displayed an erythroid, myelomonocytic or undifferentiated, lineage-marker negative phenotype. This instantaneous immortalization without any detectable crisis resembled that of the multipotent cells induced in chick bone marrow by v-Ski plus c-Kit or by v-Rel (see above).

In addition to the oncogenes introduced or activated during leukemogenesis or cell line outgrowth, there are mutations inherent to immortalization itself. Besides p53 deletions and mutations, particularly common in murine lines, the cyclin/cdk-inhibitor (CDI) p16 is mutated or deleted in the vast majority of immortalized cell lines (for review see [40, 41]). Abnormalities in expression and/or function of other CDI´s (p21, p27) were also found. (J. Smith, personal communication). Most importantly, complementation analysis revealed at least four different loci, which are deleted in various cell lines [42]. Two of them have been localized to defined regions in human chromosomes 1 and 4 [43, 44]. Interestingly, the CDI p21 was cloned as a DNA synthesis inhibitor overexpressed in senescent cells [45]. It remains to be determined, if and to what these mutations affect the cell cycle behaviour and/or differentiation capacity of immortal cell lines (see below).

Terminal Differentiation of Immortalized Hematopoietic Cell Lines ?

Normal avian erythroid progenitors can undergo terminal differentiation *in vitro*, resembling the *in vivo* differentiation program in many aspects. This included the number of cell divisions during differentiation, changes in cell cycle regulation during differentiation and loss of cell size control. Differentiating cells grew with doubling times of 11-12 hours, due to a drastically shortened G1 phase and underwent a size reduction from 350 to 70 femtoliters within the five cell divisions accompanying terminal differentiation (see [6] and Dolznig et al, manuscript submitted). In addition, a concerted reprogramming of gene expression was observed 16 hours after differentiation induction, leading to downregulation of self renewal-specific genes and concerted upregulation of numerous erythroid-specific transcription factors and late erythrocyte-specific proteins. Finally, both cell morphology and hemoglobin content corresponded to that of normal erythrocytes (Dolznig et al, manuscript submitted).

In contrast, established erythroleukemia cell lines such as the murine Friend erythroleukemia cells and the human erythroleukemia cell line K562 require nonphysiological stimuli for induction of differentiation (e.g DMSO, DMBA or hemin, usually fail to differentiate into enucleated erythrocytes (for a possible exception see [46], express erythroid proteins and functions incompletely or aberrantly and/or show aberrant growth properties (for review see [47, 48] and references therein). In particular, the phase of accelerated growth typical for normal erythroid progenitors has not been observed in any of the differentiating, immortalized erythroid cell lines, although the fact that formation of a 32-cell CFU-E colony takes between 2 and 3 days in both mouse and human bone marrow colony assays argues that normal mammalian erythroid progenitors show this property.

A limited ability to terminally differentiate along the various lineages is also exhibited by the existing multipotential cell lines. Even in the most advanced of these lines, the lymphohematopoietic progenitors induced by a dominant negative RAR, differentiation along the various lineages stops well before the cells are mature. Erythroid differentiation proceeds only to a nucleated, early reticulocyte stage, neutrophil differentiation produces cells with ring-like or segmented nuclei, but no granules are formed, the mast cells appear far from mature and the B cells are blast-like rather than resembling mature B cells [49]. In contrast, fully mature macrophages, erythrocytes and macrophages (the latter actively phagocytosing apoptotic red cells) were obtained from the non immortalized, differentiating avian ski BM cells.

Plasticity of Immortalized Hematopoietic Cell Lines

A particularly striking feature of most, if not all immortalized cell lines is their ability to alter their proliferation or differentiation characteristics upon selective pressure. This was particularly striking in the chicken system, where defined, mortal clones of erythroblasts transformed by v-ErbB Ts-mutants and capable of temperature-induced differentiation were developed into immortal erythroblast cell lines, allowing a direct comparison of mortal strains and immortal lines with identical genetic background [50]. In contrast to the primary clones, the immortalized cell populations were very heterogeneous in their ability to terminally differentiate. Repeated screening for well differentiating clones was required to obtain a cloned line differentiating as well as the mortal clone. From these cells, variants were readily selected, which were resistant to differentiation induction by temperature shift, drugs like butyric acid or both [51]. A similar plasticity has been observed in Friend cell lines [52].

The plasticity of immortalized cell lines may also facilitate lineage switching, most efficiently induced by lineage-specific transcription factors. In 416B, a primitive myeloid cell line, GATA-1 induced megakaryocyte differentiation, accompanied by reduced expression of myeloid markers [53]. An even more striking lineage switch was observed in the chicken macrophage cell line HD11. Forced expression of GATA-1 in this line induced the formation of cell clones resembling more immature myeloblasts, blasts with eosinophilic markers and peroxidase-positive granules, and immature erythroblast-like cells, that expressed the erythroid-specific histone H5 as well as low levels of hemoglobin [54]. Interestingly, globin levels could be enhanced by DMSO, a drug that is completely inactive on avian and mammalian erythroblasts transformed by tyrosine kinases [34, 51].

Conclusions and Speculations

An "ideal" *in vitro* cell system to study aspects of hematopoiesis should exhibit a long (or unlimited) life span during self renewal, should be dependent on "self renewal" factors known to act *in vivo* and should undergo terminal differentiation into fully mature cells in response to *in vivo* differentiation factors. Although no system fully meeting these requirements exists to date, the self-renewing avian erythroid progenitors approach this goal [6]. Their main disadvantage is their chicken origin. Most genes currently attracting attention, such as growth factor receptors, signal transduction intermediates, transcription factors and cell cycle regulators have been cloned in mouse and man, but not in the chicken. In addition, the *in vitro* life span of these cells (30-40 generations) allow to generate sufficient cell numbers for biochemical characterization, but is still too short to allow gene transfer experiments as in immortal cell lines. The same holds true for oncoprotein-transformed avian cell strains of myeloid origin. Like other primary cells, these chicken cell strains are also resistant to most DNA transfection protocols.

The avian multipotent progenitor cell strains circumvent this life span problem, since they are comparable in this respect to immortalized cell lines (defined as able to undergo a minimum of 150 doublings in culture). A major advantage of the v-Ski plus SCF induced cells is their ability to commit and terminally differentiate along various lineages in response to physiological combinations of growth factors and steroid hormones. However, these strains are induced by transforming oncoproteins, which may well alter their behaviour. The use of conditional oncoproteins (ts-mutants, ER-fusion proteins) may not completely solve this problem, since residual oncoprotein activities may persist even after turning off oncoprotein function. The major disadvantage of the avian system, however, is the almost total lack of recombinant cytokines and growth factors of chicken origin (for an update, see [55]). Since mammalian cytokines are usually inactive on avian cells, one way to rapidly overcome this seems to be the use of retroviral vectors expressing mammalian cytokine receptors. To date, receptors for Epo, GM-CSF and Il-3 have been expressed in primary chicken hematopoietic cells and were found to function similar than in mammalian cells [56], O. Wessely, P. Steinlein and H.B., unpublished.

With these disadvantages in mind: why do we bother to use primary cells and not just switch to established cell lines ? One recent, very speculative argument is that immortalized cell lines may not represent what we think they do. Currently, an immortalized line showing characteristic features of an erythroblast or monoblast and able to differentiate towards erythrocytes or macrophages is used as a model for the respective normal, committed progenitor. It is conceded, that the genetic changes inherent to immortalization may somewhat alter cell cycle behaviour and interfere with the very last steps during differentiation, but immortalization is not thought to cause more fundamental alterations of the progenitor phenotype. Several unrelated observations have caused us to speculate, that immortalization may induce a committed progenitor to behave like a multipotent stem cell in certain aspects. We postulate, that stem cells and multipotent progenitors may simply lack the control mechanisms enforcing a Hayflick limit. This may be due to the fact that some or all of the genes deleted or mutated during immortalization may not be expressed or functionally inactive in these totipotent or multipotent cells.

Is there any support for these speculations? Firstly, multipotent progenitors are altered in their life span behaviour. Two avian multipotent progenitor types (induced by v-Ski and v-Rel) exhibit highly enhanced or even unlimited *in vitro* life span without ever undergoing a Hayflick crisis [27, 28]. The same is true for murine Myc/Ras transformed progenitors induced in p53-/- cells [32], which may turn out to be multipotent progenitors. Similarly, totipotent mouse ES cells grow out as immortal cells without any detectable crisis. The latter two cell types have been shown to lack the chromosomal abnormalities usually associated with immortalization, the most rigorous proof for this being the fact, that ES cells give rise to whole, healthy animals.

A second argument in favour of our idea is the plasticity of cell lines. In striking contrast to primary human or avian cell clones, in which we have been unable to select for any significant phenotypical changes, established cell lines readily change their phenotype in response to selective pressure or expression of transcription factors from the "wrong" lineage (see above). Particularly the latter behaviour is the expected one for multipotential cells, which according to recent evidence coexpress lineage-specific transcription factors as well as other lineage markers and only downregulate the "wrong" transcription factors and markers upon terminal differentiation ([27], N. Iscove, personal communication). This may also explain the "lineage infidelity" seen in many human leukemias, suggesting that these leukemic cells may be arrested at the stage of a multipotent progenitor.

What could be the genes that control the Hayflick limit and are deleted or mutated during immortalization? Likely candidates are the p53 gene or genes acting upstream or downstream, such as mdm-2 or the cdk-inhibitor p21. Other cdk-inhibitors are also likely candidates (p16, see above). Finally, gene products negatively regulating the expression or function of the telomerase complex may also be considered (for review see [57]. While mortal cells of human (and probably chicken) origin lack telomerase activity, essentially all immortalized cell lines as well as the majority of human tumors express this activity [58]. Interestingly, human leukemias which have been shown to be nonimmortalized when isolated from the patient and go through a typical Hayflick crisis before eventually becoming immortalized (H. Messner, personal communication) do express telomerase [59]. Also, telomerase activity is detected in Xenopus during oogenesis and early embryogenesis [60]. Thus , it is possible that toti- and multipotent cells, as well as established cell lines escape telomer shortening leading to senescence [61] by expression of telomerase.

In conclusion, primary chicken hematopoietic cells offer a viable alternative to human cells, since they have a comparable *in vitro* life span and thus allow production of primary cells in large quantities. Due to the different Hayflick limit, mouse cells do not offer this alternative. Additional arguments in favor of chicken rather than mouse cells is their much higher genetic stability of primary clones and the lack of spontaneous immortalization with high frequency. This is particularly important in cases, where a given human cell type is not easily available or where gene transfer experiments are problematic due to biosafety- or ethical reasons. Immortalized cell lines are invaluable and indispensable tools for many biological questions, but the avian system offers the possibility to perform those types of studies, where the specific mutations or the inherent genetic instability of immortalized cell lines are likely to seriously affect the results or conclusions obtained.

References

1. Keller G (1992) Hematopoietic stem cells. Curr Opinion in Immunology 4: 133-139.
2. Sawyers CL, Denny CT,Witte ON (1991) Leukemia and the disruption of normal hematopoiesis. Cell 64: 337-50.
3. Till JE, McCulloch EA (1980) Hematopoietic stem cell differentiation. BiochimBiophys Acta 605: 431-459.
4. Daley GQ, Van Elten RA, Baltimore D (1990) Induction of chronic myelogenous leukemia in mice by the p210*bcr-abl* gene of the Philadelphia chromosome. Science 247: 824-830.
5. Kelliher MA, McLaughlin J, Witte ON, Rosenberg N (1990) Induction of a chronic myelogenous leukemia-like syndrome in mice with v-abl and BCR/ABL. Proc Natl Acad Sci USA 87: 6649-6653.
6. Beug H, Müllner EW, Hayman MJ (1994) Insights into erythroid differentiation obtained from studies on avian erythroblastosis virus. Curr Op Cell Biol 6: 816-824.
7. Spangrude GJ, Smith L, Uchida N, Ikuta K, Heimfeld S, Friedman F, Weissman IL (1991) Mouse hematopoietic stem cells. Blood 78: 1395-1402.
8. Sawada K, Krantz SB, Kans JS, Dessypris EN, Sawyer ST, Glick AD, Civin CI (1987) Purification of human erythroid colony forming units and demonstration of specific binding of erythropoietin. J Clin Invest 80: 357-366.
9. Sawada K, Krantz SB, Dai CH, Koury ST, Horn ST, Glick AD, Civin CI (1990) Purification of human blood burst-forming units erythroid and demonstration of the evolution of erythropoietin receptors. J Cell Physiol 142: 219-230.
10. Cotton EW, Means Jr. RT, Cline SM, Krantz SB (1991) Quantitation of insulin-like growth factor-I binding to highly purified human erythroid colony-forming units. Exp Hematol 19: 278-281.
11. Dai CH, Krantz SB, Zsebo KM (1991) Human burst-forming units-erythroid need direct interaction with stem cell factor for further development. Blood 78: 2493-2497.
12. Fibach E, Manor D, Oppenheim A, Rachmilewitz EA (1989) Proliferation and maturation of human erythroid progenitors in liquid culture. Blood 73: 100-103.
13. Dalyot N, Fibach E, Ronchi A, Rachmilewitz EA, Ottolenghi S, Oppenheim A (1993) Erythropoietin triggers a burst of GATA-1 in normal human erythroid cells differntiating in tissue culture. Nucl Acids Res 21: 4031-4037.
14. Hayflick L, Moorhead PS (1961) The serial cultivation of human diploid cell strains. Exp Cell Res 25: 585-621.
15. Schroeder C, Gibson L, Nordström C, Beug H (1993) The estrogen receptor cooperates with the TGFα receptor (c-*erb*B) in regulation of chicken erythroid progenitor self renewal. EMBO J 12: 951-960.
16. Hayman MJ, Meyer S, Martin F, Steinlein P, Beug H (1993) Self renewal and differentiation of normal avian erythroid progenitor cells: Regulatory roles of the c-*erb*B/TGFα receptor and the c-*kit* /SCF receptor. Cell 74: 157-169.
17. Steinlein P, Wessely O, Meyer S, Deiner EM, Hayman MJ, Beug H (1995) Primary, self-renewing erythroid progenitors develop from BFU-E through activation of both tyrosine kinase and steroid hormone receptors. Curr Biol 5: 191-204.
18. Graf T, Beug H (1978) Avian leukemia viruses: interaction with their target cells in vivo and in vitro. Biochim Biophys Acta 516: 269-99.
19. Hayman MJ, Beug H,(1992), Avian erythroblastosis: A model system to study oncogene cooperativity in leukemia. In: . Oncogenes in the Development of Leukemia, Imperial Cancer Res. Fund.: London. pp. 53-68.

20. Beug H, Graf T (1989) Co-operation between viral oncogenes in avian erythroid and myeloid leukaemia. Eur J Clin Invest 19: 491-502.
21. Beug H, von Kirchbach A, Doderlein G, Conscience JF, Graf T (1979) Chicken hematopoietic cells transformed by seven strains of defective avian leukemia viruses display three distinct phenotypes of differentiation. Cell 18: 375-390.
22. Beug H, Leutz A, Kahn P, Graf T (1984) Ts mutants of E26 leukemia virus allow transformed myeloblasts, but not erythroblasts or fibroblasts, to differentiate at the nonpermissive temperature. Cell 39: 579-588.
23. von Weizsäcker F, Beug H, Graf T (1986) Temperature-sensitive mutants of MH2 avian leukemia virus that map in the v-mil and the v-myc oncogene respectively. Embo J 5: 1521-7.
24. Graf T, McNagny K, Brady G, Frampton J (1992) Chicken "erythroid" cells transformed by the gag-myb-ets encoding E26 leukemia virus are multipotent. Cell 70: 201-213.
25. Kraut N, Frampton J, McNagny K, Graf T (1994) A functional Ets DND-binding domain is required to maintain multipotency of hematopoietic progenitors transformed by Myb-Ets. Genes Dev 8: 33-44.
26. McNiece IK, Langley KE, Zsebo KM (1991) Recombinant human stem cell factor synergized with GM-CSF; G-CSF, Il-3 and Epo to stim,ulate human progenitor cells of the myeloid and erythroid lineages. Exp Hematol 19: 226-231.
27. Beug H, Dahl R, Steinlein P, Meyer S, Deiner E, Hayman MH (1995) In vitro growth of factor-dependent multipotential cells is induced by the nuclear oncoprotein v-ski. Oncogene in press.
28. Morrison LE, Boehmelt G, Beug H, Enrietto PJ (1991) Expression of v-rel in a replication competent virus: transformation and biochemical characterization. Oncogene 6: 1657-66.
29. Beug H, Müller H, Grieser S, Doederlein G, Graf T (1981) Hematopoietic cells transformed in vitro by REVT avian reticuloendotheliosis virus express characteristics of very immature lymphoid cells. Virology 115: 295-309.
30. Boehmelt G, Madruga J, Dorfler P, Briegel K, Schwarz H, Enrietto PJ, Zenke M (1995) Dendritic cell progenitor is transformed by a conditional v-Rel estrogen receptor fusion protein v-RelER. Cell 80: 341-52.
31. Klinken SP, Nicola NA, Johnson GR (1989) In vitro-derived leukemic erythroid cell lines induced by a raf- and myc-containing retrovirus differentiate in response to erythropoietin. Proc Natl Acad Sci USA 85: 8506-8510.
32. Metz T, Harris AW, Adams JM (1995) Absence of p53 allows direct immortalization of hematopoietic cells by the *myc* and *raf* oncogenes. Cell in press.
33. Rohme D (1981) Evidence for a relationship between longevity of mammalian species and life spans of normal fibroblasts in vitro and erythrocytes in vivo. Proc Natl Acad Sci USA 78: 5009-5013.
34. Kennedy M, Beug H, Wagner EF, Keller G (1992) Factor-dependent erythroid cell lines derived from mice transplanted with hematopoietic cells expressing the v-src oncogene. Blood 79: 180-190.
35. Vennstrom B, Kahn P, Adkins B, Enrietto P, Hayman MJ, Graf T, Luciw P (1984) Transformation of mammalian fibroblasts and macrophages in vitro by a murine retrovirus encoding an avian v-myc oncogene. EMBO J 3: 3223-3229.
36. Rosenberg N (1994) *abl*-mediated transformation, immunoglobulin gene rearrangement and arrest of B lymphocyte differentiation. Sem Cancer Biol 5: 95-102.
37. Li J, D'Andrea A, Lodish H, Baltimore D (1990) Activation of cell growth by binding of Friend spleen focus-forming virus gp55 glycoprotein to the erythropoietin receptor. Nature 343: 762-644.
38. Moreau-Gachelin F, Tavitian A, Tambourin P (1988) Spi-1 is a putative oncogene in virally induced murine erythroleukaemias. Nature 331: 277-280.
39. Munroe D, Peacock J, Benchimol S (1990.) Inactivation of the cellular p53 gene is a common feature of Friend virus-induced erythroleukemia: relationship of inactivation to dominant transforming alleles. Mol Cell Biol 10: 3307-3313.
40. Hunter T (1993) Braking the cell cycle. Cell 75: 839-841.
41. Pines J (1994) Arresting developments in cell cycle control. Trends Biochem Sci 19: 143-145.

42. Pereira-Smith OM, Smith J (1988) Genetic analysis of indefinite division in human cells: Identification of four complementation groups. Proc Natl Acad Sci USA 85: 6042-6046.
43. Ning Y, Weber JL, Killary AM, Ledbetter DH, Smith JR,Pereira-Smith OM (1991) Genetic analysis of indefinite division in human cells: evidence for a cell senescence-related gene(s) on human chromosome 4. Proc Natl Acad Sci USA 88: 5635-5639.
44. Hensler PJ, Annab LA, Barrett JC, Pereira-Smith OM (1994) A gene involved in control of human cellular senescence on human chromosome 1q. Mol Cell Biol 14: 2291-2297.
45. Noda A, Ning Y, Venable SF, Pereira-Smith OM, Smith JR (1994) Cloning of senescent cell-derived inhibitors of DNA synthesis using an expression screen. Exp Cell Res 211: 90-98.
46. Patel VP, Lodish HF (1987) A fibronectin matrix is required for differentiation of murine erythroleukemia cells into reticulocytes. J Cell Biol 105: 3105-3118.
47. Alitalo R (1990) Induced differentiation of K562 leukemia cells: a model for studies of gene expression in early megakaryoblasts. Leuk Res 14: 501-514.
48. Marks PA, Rifkind RA (1989) Induced differentiation of erythroleukemia cells by hexamethylene bisacetamide: a model for cytodifferentiation of transformed cells. Environ Health Perspect 80: 181-188.
49. Tsai S, Bartelmez S, Sitnicka E,Collins SJ (1994) Lymphohematopoietic progenitors immortalized by a retroviral vector harboring a dominant-negative retinoic acid receptor can recapitulate lymphoid, myeloid and erythroid development. Genes Dev 8: 2831-2844.
50. Ulrich E, Boehmelt G, Bird A, Beug H (1992) Immortalization of conditionally transformed chicken cells: loss of normal p53 expression is an early step that is independent of cell transformation. Genes Dev 6: 876-887.
51. Beug H, Doederlein G, Freudenstein C, Graf T (1982) Erythroblast cell lines transformed by a temperature-sensitive mutant of avian erythroblastosis virus: a model system to study erythroid differentiation in vitro. J Cell Physiol Suppl 1: 195-207.
52. Marks P, Rifkind R (1978) Erythroleukemic cell differentiation. Annu Rev Biochem 47: 419-448.
53. Visvader JE, Crossley M, Hill J, Orkin SH, Adams JM (1995) The C-terminal zinc finger of GATA-1 or GATA-2 is sufficient to induce megakaryocytic differentiation of an early myeloid cell line. Mol Cell Biol 15: 634-641.
54. Kulessa H, Frampton J, Graf T (1995) GATA-1 reprograms avian myelomonocytic cell lines into eosinophils, thromboblasts and erythroblasts. Genes Dev 9: 1250-1262.
55. Beug H, Bartunek P, Steinlein P, Hayman MJ (1995), Avian hematopoietic cell culture: *In vitro* model systems to study the oncogenic transformation of hematopoietic cells. Methods in Enzymology., Academic Press Inc.: New York and London, in press.
56. Steinlein P, Deiner EM, Leutz A, Beug H (1994) Recombinant murine erythropoietin receptor expressed in avian progenitors mediates terminal erythroid differentiation *in vitro.* Growth Factors 10: 1-16.
57. Greider CW (1994) Mammalian telomere dynamics: healing, fragmentation shortening and stabilization. Curr Opin Genet Dev 4: 203-211.
58. Kim NW, Piatyszek MA, Prowse KR, Harley CB, West MD, Ho PL, Coviello GM, Wright WE, Weinrich SL,Shay JW (1994) Specific association of human telomerase activity with immortal cells and cancer. Science 266: 2011-2015.
59. Nilsson P, Mehle C, Remes K, Roos G (1994) Telomerase activity in vivo in human malignant hematopoietic cells. Oncogene 9: 3043-3048.
60. Mantell LL, Greider CW (1994) Telomerase activity in germline and embryonic cells of Xenopus. EMBO J 13: 3211-3217.
61. Harley CB, Futcher AB, Greider CW (1990) Telomeres shorten during ageing of human fibroblasts. Nature 345: 458-460.

Receptors and Signal Transduction

The Effect of C-raf Antisense Oligonucleotides on Growth Factor-Induced Proliferation of Hematopoietic Cells

J.R. Keller[1], F.W. Ruscetti[2], G. Heidecker[1], D.M. Linnekin[2], U. Rapp[3], J. Troppmair[3], J. Gooya[1], and K. W. Muszynski[1]

[1]Biological Carcinogenesis and Development Program, SAIC Frederick and, [2]Laboratory of Leukocyte Biology-BRMP and, [3]Laboratory of Viral Carcinogenesis, National Cancer Intstitute-Frederick Cancer Research and Development Program, P.O. Box B, Frederick, MD, 21702-1201.

SUMMARY

While it is well established that Raf-1 kinase is activated by phosphorylation in growth factor-dependent hematopoietic cell lines stimulated with a variety of hematopoietic growth factors, little is known about the biological effects of Raf-1 activation on normal hematopoietic cells. Therefore, we examined the requirement for Raf-1 in growth factor-regulated proliferation and differentiation of hematopoietic cells using c-raf antisense oligonucleotide. Raf-1 is required for the proliferation of growth factor dependent cell lines stimulated by IL-2, IL-3, G-CSF, GM-CSF and EPO that bind to the hematopoietin class of receptors. Raf-1 is also required for the proliferation of cell lines stimulated by growth factors that use the tyrosine kinase containing receptor class, including SLF and CSF-1. In addition, Raf-1 is also required for IL-6-, LIF- and OSM-induced proliferation whose receptors share the gp 130 subunit. In contrast to previous results which demonstrated that IL-4 could not activate Raf-l kinase, c-raf antisense oligonucleotides also inhibited IL-4-induced proliferation of T cell and myeloid cell lines. Using normal hematopoietic cells, c-raf antisense oligonucleotides completely suppressed the colony formation of murine hematopoietic progenitors in response to single growth factors, such as IL-3, CSF-1 or GM-CSF. Further, c-raf antisense oligonucleotides inhibited the growth of murine progenitors stimulated with synergistic combinations of growth factors (required for primitive progenitor growth) including two, three and four factor combinations. In comparison to murine hematopoietic cells, c-raf antisense oligonucleotides also inhibited both IL-3 and GM-CSF-induced colony formation of CD 34+ purified human progenitors. In addition, Raf-1 is required for the synergistic response of CD 34+ human bone marrow progenitors to multiple cytokines; however, this effect was only observed when additional antisense oligonucleotides were added to the cultures at day 7 of a 14 day assay. Finally, Raf-1 is required for the synergistic response of human Mo-7e cells and of normal human fetal liver cells to five factor combinations. Thus, Raf-1 is required to transduce growth factor-induced proliferative signals in factor-dependent progenitor cell lines for all known classes of hematopoietic growth factor receptors, and is required for the growth of normal murine and human bone marrow-derived progenitors.

Introduction

The growth and differentiation of hematopoietic cells has been shown to be regulated *in vitro* and *in vivo* by a diverse group of hematopoietic growth factors (HGFs) (1). Hematopoietic growth factors bind to specific growth factor receptors resulting in hetero- and homo-dimerization of receptor complexes, activation of intrinsic or associated tyrosine kinases and phosphatases,and phosphorylation of intracellular substrates initiating a cascade of events which results in DNA synthesis and differentiation (2-4). In this regard, it has been shown that HGFs induce the phosphorylation of a similar but not identical set of intracellular and membrane associated proteins (3-4). One of the proteins identified is Raf-1, a cytoplasmic serine/threonine kinase that is activated by phosphorylation in response to many cytokines including HGFs (5-12). In hematopoietic progenitor cell lines, biochemical studies have shown that Raf-1 kinase activity is activated by phosphorylation in response to Interleukin-2 (IL-2), IL-3, granulocyte-colony stimulating factor (G-CSF), granulocyte/ macrophage- CSF (GM-CSF), CSF-1, erythropoietin (EPO) and steel factor (SLF) (13-17). Since Raf-1 kinase was activated in these cells by HGF stimulation, we sought to determine what the biological consequences of Raf-1 activation were on normal hematopoietic cell growth. Therefore, we studied the effect of inhibiting Raf-1 expression using antisense oligonucleotides (ODNs) on HGF-induced proliferation of hematopoietic progenitor cell lines using ligands which activate several different classes of growth factor receptors, and next, on HGF-induced proliferation and growth of normal murine and human bone marrow progenitor cells.

Results

The Effect of c-raf Antisense Oligonucleotides on Growth Factor-Induced Proliferation of Factor-Dependent Cell Lines.

To evaluate the role of Raf-1 in hematopoietic cell growth we examined the effect of c-raf antisense oligonucleotides on the proliferation of growth factor dependent cell lines. Using standard phosphoramadite chemistry, we synthesized ODNs consisting of an 18 base pair sequence complimentary (antisense) or identical (sense) to the translation start site of the c-raf mRNA gene, and oligomers of nonsense sequences with the same overall base composition. The c-raf specific sequences correspond to codons 1 through 6 and are identical in the murine and human genes (18). To establish the dose response of the ODNs, FDC-P1 cells were incubated with one half the final concentration of ODNs overnight in the absence of IL-3 after which IL-3 and the other half of the ODNs were added. Cell proliferation was measured 48 hours later using ^{3}H-thymidine incorporation assays. FDC-P1 cells and the other cell lines examined show an absolute requirement for the growth factor examined and are not viable after 24-48 hr in medium alone. The c-raf antisense ODNs completely inhibited IL-3-induced proliferation in a dose dependent manner with maximum inhibition (>95%) achieved at doses between 5 and 10 μM while sense and nonsense ODNs showed little or no effect (figure 1). In additional experiments, a 7.5 μM concentration of ODNs was shown to be optimal, a concentration where the cell viability was not significantly

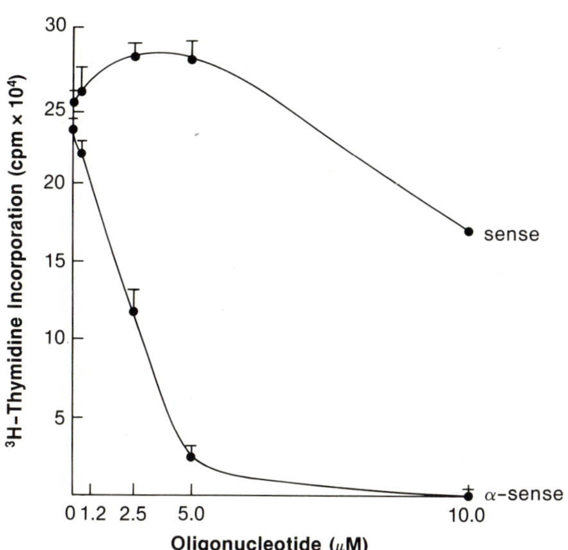

Fig. 1 Dose-dependence of c-raf antisense oligonucleotide inhibition of IL-3-induced proliferation. FDC-P1 cells were treated with c-raf antisense oligonucleotides in the presence of 30 ng/ml IL-3. Effects on proliferation were evaluated by ^{3}H-thymidine incorporation 48 hrs after the addition of IL-3. The results are reported as the mean ± SD of triplicate wells.

affected by treatment with either the sense or antisense ODNs. The effect of c-raf ODNs was further examined on the proliferation of cell lines stimulated with growth factors whose receptors are also members of the hematopoietic receptor family including GM-CSF, G-CSF and EPO. Similar to the effect of c-raf ODNs on IL-3-induced proliferation of FDC-P1 cells, c-raf antisense ODNs completely inhibited (>95%) G-CSF-, GM-CSF- and EPO-induced proliferation of the cell lines, while sense and nonsense ODNs showed no effect (Table 1). In addition, c-raf antisense ODNs but not sense or nonsense ODNs inhibited the proliferation of factor-dependent cell lines in response to SLF and CSF-1 whose receptors are members of the tyrosine kinase receptor family (Table 1). Also, c-raf antisense ODNs inhibited the proliferation of cell lines stimulated by IL-6, LIF and OSM whose receptor complexes include the gp130 subunit and are also a subclass of the hematopoietic receptor family.

Table 1. Summary of the Effects of Antisense c-raf Oligonucleotides on the Proliferation of Growth Factor Dependent Cell Lines

Growth Factor	Cell Line	Origin	Percent Inhibition of ^{3}H-Thymidine Incorporation
IL-3	FDC-P1	Myeloid	> 95%
GM-CSF	DA-3	Myeloid	> 95%
G-CSF	$32D\text{-}C1_3$	Myeloid	> 95%
EPO	HCD-57	Erythroid	> 95%
CSF-1	32D-CSF	Myeloid	> 95%
SLF	FDC-P1	Myeloid	> 95%
IL-6	B-9	B-lymphoid	> 95%
IL-6	DA-la	Myeloid	> 95%
OSM	DA-la	Myeloid	> 95%
LIF	DA-la	Myeloid	> 95%
IL-2	CTLL	T-lymphoid	> 95%
IL-4	CTLL	T-lymphoid	> 95%
IL-4	$32D\text{-}Cl_{23}$	Myeloid	> 95%

Murine cell lines responsive to specific growth factors were treated with c-raf sense and antisense oligonucleotides and were assayed for ^{3}H-thymidine incorporation. C-raf sense oligonucleotides did not effect growth factor stimulated proliferation (≥5%) for any of the cell lines tested.

Thus, antisense c-raf ODNs inhibit the proliferation of growth factor dependent hematopoietic progenitor cell lines regardless of the growth factor used to stimulate proliferation.

Specificity of c-raf Antisense Oligonucleotides.
To demonstrate that the effects of c-raf antisense treatment on HGF-induced proliferation of factor-dependent cell lines was specifically related to loss of c-raf gene expression, cell lysates from FDC-P1 cells were analyzed by western blot for the effect of antisense c-raf ODNs on Raf-1 protein expression. As previously demonstrated and shown here for comparison (19), antisera to Raf-1 specifically detected a 74-kD Raf protein band in whole cell lysates from sense treated but not from antisense treated FDC-P1 cells after 36 hr (Figure 2, Panel A). In comparison, using the same cell lysates and antisera to A-Raf which recognizes a closely related protein also expressed in FDC-P1 cells, demonstrate that A-Raf was not affected by antisense c-raf ODNs (Figure 2, Panel B). Thus, c-raf antisense ODNs specifically inhibit Raf-1 protein expression and they do not affect the expression of A-Raf, a closely related protein.

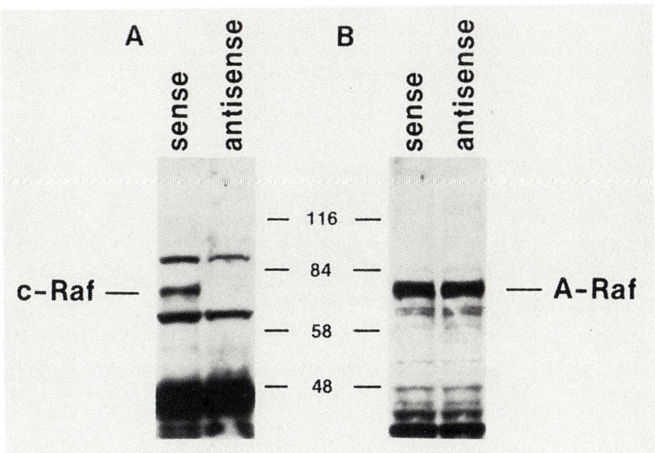

Fig. 2 c-raf antisense oligonucleotides specifically inhibit Raf-1 protein expression. FDC-P1 cells treated with c-raf antisense or sense oligonucleotides were harvested 36 hrs after the addition of IL-3. Whole cell lysates (45 μg) were loaded onto 7.5% gels and separated by SDS-PAGE. Gels were blotted onto nitrocellulose and probed with anti-SP63 antisera (c-raf specific) or anti-A-Raf antisera. Protein was detected using the ECL system and exposed to X-ray film for 5 minutes.

Since c-raf antisense ODNs inhibited the proliferation of factor-dependent cell lines regardless of the factor used to promote proliferation, we examined the effect of ODNs on the proliferation of NFS-60 cells infected with a retrovirus expressing the v-raf gene (v-raf does not abrogate IL-3 dependence). Since the v-raf mRNA does not contain sequences that are complimentary to the c-raf antisense ODNs, antisense ODNs should not affect v-raf expression. The c-raf antisense ODNs inhibited IL-3-induced proliferation of uninfected NFS-60 cells by 81% (Figure 3, Panel A). In contrast, c-raf antisense inhibited v-raf infected NFS-60 cells by only 25% at the same concentration of ODNs (Figure 3, Panel B). Thus, v-raf protein expression can substitute for Raf-1 whose expression in NFS-60 cells is inhibited by c-raf antisense ODNs and indicate that the Raf-1 protein is specifically required for growth factor-induced proliferation.

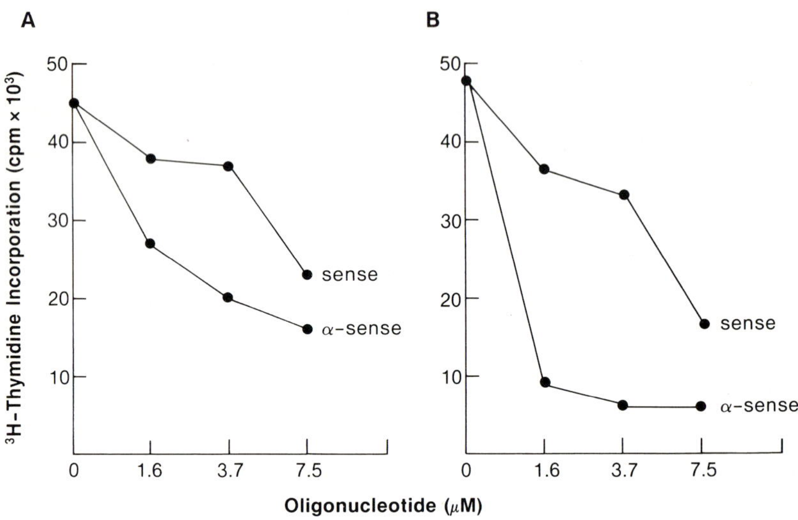

Fig. 3 Expression of v-raf partially blocks c-raf antisense oligonucleotide inhibition of IL-3-induced proliferation. NFS-60 cells infected with retrovirus containing v-raf (panel A) or parental cells infected with control vector (panel B) were treated with c-raf antisense of sense oligos at the concentrations indicated and IL-3 at 30 ng/ml. The effects on proliferation were evaluated by ^{3}H-thymidine incorporation and the results are the mean ± SD of triplicate wells.

The Effect of c-raf Antisense Oligonucleotides on IL-4-Induced Proliferation of Myeloid and Lymphoid Cells.

It has previously been shown that IL-2, but not IL-4, induces the phosphorylation of Raf-1 and activates Raf-1 kinase activity in T cells (13). In contrast, we found that c-raf antisense ODNs inhibited IL-4 induced proliferation of two myeloid cell lines and also inhibited IL-4-induced proliferation of the T cell line CTLL-2 indicating that Raf-1 expression is required for IL-4-induced proliferation of both myeloid and lymphoid cell lines (Table 1). To resolve the discrepancy between the previous biochemical data and our results using c-raf antisense ODNs, we examined the effect of IL-4 on Raf-1 activation in FDC-P1 cells using FDC-P1 cells maintained in IL-4. Raf-1 kinase activity was activated in FDC-P1 cell lysates within 5 to 10 minutes after IL-4 stimulation using myelin basic protein and MEK as *in vitro* kinase substrates for Raf-1 (data not shown). Furthermore, increased Raf-1 kinase activity directly correlated with a phosphorylation induced shift in the mobility of the Raf-1 on SDS-PAGE gels. Thus, Raf-1 is a component of IL-4-activated signal transduction pathways in myeloid cells.

The Effect of c-raf Antisense Oligonucleotides on Hematopoietic Growth Factor-Induced Colony Formation of Normal Human and Mouse Bone Marrow Cells.

To determine the effect of inhibiting Raf-1 gene expression on the growth of normal bone marrow cells, the effect of c-raf antisense ODNs on HGF-induced colony formation of murine and human bone marrow cells (BMCs) was examined. Murine and human BMCs were partially purified (lineage negative (Lin^{neg}) for murine and CD-34+ for human) to enrich for hematopoietic progenitors (20,21,22). Antisense c-raf ODNs inhibited >95% of the colony formation of Lin^{neg} murine BMCs stimulated by single growth factors (IL-3, GM-CSF and CSF-1) that promote the growth of committed bone marrow progenitors (Table 2) (23). Also, antisense c-raf ODNs inhibited the growth of Lin^{neg} BMCs stimulated by synergistic combinations of HGFs (IL-3 +

Table 2. Effect of Antisense Oligonucleotides on the Colony Formation of Purified Human and Mouse Bone Marrow Progenitor Cells

	Colony Forming Units-Culture (% Inhibition)	
Growth Factor	Murine Lin^{neg}	Human CD-34+
IL-3	> 95%	> 95%
GM-CSF	> 95%	> 95%
CSF-1	> 95%	N.A.
IL-3 + SLF	> 95%	40-50%
GM-CSF + SLF	> 95%	40-50% (14 day)
		70-80% (14 day + pulse)
		> 95% (7 day)

Normal murine and human bone marrow cells were purified as indicated in the text. Bone marrow cells were treated with oligonucleotides overnight in the presence of growth factors. The cells were plated in soft agar colony assays and scored for colony formation on day 7 for murine Lin^{neg} cells and on day 14 for CD 34+ cells unless otherwise indicated in the text. Treatment with sense oligonucleotides did not effect growth factor-stimulated colony formation >5% in the assays reported above.

SLF or GM-CSF + SLF) that promote the growth of more primitive progenitor cells (24-25). In comparison, similar to Lin^{neg} murine BMCs, antisense c-raf inhibited >95% of the colony formation of human CD-34+ cells stimulated by either GM-CSF or IL-3. In contrast, c-raf antisense ODNs only inhibited 40-50% of the colony formation of CD-34+ cells stimulated by synergistic combinations of HGFs. Unlike murine BMCs which gives rise to colonies after 7 days in culture, optimal human bone marrow colony formation requires a 14 day incubation period. Furthermore, human BMCs contain committed progenitors which give rise to colonies on day 7 while the more primitive progenitors give rise to colonies after 14 days. Comparing the effects of c-raf antisense ODNs on day 7 versus day 14 colony formation of CD-34+ cells stimulated with GM-CSF plus SLF showed that c-raf antisense inhibited 90-95% of day 7 colonies (Table 2). These results suggest that the more primitive CD-34+ progenitors stimulated by the combination of GM-CSF plus SLF do not require c-raf for their proliferation and growth. However, ODN instability could also account for the observed effects, therefore, we determined whether we could increase the inhibitory effects of c-raf antisense ODNs on day 14 colony formation by pulsing the cultures with an additional dose of ODNs on day 7 (Table 2). In experiments where cultures were pulsed with an additional dose of c-raf antisense ODNs, inhibition of GM-CSF plus SLF-induced day 14 colony formation was further inhibited to 70-75%. Thus, the partial effect of c-raf antisense ODNs on day 14 human colony formation is due, in part, to the degradation of the ODNs.

Effect of c-raf Oligonucleotides on the Proliferation of Cells Stimulated by Synergistic Combinations of Growth Factors in Short-Term Proliferation Assays. Since complete inhibition of day 14 colony formation was not observed in the colony assays regardless of the protocol used, we next examined the growth of progenitors stimulated with synergistic combinations of cytokines in short-term proliferation assays. The growth factor dependent CD-34+ progenitor cell line MO-7e proliferates in response to IL-3 or SLF but maximally proliferates in response to synergistic combinations of cytokines. The proliferation of MO-7e cells was completely inhibited by c-raf antisense ODNs in response to IL-3 or GM-CSF as well as to the synergistic combinations of SLF plus IL-3 or GM-CSF plus IL-3 (Table 3). In addition, since it is difficult to obtain sufficient quantities of CD-34+ cells for use in proliferation assays, we examined the effect of c-raf antisense ODNs on the growth of normal human fetal liver cells in 5 day proliferation assays and in 14 day colony assays. Similar to the results with MO-7e cells, c-raf antisense ODNs completely inhibited the proliferation of fetal liver cells in ^{3}H-thymidine incorporation assays stimulated with single and synergistic combinations of growth factors (Table 3). However, in colony assays, c-raf antisense ODNs only partially inhibited colony formation of fetal liver cells stimulated with single or synergistic combinations of cytokines (Table 3). Thus, while c-raf antisense ODNs completely inhibit the HGF-induced proliferation of fetal liver cells in proliferation assays, they only partially inhibited the colony formation of fetal liver cells.

Table 3. Effect of c-raf Antisense Oligonucleotides on the Proliferative Response of MO7e Cells and Fetal Liver Cells Stimulated with Synergistic Combinations of Cytokines

	Percent Inhibition (%)		
	MO-7e Cells	Fetal Liver Cells	
Growth Factor	^{3}H-Thymidine (3 day)	^{3}H-Thymidine (5 days)	CFU-C (14 day)
IL-3	> 95%	> 95%	74%
IL-3 + SLF	> 95%	> 95%	56%
EPO	N.A.	> 95%	68%
EPO + SLF	N.A.	> 95%	62%
GM-CSF	> 95%	> 95%	72%
GM-CSF + SLF	> 95%	> 95%	72%

Growth factor dependent MO-7e cells or freshly harvested human fetal liver cells were treated with sense or antisense oligonucleotides in the presence of the indicated growth factors. MO-7e cells and fetal liver cells were assayed for ^{3}H-Thymidine incorporation after 96 hrs. and 5 days respectively after the addition of growth factors. The assay for colony formation (CFU-c) of fetal liver was the same as for human bone marrow described in table 2.

Discussion

To determine the requirement for Raf-1 in the proliferation of hematopoietic cells, we examined the effect of c-raf antisense ODNs on the growth of myeloid cell lines and normal bone marrow cells. Treatment of factor-dependent cell lines with c-raf antisense ODNs inhibited IL-2, IL-4, IL-3, GM-CSF, EPO, G-CSF, CSF-1, SLF, LIF, IL-6, and OSM stimulated proliferation. This effect was specifically related to the loss of c-raf gene expression. Thus, Raf-1 is required for the proliferation of hematopoietic cell lines induced by a wide range of HGFs that utilize both hematopoietin and tyrosine kinase receptor classes.

Similar to the effect of c-raf antisense ODNs on the proliferation of factor-dependent progenitor cell lines, c-raf antisense ODNs inhibited the growth of partially purified mouse (Lin^{neg}) and human (CD-34+) bone marrow progenitor cells in colony assays in response to IL-3, GM-CSF and CSF-1 which promote the growth of the most committed progenitor cells. Furthermore, c-raf antisense ODNs inhibited the growth of murine Lin^{neg} cells in response to synergistic combinations of HGFs that stimulate the growth of the more primitive progenitor cells. In contrast, c-raf antisense ODNs only partially inhibited the growth of CD-34+ cells in response to synergistic combinations of cytokines. The partial inhibition observed with human progenitor cells could be explained by a differential sensitivity of committed versus more primitive progenitors to the c-raf antisense ODNs, or that ODN instability could result in partial effects in a 14 day colony formation assay. In this regard, a second dose of ODNs at day 7 further inhibited day 14 colony formation of CD-34+ cells stimulated with the combination of GM-CSF plus SLF suggesting that degradation of the ODNs might account for the partial effects. Furthermore, c-raf antisense ODNs completely inhibited the proliferation of MO-7e cells (3 day assay) and fetal liver cells (5 day assay) in response to synergistic combinations of cytokines in ^{3}H-thymidine incorporation assays. In comparison, 14 day colony formation of fetal liver cells in parallel assays was only partially inhibited. Taken together, partial rather than complete inhibition of HGF-induced colony formation of human cells by c-raf antisense ODNs is most likely a result of ODN instability in the colony assays. Furthermore, any study that uses antisense ODNs with human bone marrow progenitors in a 14 colony assay should be interpreted in view of potential degradation of ODNs over the length of the assay.

References

1. Metcalf D, (1984) The Hematopoietic Colony-Stimulating Factor. Elsevier, Amsterdam. pp 493 .
2. Ullrich A, Schlessinger J (1990). Signal Trnsduction by receptors with tyrosine kinase activity. Cell 61:203-212.
3. Isfort RJ, Ihle J (1990) Multiple hematopoietic growth factors signal through tyrosine phosphorylation. Growth Factors 2:213-220.
4. Miyajima A, Kitamura T, Harada N, Yokoto T, Arai K (1992) Cytokine receptors and signal transduction. pp 295-331 (Annu. Rev. Immunol. vol 10)
5. Miyashita T, Hovey L, Torigoe T, Krajewsky S, Troppmair J, Rapp UR, Reed JC (1993) Novel form of oncogene cooperation: synergistic suppression of apoptosis by combination of bcl-2 and raf oncogenes. Oncogene 9:2751-2756.
6. Morrison DK, Kaplan DR, Rapp U, Roberts TM (1988) Signal Transduction from membrane to cytoplasm: growth factors and membrane-bound oncogene products increase Raf-1 phosphorylation and associated protein kinase activity. Proc. Natl. Acad. Sci., USA 85:8855-8859.
7. Kovacina KS, Yonezawa K, Brautigan DL, Tonks NK, Rapp UR, Roth RA (1990) Insulin acitvates the kinase activity of the Raf-1 proto-oncogene by increasing its serine phosphorylation. J. Biol. Chem. 265: 12115-12118.

8. App H, Hazan R, Zilbertstein A, Ullrich A, Schlessinger J, Rapp UR (1991) Epidermal growth factor (EGF) stimulates association and kinase activity of Raf-1 with the EGF receptor. Mol. Cell. Biol. 11:913-919.
9. Kolch W, Heidecker G, Lloyd P, Rapp U (1991) Raf-1 protein kinase is required for growth of induced NIH-3T3 cells. Nature (Lond.) 349:426-428.
10. Wood KW, Sarnecki C, Roberts TM, Blenis J (1992) ras mediates nerve growth factor receptor modulation of three signal-transducing protein kinases: MAP kinase, raf-1, and RSK. Cell 68:1041-1050.
11. Sozeri O, Vollmer K, Liyanage M, Frith D, Kour G, Mark GE, Stabel S (1992) Activation of the c-raf protein kinase by protein kinase C phosphorylation.Oncogene 7:2259-2262.
12. Kyriakis JM, App H, Zhang X, Banerjee P, Brautigan DL, Rapp UR, Avruch J (1992) Raf-1 activates MAP kinase-kinase. Nature (Lond.) 358:417-121.
13 Turner B, Rapp U, App H, Greene M, Dobashi K, Reed J (1991) Interleukin 2 induces tyrosine phosphorylation and activation of p72-74 Raf-1 kinase in a T-cell line. Proc. Natl. Acad. Sci. USA. 88:1227-1231.
14. Miyazawa K, Hendrie PC, Mantel C, Wood K, Ashman LK, Broxmeyer HE (1991) Comparative analysis of signaling pathways between mast cell growth factor (c-kit ligand) and granulocyte-macrophage colony-stimulating factor in a human factor-dependent myeloid cell line involves phosphorylation of Raf-1, GTPase-activating protein and mitogen-activated protein kinase. Exp. Hematol. 19:1110-1123.
15. Carroll MP, Spivak JL, McMahon M, Weich N, Rapp UR, May WS (1991) Erythropoietin induces Raf-1 activation and Raf-1 is required for erythropoietin-mediated proliferation. J. Biol. Chem. 266:14964-14969.
16. Carroll MP, Clark-Lewis I, Rapp UR, May WS (1990) Interleukin-3 and granulocyte-macrophage colony-stimulating factor mediate rapid phosphorylation and activation of cytostolic-Raf. J. Biol. Chem. 265:19812-19817.
17. Baccarini M, Sabitini DM, App H, Rapp UR, Stanley ER (1990) Colony Stimulating factor-1 (CSF-1) stimulates temperature dependent phosphorylation and activation of the Raf-1 proto-oncogene product. EMBO J. 9:3649-3657.
18. Rapp UR, Heidecker G, Huleihel M, Cleveland JL, Choi WC, Pawson T, Ohle JN, Anderson WB (1988) Raf family serine/threonine protein kinases in mitogen signal transduction. Proc. of Cold Spring Harbor Symp. on quan. Biol: Signal Transduction 53:173-188.
19. Muszynski KW, Ruscetti FW, Heidecker G, Rapp U, Troppmair J, Gooya JM, Keller JR (1995) Raf-1 protein is required for growth factor-induced proliferation of hematopoietic cells. J. Exp. Med. 181:2189-2199.
20. Spangrude GJ, Heimfeld S, Weissman IL (1988) Purification and characterization of mouse hematopoietic stem cells. Science 241:58-62.
21. Andrews RG, Singer JW, Bernstein ID (1989) Precursors of colony-forming cells in humans can be distinguished from colony-forming cells by expression of the CD33 and CD34 antigens and light scatter properties. J. Exp. Med. 169:1721-1731.
22. Tindle RW, Nichols RAB, Chan L, Campana D, Catovsky D, Birnie GD (1985) A novel monoclonal antibody BL-3C5 recognizes myeloblasts and non-B and T lymphoblasts in acute leukaemias and CGL blast crises, and react with inmature cells in normal bone marrow. Leukemia Res. 9:1-9.
23. Metcalf D (1987) The molecular control of normal and leukaemic granulocytes and macrophages. pp 389 (Proc R Soc Lond vol 230)
24. Ruscetti FW, Dubois CM, Jacobsen SEW, Keller JR (1991) Transforming growth factor B and interleukin-1, in Lord B, Dexter TM (eds): Growth factors in haemopoiesis. W.B. Saunders Company, London
25. Ogawa M (1993) Differentiation and proliferation of hematopoietic stem cells. Blood 81:2844-2849.

Protein Kinase C-δ , an Important Signaling Molecule in the Platelet-Derived Growth Factor β Receptor Pathway

W. Li and J. H. Pierce
Laboratory of Cellular and Molecular Biology, National Cancer Institute, Bethesda, MD 20892

Introduction

The protein kinase C (PKC) family comprises of a group of serine/threonine kinases that are thought to play a central role in cell growth, differentiation and cellular transformation (for reviews see [1, 2]). To date, more than 10 different PKC isoenzymes have been identified. The differential expression pattern of these isoenzymes in different tissues and cell types implicates that they may be involved in different signaling pathways. This prompted us to individually overexpress different PKC isoenzymes (α, β, δ, ε, ζ, and η) in an interleukin-3 (IL-3)-dependent murine myeloid progenitor cell line, 32D, in order to investigate their possible involvement in mitogenic and differentiation pathways of hematopoietic cells (3). 32D cells express low levels of PKC-δ, -α and -η, as detected by immunoblot analysis using isoenzyme-specific antibodies (3). After overexpression of different PKCs in 32D cells, none of these transfectants demonstrated any morphological changes in the presence of IL-3, and all remained growth factor dependent. However, treatment of 32D cells overexpressing either PKC-α or PKC-δ (32D/PKC-δ) with 12-O-tetradecanoylphorbol-13-acetate (TPA) resulted in pronounced monocytic differentiation, while the parental 32D cells and other PKC isoenzyme transfectants did not readily undergo monocytic differentiation in response to TPA stimulation (3). These data indicate that PKC-α and PKC-δ play a pivotal role in monocytic differentiation and that the activation of certain substrates involved in differentiation can only occur when PKC-α and PKC-δ are activated in the 32D cell system.

Platelet-derived growth factor (PDGF) is a growth factor for cells of connective tissue origin. Two related receptor molecules, designated as α and β PDGF receptor (PDGFR), have been isolated and both bind PDGF-BB with high affinity (for review see [4]). PDGF has been linked to many physiological and pathological processes, such as wound repair, atherosclerosis, and neoplasia (for review see [5]). 32D cells do not endogenously express either PDGF receptor (6). When human PDGF-βR was overexpressed in 32D cells (32D/PDGF-βR), PDGF-BB was able to elicit many cellular responses including receptor autophosphorylation, activation of intracellular enzymes such as phospholipase C-γ (PLC-γ) and phosphatidylinositol (PI)-3 kinase, Ca^{2+} mobilization, and hydrolysis of PI and phosphatidylcholine (6). Hydrolysis of these phospholipids results in increased levels of diacylglycerol (DAG) (2), the native activator of PKC which may induce mitogenic signals in these transfected cells in response to PDGF stimulation.

Although many experiments have been performed to attempt to dissect the role played by PKC in PDGF-βR signaling pathway (7, 8, 9), the isoenzymes involved have not been fully investigated. We have utilized 32D/PDGF-βR and NIH-3T3 cells to elucidate the role of PKC-δ in signal transduction induced by this receptor tyrosine kinase. Our findings demonstrate that PKC-δ undergoes ligand-dependent translocation from the cytosol to the membrane, tyrosine phosphorylation, and activation in response to PDGF exposure in both 32D and NIH-3T3 cell systems. PKC-δ associates with PDGF-βR in PDGF-BB-dependent manner. Moreover, activation of PKC-δ through PDGF stimulation of the reconstituted PDGF-βR signaling pathway is involved in mediating 32D monocytic differentiation.

PKC-δ Is Phosphorylated on Tyrosine Residue(S) upon PDGF-βR Activation. We recently reported that PKC-δ becomes phosphorylated on tyrosine in response to TPA stimulation of either 32D or NIH-3T3 cells transfected with PKC-δ (10). In order to determine whether tyrosine phosphorylation of PKC-δ would occur in response to activation of a receptor tyrosine kinase, we utilized 32D cells that had been previously transfected with expression vectors containing either PDGF-αR (32D/PDGF-αR) or PDGF-βR (32D/PDGF-βR) (6). As shown in Fig. 1A, PDGF-BB stimulation of either 32D/PDGF-αR or 32D/PDGF-βR cells resulted in tyrosine phosphorylation of endogenous PKC-δ as detected by immunoprecipitation of cell lysates with anti-PKC-δ serum and subsequent immunoblot analysis with anti-phosphotyrosine antibody (anti-pTyr). TPA stimulation also caused tyrosine phosphorylation of endogenous PKC-δ in both 32D/PDGF-αR and 32D/PDGF-βR cells (Fig. 1A).

To amplify the detection of tyrosine-phosphorylated PKC-δ, a PKC-δ expression vector was introduced in each 32D/PDGFR line and these transfectants were designated as 32D/PDGF-βR/PKC-δ and 32D/PDGF-αR/PKC-δ. As shown in Fig. 1B, the levels of PKC-δ were increased by 10.5 fold in the 32D/PDGF-αR/PKC-δ and 10.2 fold in the 32D/PDGF-βR/PKC-δ transfectants, as determined by immunoblot analysis with anti-PKC-δ serum. After PDGF-BB stimulation, tyrosine phosphorylation of PKC-δ was increased in 32D/PDGF-βR/PKC-δ and 32D/PDGF-αR/PKC-δ cells when compared to 32D/PDGF-βR and 32D/PDGF-αR cells, respectively (Fig. 1A). The level of tyrosine phosphorylation of PKC-δ induced by PDGF-αR activation was 6.4 fold lower for endogenous PKC-δ and 2.6 fold lower for overexpressed PKC-δ than that caused by PDGF-βR activation. However, the numbers of PDGF-αR and PDGF-βR binding sites per cell expressed on the respective transfectants were previously shown to be similar as determined by Scatchard analysis (6). Moreover, the levels of PKC-δ protein (Fig. 1B) and TPA-induced tyrosine phosphorylation of PKC-δ (Fig. 1A) were comparable between 32D/PDGF-αR and 32D/PDGF-βR and between 32D/PDGF-βR/PKC-δ and 32D/PDGF-αR/PKC-δ transfectants. Thus, the difference in levels of tyrosine-phosphorylated PKC-δ induced by activation of PDGF-αR and PDGF-βR may reflect an inferior capacity of PDGF-αR to directly tyrosine phosphorylate PKC-δ or to couple with downstream tyrosine kinases which mediate PKC-δ phosphorylation.

PKC-δ tyrosine phosphorylation was also detected in NIH-3T3/PKC-δ cells after PDGF-BB, PDGF-AA or TPA stimulation (Fig. 1C). PDGF-BB induced the strongest PKC-δ tyrosine phosphorylation among the three agonists (Fig. 1C). Since NIH-3T3 cells express both PDGF-αR and -βR and PDGF-BB binds to

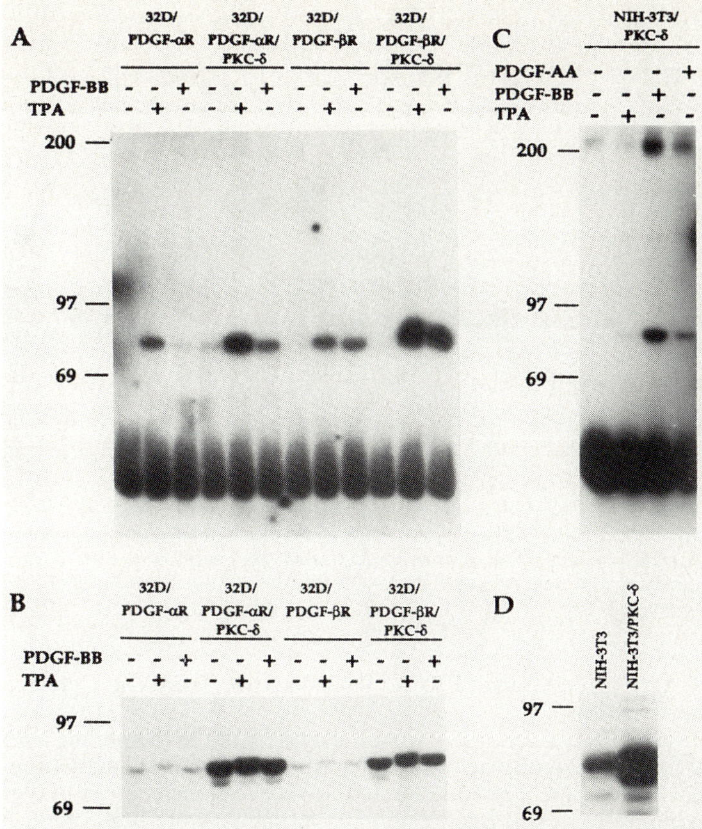

Fig. 1. PDGFR activation results in PKC-δ tyrosine phosphorylation in 32D and NIH-3T3 cell systems. (A) 32D/PDGF-αR, 32D/PDGF-αR/PKC-δ, 32D/PDGF-βR and 32D/PDGF-βR/PKC-δ cells were treated with PDGF-BB or TPA. Cell lysates were immunoprecipitated with anti-PKC-δ serum. Proteins were separated by SDS-PAGE, transferred, and immunoblotted with anti-pTyr. (B) Proteins from 32D cell lysates were separated by SDS-PAGE, transferred, and immunoblotted with anti-PKC-δ serum. (C) NIH-3T3/PKC-δ cells were stimulated with PDGF-BB, PDGF-AA or TPA. Cells were stimulated with 100 ng/ml of each agonist for 10 min in all experiments unless otherwise specified. Immunoprecipitation and immunoblot analysis were performed as in (A). (D) Direct immunoblot analysis of proteins from untreated NIH-3T3 and NIH-3T3/PKC-δ cell lysates was performed using anti-PKC-δ serum. Marker proteins are given in KDa.

both receptors, it is likely that activation of both receptors contributed to PKC-δ tyrosine phosphorylation. PKC-δ overexpression in NIH-3T3 cells was confirmed by direct immunoblot analysis with anti-PKC-δ serum (Fig. 1D).

To further investigate the role played by PDGF-βR in inducing PKC-δ phosphorylation, we immunoprecipitated [^{32}P]-labeled NIH-3T3/PKC-δ cell lysates with either anti-PKC-δ or anti-pTyr. As shown in Fig. 2, PKC-δ phosphorylation levels were strikingly increased after either PDGF-BB or TPA stimulation after immunoprecipitation with either antibody. Similar results were also obtained with 32D/PDGF-βR/PKC-δ cells (data not shown). Taken together, the above results demonstrate that PKC-δ phosphorylation, that presumably reflects its activation (11), occurs after PDGF-βR and -αR activation in both 32D and NIH-3T3 cell systems.

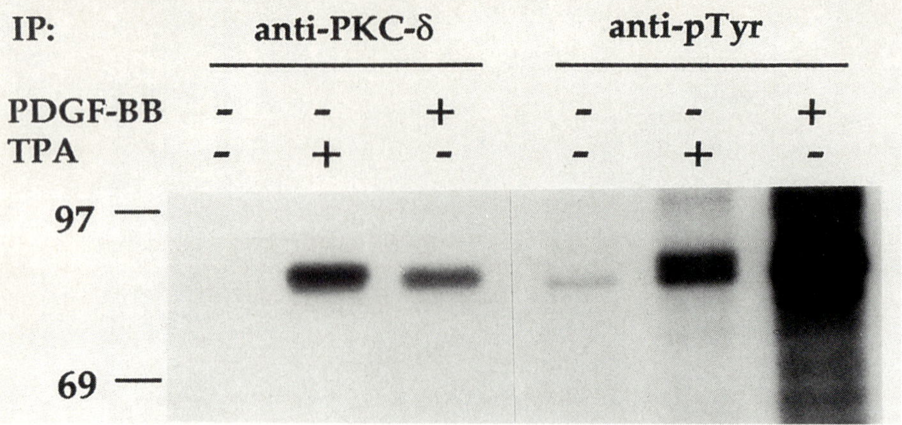

Fig. 2. PKC-δ is phosphorylated *in vivo* after PDGF-BB stimulation. NIH-3T3/PKC-δ cells were serum starved, labeled with [^{32}P] orthophosphate, and either untreated or stimulated with PDGF-BB or TPA. Cell lysates were immunoprecipitated (IP) with anti-PKC-δ or anti-pTyr. Immunoprecipitates were resolved by SDS-PAGE and the dried gel was autoradiographed. The markers are given in KDa.

PDGF-βR Activation Causes PKC-δ Translocation from the Cytosolic to the Membrane Fraction.

Translocation of PKC isoenzymes from the cytosol to the membrane after phorbol ester treatment is generally accepted as an indicator of PKC activation (12). To determine whether PDGF stimulation led to PKC-δ translocation in the 32D and NIH-3T3 cell systems, cell fractionation experiments were performed. As shown in Fig. 3A and B, PDGF-BB stimulation initiated translocation of both endogenous (32D/PDGF-βR, 15% of the total amount) and overexpressed (32D/PDGF-βR/PKC-δ, 11% of the total amount) PKC-δ from the cytosol to the membrane as determined by immunoblot analysis with anti-PKC-δ serum. Time kinetics of PKC-δ translocation in 32D/PDGF-βR/PKC-δ cells demonstrated that translocation was maximal at 0.5 min after PDGF-BB stimulation (Fig. 3B). TPA induced 8-11 fold more PKC-δ translocation than that induced by PDGF-BB at the different time points (Fig. 3B). PDGF-BB also induced PKC-δ translocation in NIH-3T3/PKC-δ cells (Fig. 3C). Although some PKC-δ was found in the membrane fraction prior to stimulation, translocation after PDGF-BB stimulation was more evident in NIH-3T3/PKC-δ (25% of the total amount after subtracting the amount of PKC-δ already present in the membrane fraction before stimulation) than in 32D/PDGF-βR/PKC-δ cells. This may reflect PDGF-αR involvement after PDGF-BB treatment of NIH-3T3/PKC-δ cells. Thus, PDGF-βR activation results in PKC-δ translocation in two different cell systems.

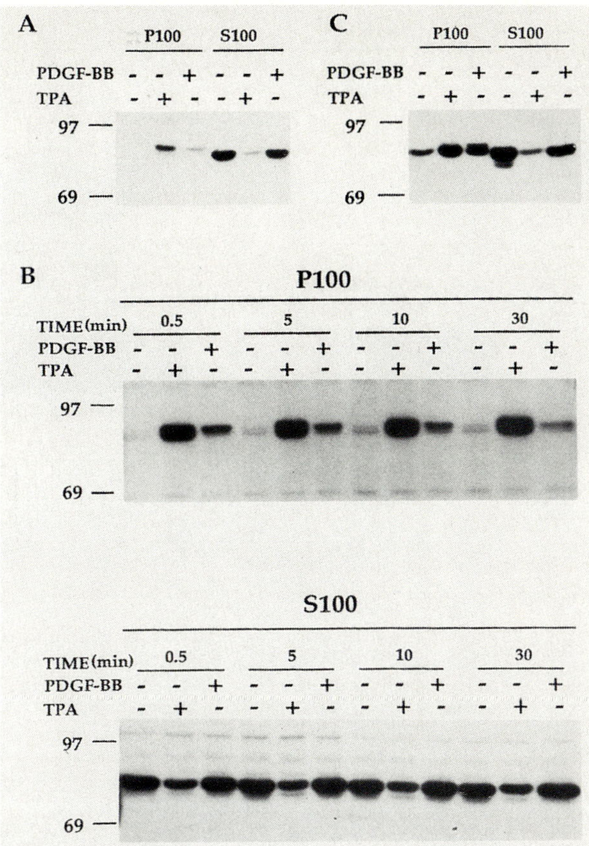

Fig. 3. PDGF-βR activation leads to PKC-δ translocation from the cytosolic to the membrane fraction. (A) 32D/PDGF-βR cells, (B) 32D/PDGF-βR/PKC-δ cells or (C) NIH-3T3/PKC-δ cells were treated with PDGF-BB or TPA for 10 min (A and C) or for 0.5, 5, 10, 30 min (B). The P100 fraction was separated from the S100 fraction according the protocol published (17). Equal amounts of protein from each cell fraction were separated by SDS-PAGE and transferred proteins were immunoblotted with anti-PKC-δ serum. Marker proteins are indicated in KDa.

Tyrosine-Phosphorylated PKC-δ Can Be Detected Only in the Membrane Fraction.

Since TPA and PDGF-BB were both able to induce tyrosine phosphorylation and translocation of PKC-δ, the localization of tyrosine-phosphorylated PKC-δ was investigated. As shown in Fig. 4A and C, tyrosine-phosphorylated PKC-δ could be observed only in the membrane fraction after PDGF-BB or TPA treatment of 32D/PDGF-βR or NIH-3T3/PKC-δ cells as determined by immunoprecipitation with anti-PKC-δ serum followed by immunoblot analysis with anti-pTyr. Time kinetics revealed that maximal PKC-δ tyrosine phosphorylation in the membrane fraction occurred at 0.5 min after PDGF-BB stimulation of 32D/PDGF-βR/PKC-δ cells (Fig. 4B), whereas no tyrosine phosphorylation of PKC-δ could be detected in the cytosolic fraction at any time point used for stimulation (data not shown). TPA-induced PKC-δ translocation was 8-11 fold higher than that induced by

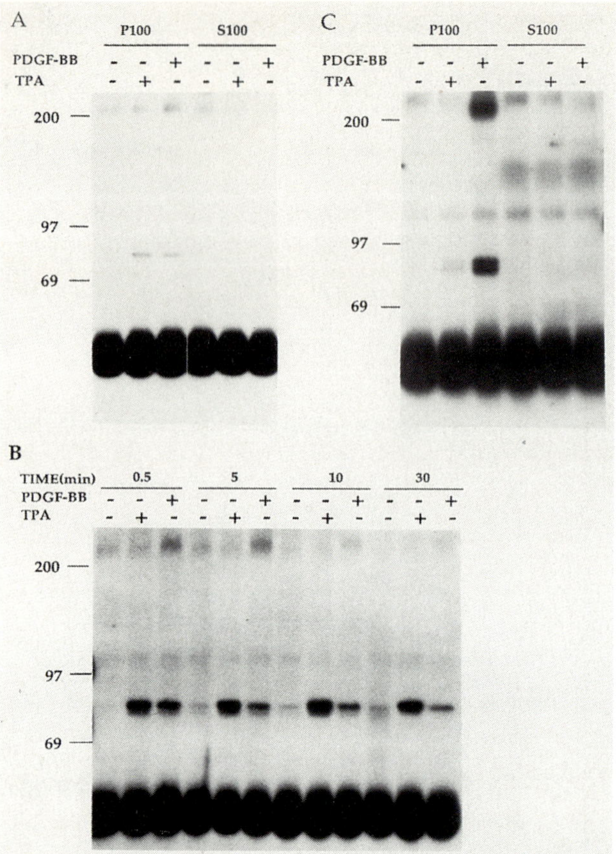

Fig. 4. Tyrosine-phosphorylated PKC-δ can be detected only in the membrane fraction after PDGF-βR activation. (A) 32D/PDGF-βR cells, (B) 32D/PDGF-βR/PKC-δ cells or (C) NIH-3T3/PKC-δ cells were treated with PDGF-BB or TPA for 10 min (A and C) or for 0.5, 5, 10, 30 min (B). The P100 fraction was separated from the S100 fraction as previously described (17). Equal amounts of protein from each cell fraction were immunoprecipitated with anti-PKC-δ serum (only the P100 fraction was included in 4B). The immunoprecipitates were separated by SDS-PAGE and transferred proteins were immunoblotted with anti-pTyr. Marker proteins are given in KDa.

PDGF-BB at different time points (see Fig. 3B). However, tyrosine phosphorylation of PKC-δ generated by TPA stimulation was only 1.5-4 fold higher than that detected after PDGF-BB stimulation (Fig. 4B), suggesting that PDGF-βR might directly phosphorylate PKC-δ. Although anti-PKC-δ serum efficiently immunoprecipitated abundant levels of PKC-δ remaining in the cytosolic fraction after factor treatment as detected by immunoblot analysis with anti-PKC-δ serum (data not shown; also see direct immunoblot analysis with anti-PKC-δ in Fig. 3), tyrosine-phosphorylated PKC-δ could not be detected in this fraction. Thus, tyrosine-phosphorylated PKC-δ was exclusively located in the membrane fraction in both cell systems.

PDGF-BB Stimulation Increases PKC Activity in the Membrane Fraction. To directly address the question of whether PKC activity was increased after its translocation to the membrane, we enriched for PKC-δ from the membrane and cytosolic fractions of 32D/PDGF-βR/PKC-δ or NIH-3T3/PKC-δ cells stimulated with or without agonists by DE 52 ion exchange chromatography for subsequent PKC activity assay. As shown in Table 1, PKC activities in the membrane of 32D/PDGF-βR/PKC-δ cells were increased 1.6 and 2.3 fold after PDGF-BB and TPA stimulation, respectively. Increased PKC activity was also observed in NIH-3T3/PKC-δ cells after PDGF-BB or TPA stimulation. Higher kinase activities were found after TPA stimulation than after PDGF-BB stimulation in both cell types (Table 1). The simultaneous reduction of cytosolic PKC-δ activity was also observed in both transfectants (Table 1). Thus, PDGF stimulation increases membrane PKC activity in two cell systems.

Table 1. PDGFR Activation Results in Increased PKC Activity in the Membrane Fraction of 32D/PDGF-βR/PKC-δ and NIH-3T3/PKC-δ Cells

Cell Line	Stimulation	PKC Activity * (fold increase)	
		P100	S100
32D/PDGF-βR/PKC-δ	None	7.7	150.7
	TPA	18.0 (2.3)	118.3
	PDGF-BB	12.6 (1.6)	149.0
NIH-3T3/PKC-δ	None	23.1	539.7
	TPA	63.8 (2.8)	164.9
	PDGF-BB	42.9 (1.9)	438.7

* PKC activity was measured as described (17) and was represented as 10^3 cpm/mg protein/min. The background phosphorylation (without kinase source, 414 cpm) was subtracted from all samples and results in Table 1 represent the mean values of three individual samples. The variation was less than 5% of the mean value.

Activated PDGF-βR Phosphorylates PKC-δ on Tyrosine *in vitro* and Directly Associates with PKC-δ *in vivo*.
To further confirm that PKC-δ is a direct substrate of PDGF-βR, we attempted to phosphorylate purified PKC-δ *in vitro* by purified PDGF-βR, both derived from baculovirus expression system. As shown in Fig. 5A, the activated PDGF-βR was able to efficiently phosphorylate PKC-δ as determined by immunoblot analysis with anti-pTyr. Autophosphorylation of PDGF-βR was also observed. The subsequent PKC-δ kinase assay utilizing a PKC-δ pseudosubstrate region-derived peptide as the substrate demonstrated that PKC-δ activity was increased by 18 % after PKC-δ was phosphorylated by PDGF-βR (10).

Many studies have shown that a tyrosine kinase which phosphorylates a substrate can be found physically associated with it (for review see [13]). Since PDGF-βR was shown to phosphorylate PKC-δ *in vitro* (Fig. 5A), we investigated whether PKC-δ associated directly with PDGF-βR *in vivo* by performing coprecipitation experiments. As shown in Fig. 5B, PKC-δ could be detected after anti-PDGF-βR immunoprecipitation followed by immunoblot analysis with anti-PKC-δ serum. Immunoprecipitation with anti-PKC-δ serum followed by immunoblot analysis with anti-PDGF-βR serum also revealed that these two proteins were coprecipitable (data not shown). Although some association could be detected before PDGF-BB stimulation, a 1 - 2 fold increase in association was observed after PDGF-BB stimulation, but not after TPA stimulation. Taken together, these

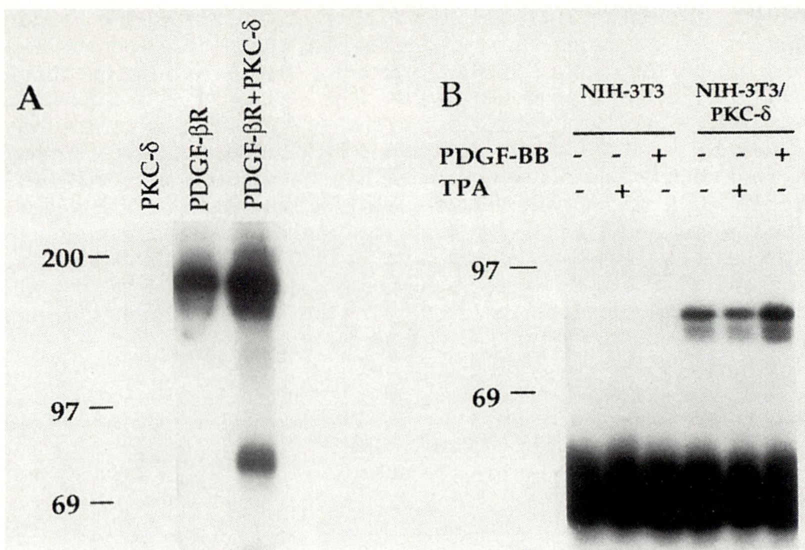

Fig. 5. PKC-δ is phosphorylated on tyrosine by purified PDGF-βR *in vitro* and associates with activated PDGF-βR *in vivo*. (A) PDGF-βR was incubated with purified PKC-δ in an *in vitro* tyrosine kinase assay (10). The reaction was stopped by addition of sample buffer and was subjected to SDS-PAGE. Transferred proteins were immunoblotted with anti-pTyr. (B) NIH-3T3 cells or NIH-3T3/PKC-δ cells were stimulated with TPA or PDGF-BB for 10 min. Equal amounts of protein were immunoprecipitated with anti-PDGF-βR serum. Transferred proteins were immunoblotted with anti-PKC-δ serum . Marker proteins are given in KDa.

results suggest that PKC-δ is a direct substrate of PDGF-βR and associates with PDGF-βR in a ligand-dependent manner.

PDGF-βR Activation Leads to Monocytic Differentiation of 32D Cells when PKC-δ Is Sufficiently Expressed and Activated.

Having demonstrated that PDGF-βR activation induced PKC-δ translocation, tyrosine phosphorylation, association, and activation, we were interested to determine if PDGF stimulation would affect 32D cell differentiation. Therefore, 32D/PDGF-βR and 32D/PDGF-βR/PKC-δ cells were incubated with 100 ng/ml of PDGF-BB or TPA overnight and were subjected to flow cytometry using antibodies directed against two cell surface antigens, Mac-1 and FcγRII, which are known to increase during monocytic differentiation (14, 15). As shown in Fig. 6, PDGF-βR activation led to obvious increases in Mac-1 (Fig. 6A) and FcγRII (Fig. 6B) expression on the surface of 32D/PDGF-βR/PKC-δ cells, whereas only negligible increases in antigen expression were observed in 32D/PDGF-βR cells under the same conditions (Fig. 6A and B). As expected, TPA stimulation of 32D/PDGF-βR/PKC-δ cells also resulted in increased mean fluorescence intensities for Mac-1 and FcγRII (Fig. 6A and B). After PDGF treatment, 32D/PDGF-βR/PKC-δ cells became strongly adherent and displayed features common to mature macrophages such as a decreased nucleus to cytoplasm ratio, membrane ruffling and cytoplasmic vacuolation as determined by Wright-Giemsa staining, whereas less apparent changes were observed for 32D/PDGF-βR cells

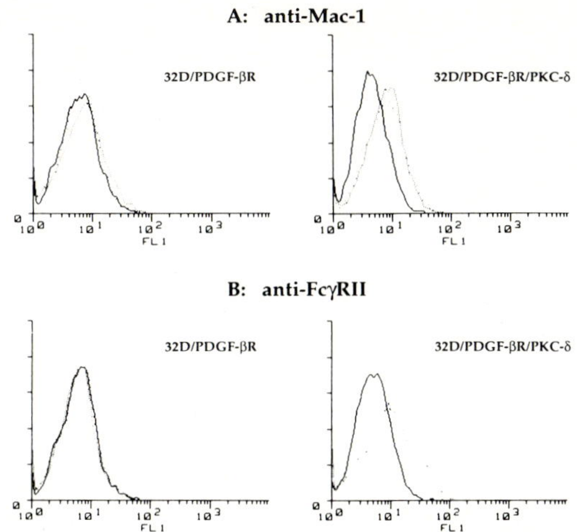

Fig. 6. Overnight PDGF-BB treatment increases expression of monocytic differentiation markers on the 32D/PDGF-βR/PKC-δ transfectants. 32D/PDGF-βR or 32D/PDGF-βR/PKC-δ cells were untreated (——) or exposed to either PDGF-BB (·········) or TPA (· · · ·) overnight. The cells were incubated with (A) FITC-conjugated anti-Mac-1 or (B) FITC-conjugated anti-FcγRII and subjected to flow cytometry. The X axis represents log fluorescence intensity (FL1 represents FITC fluorescence) and the Y axis represents relative cell number.

(data not shown). These results indicate that PDGF is able to induce monocytic differentiation as long as PKC-δ is sufficiently expressed and activated.

Summary

In the present study, we have investigated the consequences of PDGFR-mediated signaling on PKC-δ activation in both murine hematopoietic and fibroblast cell systems. Our results demonstrate that PDGF-βR activation leads to activation, tyrosine phosphorylation, translocation of PKC-δ, and that association of PDGF-βR with PKC-δ takes place in a ligand-dependent manner. Although the overexpression of PKC-δ in 32D/PDGF-βR cells or NIH-3T3 cells did not affect mitogenic responses induced by PDGF-BB stimulation (W. Li and J. H. Pierce, unpublished observation), PKC-δ appeared to function in cell differentiation induced by PDGF in hematopoietic cells. It was recently reported that human monocytes derived from blood were able to differentiate towards macrophages in the presence of PDGF-BB (16). During the induction period, the human PDGF-βR was upregulated and this upregulation was considered to be PKC dependent. The authors also demonstrated that mouse peritoneal macrophages were able to bind PDGF-BB after being elicited by thioglycolate *in vivo*. Our results suggest that PDGF stimulation of 32D/PDGF-βR/PKC-δ cells induced monocytic differentiation of these myeloid precursor cells. It was necessary to increase PKC-δ expression to observe convincing differentiation in response to overnight PDGF stimulation. Interestingly, upregulation of PKC-δ protein and RNA levels was observed after long-term PDGF-BB treatment of 32D/PDGF-βR cells (17). Thus, the development of circulating monocytes into tissue macrophages may be partially due to the induction of PDGF-βR expression, PDGF exposure, and subsequent PKC-δ upregulation and activation. Since 32D cells were also able to undergo monocytic differentiation after activation of either EGF or CSF-1 signaling pathways (18, 19), it will be interesting to investigate the involvement of PKC-δ in the differentiation process initiated by these two molecules. Interestingly, EGF and CSF-1 stimulation lead to increased DAG production (20, 21) and both induce PKC-δ tyrosine phosphorylation (W. Li and J. H. Pierce, unpublished observation). Thus, the 32D system should prove to be a good model to dissect the role played by a single PKC isoenzyme in a specific signal transduction pathway. Currently, we are investigating whether other PKCs are activated by stimulation of the PDGFR signaling pathway and whether this activation leads to enhanced monocytic differentiation. We are particularly interested in PKC-α which can also transduce differentiation signal after its overexpression in 32D cells and activation by TPA (3). However, the high expression pattern of PKC-δ in hematopoietic cells (22) strengthen its possible physiological involvement in differentiation when compared to other PKCs.

In this report, we characterize tyrosine phosphorylation of PKC-δ induced by PDGF-βR activation in two different cell systems. The evidence presented in this study is not sufficient to propose a direct relationship between tyrosine phosphorylation of PKC-δ and its activation or role in differentiation. Definitive prove that this phosphorylation affects PKC-δ structure, activity, and subsequent biological effects awaits phosphopeptide mapping and site-directed mutagenesis. However, the following observations suggest a positive correlation between tyrosine phosphorylation of PKC-δ and its kinase activity. After PKC-δ was phosphorylated on tyrosine *in vitro* by different tyrosine kinases, its kinase activity was increased towards its substrate derived from the PKC-δ pseudosubstrate region (10). Tyrosine phosphorylation of PKC-δ could be detected only in the membrane fraction where the enzymatic activity was increased (Fig. 4 and Table 1). After PDGF-BB or TPA stimulation, tyrosine phosphorylation of PKC-δ coincided with its autophosphorylation on serine (Fig. 2 and W. Li, unpublished observation) which also reflects the active status of the enzyme (11). No matter how tyrosine phosphorylation affects its kinase activity, our data suggest that PKC-δ tyrosine phosphorylation is an excellent indicator of PKC-δ translocation and activation.

Our data suggest that PKC-δ is a direct substrate of the PDGF-βR *in vivo*. The direct association of PKC-δ with PDGF-βR was detected by coprecipitation assays utilizing a combination of antibodies against PDGF-βR and PKC-δ (Fig. 5). This finding is very interesting because PKC-δ does not possess any SH-2 domain which is utilized by most of tyrosine kinase substrates, such as PLC-γ, p85 of PI-3 kinase, Syp, Nck, Shc, to associate with phosphorylated tyrosine kinase receptors (for review see [23]). The mechanism that directs PKC-δ to associate with PDGF-βR in a ligand-dependent manner will be an important issue to address in the future.

References

1. Huang K-P (1990) Role of protein kinase C in cellular regulation. BioFactors 2:171-178.
2. Nishizuka Y (1992) Intracellular signaling by hydrolysis of phospholipids and activation of protein kinase C. Science 258:607-614.

3. Mischak H, Pierce JH, Goodnight, J, Kazanietz MG, Blumberg PM, Mushinski JF (1993) Phorbol ester-induced myeloid differentiation is medicated by protein kinase C-α and -δ and not protein kinase C-βII, ε, ζ, and η. J Biol Chem 268:20110-20115.
4. Claesson-Welsh L (1994) Platelet-derived growth factor receptor signals. J Biol Chem 51:32023-32026.
5. Ross R, Raines EW, Bowen-Pope DF (1986) The biology of platelet-derived growth factor. Cell 46:155-169.
6. Matsui T, Pierce JH, Fleming TP, Greenberger JS, LaRochelle WJ, Ruggiero M, Aaronson SA (1989) Independent expression of human α or β platelet derived growth factor receptor cDNA in a naive hematopoietic cell leads to functional coupling with mitogenic and chemotactic signaling pathways. Proc Natl Acad Sci USA 86:8314-8318.
7. Coughlin SR, Lee WM, Williams PW, Giels GM, Williams LT (1985) c-myc gene expression is stimulated by agents that activate protein kinase C and does not account for the mitogenic effect of PDGF. Cell 43:243-251.
8. Kihara M, Robinson PJ, Buck SH Dage RC (1989) Mitogenesis by serum and PDGF is independent of PI degradation and PKC in VSMC. Am J Physiol 256:C886-C892.
9. Sharma R, Bhalla R (1993) PDGF-induced mitogenic signaling is not mediated through protein kinase C and c-fos pathway in VSM cells. Am J Physiol 264:C71-C79.
10. Li W, Mischak H, Yu J-C, Wang L-M, Mushinski JF, Heidaran MA, Pierce JH (1994) Tyrosine phosphorylation of protein kinase C-δ in response to its activation. J Biol Chem 269:2349-2352.
11. Mitchell F, Marais ERM, Parker PJ (1989) The phosphorylation of protein kinase C as a potential measure of activation. Biochem J 261:131-136.
12. Kraft AS, Anderson WB (1983) Phorbol esters increase the amount of Ca^{2+}, phospholipid-dependent protein kinase associated with plasma membrane. Nature (London) 301:621-623.
13. Cantley L, Auger CKR, Carpenter C, Duckworth B, Graziani A, Kapeller R, Soltoff S (1991) Oncogenes and signal transductions. Cell 64:281-302.
14. Springer T, Galfre G, Secher DS, Milstein C (1979) Mac-1: a macrophage differentiation antigen identified by monoclonal antibody. Eur J Immunol 9:301-306.
15. Unkless J (1979) Characterization of a monoclonal antibody directed against mouse macrophage and lymphocyte Fc receptors. J Exp Med 150:580-584.
16. Inaba T, Shimano H, Gotoda T, Harada K, Shimada M, Ohsuga J-L, Watanabe Y, Kawamura M, Yazaki Y Yamada N (1993) Expression of platelet-derived growth factor β receptor on human monocyte-derived macrophage and effects on platelet-derived growth factor BB dimer on the cellular function. J Biol Chem 268:24353-24360.
17. Li W, Yu J-C, Michieli P, Beeler JF, Ellmore N, Heidaran MA, Pierce JH (1994) Stimulation of the platelet-derived growth factor β receptor signaling pathway activates protein kinase C-δ. Mol Cell Biol 10:6727-6735.
18. Pierce JH, DiMarco E, Cox GW, Lombardi D, Ruggiero M, Varesio L, Wang L-M, Choudhury GG, Sakaguchi A, DiFiore PP, Aaronson SA (1990) Macrophage-colony-stimulating factor (CSF-1) induces proliferation, chemotaxis, and reversible monocytic differentiation in myeloid progenitor cells transfected with the human c-fms/CSF-1 receptor cDNA. Proc Natl Acad Sci USA 87:5613-5617.
19. Pierce JH, Ruggiero M, Fleming TP, DiFiore PP, Greenberger JS, Varticovski L, Schlessinger J, Rovera G, Aaronson SA (1988) Signal transduction through the EGF receptor transfected in IL-3 dependent cells. Science 239:628-631.
20. Imamura K, Dianoux A, Nakamura T, Kufe D (1990) Colony-stimulating factor 1 activates protein kinase C in human monocytes. EMBO J 9:2423-2429.
21. Pike LJ, Eakes AT (1987) Epidermal growth factor stimulates the production of phosphatidylinositol monophosphate and the breakdown of polyphosphoinositides in A431 cells. J Biol Chem 262:1644-1651.
22. Mischak H, Pierce JH, Goodnight J, Kazanietz MG, Blumberg PM, Mushinski JF (1993) Expression of protein kinase C genes in hemopoietic cells is cell-type- and B cell-differentiation stage specific. J Immunol 147:3981-3987.
23. Pawson T (1995) Protein modules and signalling networks. Nature 373:573-578.

Novel Adapter Proteins that Link the Human GM-CSF Receptor to the Phosphatidylino-sitol 3-Kinase and Shc/Grb2/ras Signaling Pathways

M. Jücker[1] and R. A. Feldman[1,2]
[1]Department of Microbiology and Immunology, University of Maryland School of Medicine, Baltimore, MD 21201; [2]Medical Biotechnology Center, University of Maryland Biotechnology Institute, Baltimore, MD 21201, USA.

Summary

We have used a human GM-CSF-dependent hematopoietic cell line that responds to physiological concentrations of hGM-CSF to analyze a set of signaling events that occur in normal myelopoiesis and whose deregulation may lead to leukemogenesis. Stimulation of these cells with hGM-CSF induced the assembly of multimeric complexes that contained known and novel phosphotyrosyl proteins. One of the new proteins was a major phosphotyrosyl substrate of 76-85 kDa (p80) that was directly associated with the p85 subunit of phosphatidylinositol (PI) 3-kinase through the SH2 domains of p85. p80 also associated with the β subunit of the activated hGM-CSF receptor, and assembly of this complex correlated with activation of PI 3-kinase. A second phosphotyrosyl protein we identified, p140, associated with the Shc and Grb2 adapter proteins by direct binding to a novel phosphotyrosine-interacting domain located at the N-terminus of Shc, and to the SH3 domains of Grb2, respectively. The Shc/p140/Grb2 complex was found to be constitutively activated in acute myeloid leukemia cells, indicating that activation of this pathway may be a necessary step in the development of some leukemias. The p80/p85/PI 3-kinase and the Shc/Grb2/p140 complexes were tightly associated with Src family kinases, which were prime candidates for phosphorylation of Shc, p80, p140 and other phosphotyrosyl substrates pesent in these complexes. Our studies suggest that p80 and p140 may link the hGM-CSF receptor to the PI 3-kinase and Shc/Grb2/ras signaling pathways, respectively, and that abnormal activation of hGM-CSF-dependent targets may play a role in leukemogenesis.

Introduction

GM-CSF is a hematopoietic growth factor that regulates proliferation, differentiation, and effector functions of monocytes, macrophages, granulocytes, and other cell types [23]. The high affinity receptor for human GM-CSF (hGM-CSF) is a heterodimer consisting of two subunits, termed α and β. The α subunit bears the specificity for ligand recognition and can bind hGM-CSF with low affinity, whereas the β subunit can not bind hGM-CSF by itself, but is required for high affinity binding of hGM-CSF [9, 16]. In the human, the same β subunit is utilized by the receptors for GM-CSF, IL-3, and IL-5, providing an explanation for the overlapping biological activities of these cytokines [23]. Deregulation of the human GM-CSF receptor (hGMR) signaling pathway has been implicated in leukemogenesis by mechanisms involving autocrine and paracrine stimulation of hGMR [22, 28]. Direct support for this hypothesis was obtained by demonstration that coexpression of the α and β subunits of hGMR in established murine fibroblasts is sufficient to reconstitute a functional receptor, which is capable of causing ligand-dependent transformation [1]. These conditional transformants expressed 1,500-3,800 high affinity receptors per cell, formed foci of transformed cells in the presence of pM levels of hGM-CSF, and were able to form anchorage-independent colonies in soft agar in the presence but not in the absence of hGM-CSF. Thus, in the proper cellular environment, the normal hGMR can generate a ligand-dependent oncogenic signal [1]. hGMR was quite potent as an oncogenic agent. Its transforming efficiency was 5-10 times lower than that obtained with v-*fps/fes*, a retroviral oncogene [1].

Since deregulation of the hGMR signaling pathway may be a contributing factor to leukemogenesis we undertook a detailed characterization of some of the events that occur during hGM-CSF signaling. To this end we used as a model system TF-1, a hGM-CSF-dependent cell line that is responsive to physiological concentrations of the ligand [15]. We found that hGM-CSF induced tyrosine phosphorylation and activation of a number of signaling proteins and their assembly into multi-subunit complexes. Among these signaling molecules we identified two novel phosphotyrosyl proteins that may link hGMR to the PI 3-kinase and Shc/Grb2/ras pathways, and found that one of these complexes is activated in acute myeloid leukemias.

Results and Discussion

An 80 kDa Protein that is Tyrosine Phosphorylated in hGM-CSF-Stimulated Cells Associates with p85/PI 3-Kinase and with the Activated hGMR.

The mitogenic PI 3-kinase pathway is activated by oncogenic tyrosine kinases and by virtually every growth factor or cytokine that has been examined, including GM-CSF [2, 5, 10, 11, 13, 17]. PI 3-kinase is a heterodimer consisting of two subunits: an 85 kDa protein (p85) containing src homology 2 (SH2) and SH3 domains and a 110 kDa catalytic subunit (p110) [4]. p85 functions as an adapter molecule that targets p110 to activated, tyrosine phosphorylated growth factor receptors [10, 13, 21]. When a tyrosine kinase receptor such as PDGFR is activated by its ligand, tyrosine phosphorylation of the receptor creates a binding site for the SH2 domains of p85, which brings PI 3-kinase to the activated receptor. In other cases, binding to an activated receptor is indirect. For instance, insulin induces binding of p85 to IRS-1, a phosphotyrosyl substrate of the insulin receptor [30, 31], and IL-4 induces binding of p85 to IL-4 receptor via a tyrosine-phosphorylated adapter protein designated 4PS [14]. GM-CSF induces the association of p85/PI 3-kinase with tyrosine-phosphorylated proteins present in the activated hGMR complex, but binding of p85/PI 3-kinase to hGMR is likely to be indirect because this receptor does not have the optimal consensus sequence for p85 binding [7]. As described below, we identified a new tyrosine phosphorylated protein that may connect PI 3-kinase to the hGMR.

In TF-1 cells, hGM-CSF induces tyrosine phosphorylation of multiple cellular proteins. The major phosphotyrosyl protein is a broad band of 76 to 85 kDa (p80)

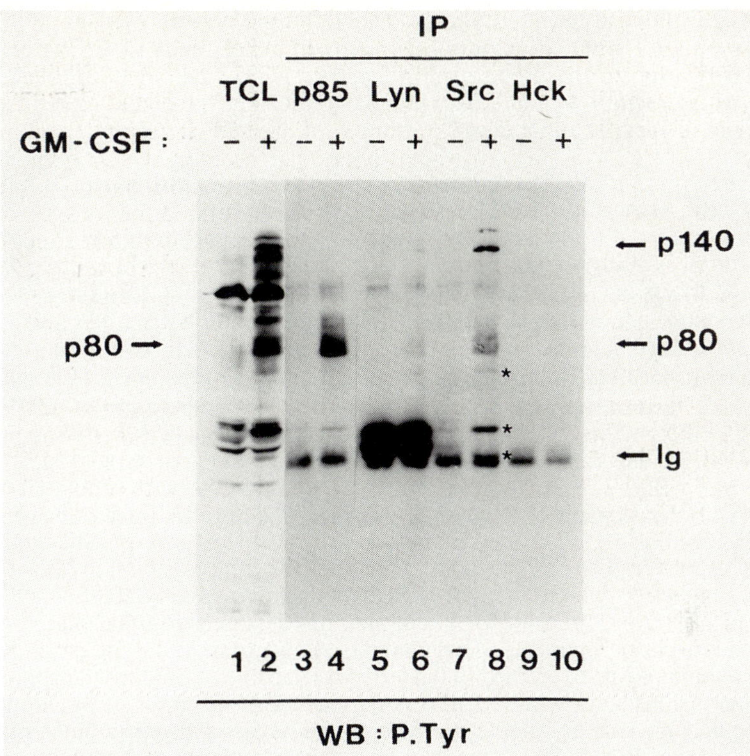

Fig. 1. Association of tyrosine-phosphorylated p80 with the p85 subunit of PI 3-kinase and Src family kinases. After starvation for 18 h TF-1 cells were either not stimulated (-) or stimulated with 10 ng/ml hGM-CSF for 2 min (+). Total cellular lysates (TCL) or cell lysates immunoprecipitated (IP) with antibodies directed against p85, Lyn, Src or Hck were analysed by phosphotyrosine immunoblotting (P.Tyr). The positions of p80, p140 and the immunoglobulin heavy chain (Ig) are indicated. Stars indicate the positions of $p46^{Shc}$, $p52^{Shc}$ and $p66^{Shc}$.

(Fig. 1, lane 2). Phosphotyrosyl p80 was present in immunoprecipitates of the p85 subunit of PI 3-kinase (Fig. 1, lane 4), but p80 and p85 were shown to be different proteins because p80 was not directly recognized by anti-p85 antibody, and the two proteins had different electrophoretic mobilities [11]. To analyze the interactions between these two proteins we have expressed the C-terminal and N-terminal SH2 domains of p85 as bacterial glutathione S-transferase (GST) fusion proteins, and used the GST fusions as affinity reagents. We found that the C-terminal and to a lesser extent the N-terminal SH2 domains of the p85 subunit bound phosporylated p80 [11]. hGM-CSF also induced the association of phosphotyrosyl p80 to the β subunit of the activated hGMR and this correlated with activation of PI 3-kinase [11]. The association of p80 with both p85/PI 3-kinase and with the β subunit of hGMR suggests that p80 is a good candidate for connecting the activated hGMR to the PI 3-kinase pathway.

So far, the functional role we described for p80 appears to be restricted to signaling through the hGMR β subunit. IL-3, whose receptor shares this common β subunit also induced phosphorylation of p80, but no phosphotyrosyl p80 protein was observed after stimulation with other growth factors such as CSF-1 and EGF [11]. Clarification of the precise role of p80 in hGM-CSF signaling, and of the nature of its interaction with hGMR will have to await its molecular cloning and the development of specific antibodies.

Phosphotyrosyl p140 Associates with the Shc and Grb2 Adapter Proteins and May Link hGMR to the ras Pathway.

Binding of GM-CSF to its receptor activates the ras pathway [11, 18, 23], most likely through binding of the Shc and Grb2 adapter proteins [19, 25]. In cells transformed by oncogenic tyrosine kinases and after stimulation of cells with growth factors, the three alternatively spliced Shc proteins $p46^{Shc}$, $p52^{Shc}$ and $p66^{Shc}$ become phosphorylated on tyrosine [20]. This triggers their association with the SH2 domain of Grb2 [27], an adapter protein that binds to the nucleotide exchange factor Sos [6], resulting in an increase in the active GTP-bound form of $p21^{ras}$.

In TF-1 cells, hGM-CSF also induces tyrosine phosphorylation of Shc proteins (Fig. 2, lane 3) and their association to the SH2 domain of Grb2 [11, 18]. In addition, Shc (Fig. 2, lane 3) and Grb2 [11, 18] associate with a novel 140 kDa protein (p140) that is phosphorylated on tyrosine. The relevant interactions between Shc, Grb2, and p140 were studied using different domains of Shc and Grb2 expressed as bacterial fusion proteins. Although the tight association between Shc and p140 suggested the involvement of SH2-phosphotyrosine interactions, the SH2 domains of Shc and Grb2 failed to recognize p140, suggesting that other types of interactions might be involved. Our analysis showed this to be the case. The N-terminal domain of Shc expressed in bacteria was capable of binding phosphorylated p140. This is consistent with recent reports that the N-terminus of Shc contains a non-SH2 phosphotyrosine-interacting domain that is also present in other signaling molecules such as IRS-1 [8]. This novel phosphotyrosine-interacting domain (PID) specifically recognizes a phosphorylated NPXY motif in target proteins such as the insulin and EGF receptors [3, 8]. Thus, we predict that when p140 is cloned, it will be shown to have an NPXY motif. p140 may be related or identical to a 145 kDa protein that is tyrosine-phosphorylated in response to growth factors and other stimuli and can also bind to the PID of Shc [12].

Since Grb2 immunoprecipitates also contained p140 we wanted to determine if p140 could bind directly to Grb2. Using bacterially expressed GST fusion proteins we found that the C-terminal and to a lesser extent the N-terminal SH3 domains of Grb2 recognized p140 [11, 18]. Thus, p140 may have proline-rich sequences that serve to anchor p140 to the SH3 domains of Grb2. The SH2 domain of Grb2 recognized the three Shc proteins but as mentioned above, it did not recognize p140 [11].

Our results suggest that p140 is a multi-functional adapter protein that can associate directly with both Shc and Grb2. It remains to be determined whether p140 forms a single complex with these two proteins or whether it binds to each of these adapters separately. Nevertheless, the association of p140 with Shc and Grb2 strongly suggests a role of p140 in the Shc/Grb2/ras signaling pathway. Further analysis will be required to determine the precise role of p140 in activation of the Shc/Grb2/ras pathway.

Shc and p140 Are Constitutively Tyrosine Phosphorylated and Form a Complex in Acute Myeloid Leukemia Cells.

The Shc proteins are constitutively tyrosine phosphorylated in cells transformed by v-*src* and v-*fps* [20] and their overexpression can cause transformation in NIH

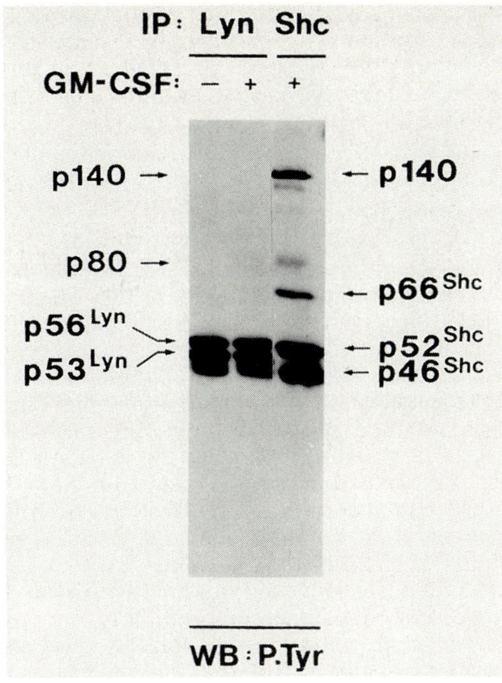

Fig. 2. hGM-CSF-dependent association of Lyn tyrosine kinase with p80 and p140. Cell lysates from unstimulated (-) or hGM-CSF-stimulated (+) TF-1 cells were immunoprecipitated (IP) with anti-Lyn or anti-Shc antibodies and analyzed by anti-phosphotyrosine immunoblotting (P.Tyr). The positions of tyrosine phosphorylated p140, p80, p53/p56^{Lyn} and the three isoforms of Shc (p46, p52 and p66) are indicated.

3T3 cells [25], suggesting that Shc activation may be a step in neoplastic transformation. A preliminary analysis involving a few samples of primary leukemia cells showed that the Shc proteins and p140 were constitutively tyrosine-phosphorylated and formed a complex in blood cells from acute myeloid leukemia patients, but not in samples from chronic lymphocytic leukemia patients or in peripheral blood mononuclear cells from healthy donors [11]. These results suggest that activation of the Shc/p140 phosphotyrosyl complex may be a necessary step in the pathogenesis of some leukemias.

p80 and p140 Are Tightly Associated With Src Family Kinases.
Phosphorylation of the multiple hGM-CSF-dependent targets requires the concerted action of several different protein tyrosine kinases. After GM-CSF stimulation, Jak-2 kinase, which is normally associated with the β subunit of hGMR becomes phosphorylated on tyrosine [29]. The activation of Jak-2 results in phosphorylation of the β subunit and probably one or more members of the STAT family of transcription factors [26]. STAT proteins are rapidly translocated to the nucleus where they activate a set of responsive genes [26]. Lyn and Yes tyrosine kinases have also been proposed to be involved in hGM-CSF signaling [5, 32], but their targets have not been identified.

To identify the kinases involved in phosphorylation of the hGM-CSF-dependent substrates we described, we analyzed immunoprecipitates of several different tyrosine kinases for the presence of these phosphotyrosyl proteins. Anti-Lyn immunoprecipitates from hGM-CSF-stimulated cells contained tyrosine phosphorylated p80 and p140 (Fig. 1, lane 6, and Fig. 2, lane 2). In some experiments, $p66^{Shc}$ was also detectable in anti-Lyn immunoprecipitates, whereas the other two Shc proteins, $p46^{Shc}$ and $p52^{Shc}$ were masked by $p53^{Lyn}$ and $p56^{Lyn}$, which were already tyrosine phosphorylated before hGM-CSF stimulation (Fig. 1, lanes 5 and 6). The association of Lyn with the p80/p85/PI 3-kinase complex is consistent with the observation that hGM-CSF induces the appearance of PI 3-kinase activity in anti-Lyn immunoprecipitates [32]. The presence of p140 in these precipitates suggests that Lyn is also part of the Shc/Grb2/p140 complex and that this kinase may phosphorylate some of the proteins in this complex as well.

Similar immunoblot analysis using anti-Src antibodies, which recognize both Src and Yes in TF-1 cells [11], showed that anti-Src immunoprecipitates from hGM-CSF-stimulated cells contained p80, the three Shc proteins, p140, and an unidentified phosphotyrosyl protein of 160 kDa (Fig. 1, lane 8). Further analysis showed that Src and/or Yes were already associated with the p85 subunit of PI 3-kinase before hGM-CSF stimulation [11]. This observation is consistent with reports that the SH3 domain of Src kinases can bind directly to a proline-rich region present in the p85 subunit of PI 3-kinase [24]. Thus, Src/Yes are good candidates for phosphorylation of p80 because these kinases were already part of the PI 3-kinase complex before hGM-CSF stimulation. Similarly, the presence of Shc and p140 in Src/Yes precipitates suggests that Src kinases may also be involved in phosphorylation of proteins in the p140/Grb2/Shc complex. Hck, another Src family tyrosine kinase did not associate with any hGM-CSF-dependent substrates, suggesting that this kinase may not phosphorylate any of the targets we described (Fig. 1, lanes 9 and 10).

These data suggest that at least two members of the Src family i.e. Lyn, and Src and/or Yes are involved in hGM-CSF signaling and may directly phosphorylate p80, Shc, p140, and other proteins present in the complexes of phosphotyrosyl proteins induced by hGM-CSF. Thus, the Src family of tyrosine kinases appears to play a major role in the mechanism of action of hGM-CSF.

The signaling events discussed here are illustrated in Fig. 3. hGM-CSF activates at least three different signaling pathways, namely the p85/PI 3-kinase, the Shc/Grb2/ras, and the Jak-2/Stat pathways. hGM-CSF treatment results in phosphorylation of a number of cellular proteins by different tyrosine kinases. Jak-2 may phosphorylate the β subunit of hGMR, triggering its association with other signaling molecules into one or more multimeric complexes. p80, a novel adapter protein that is phosphorylated on tyrosine by one or more src family kinases binds to the β subunit of hGMR. p80 also binds to a pre-existing complex of p85/PI 3-kinase and Src family kinases through the SH2 domains of p85. Thus, p80 may participate in linking the activated hGMR to the mitogenic PI 3-kinase pathway. In the diagram we have drawn a hypothetical SH2 domain in p80 which might recognize a phosphotyrosine site in the β subunit. hGMR activation also results in the assembly of a complex involved in activation of the ras signaling pathway. Tyrosine phosphorylation of the Shc proteins induces their association with the SH2 domain of Grb2, which participates in conversion of the inactive GDP-bound form of ras into the active GTP-ras through its association with the nucleotide exchange factor Sos [6]. The Shc/Grb2 complex contains another multi-functional adapter protein,

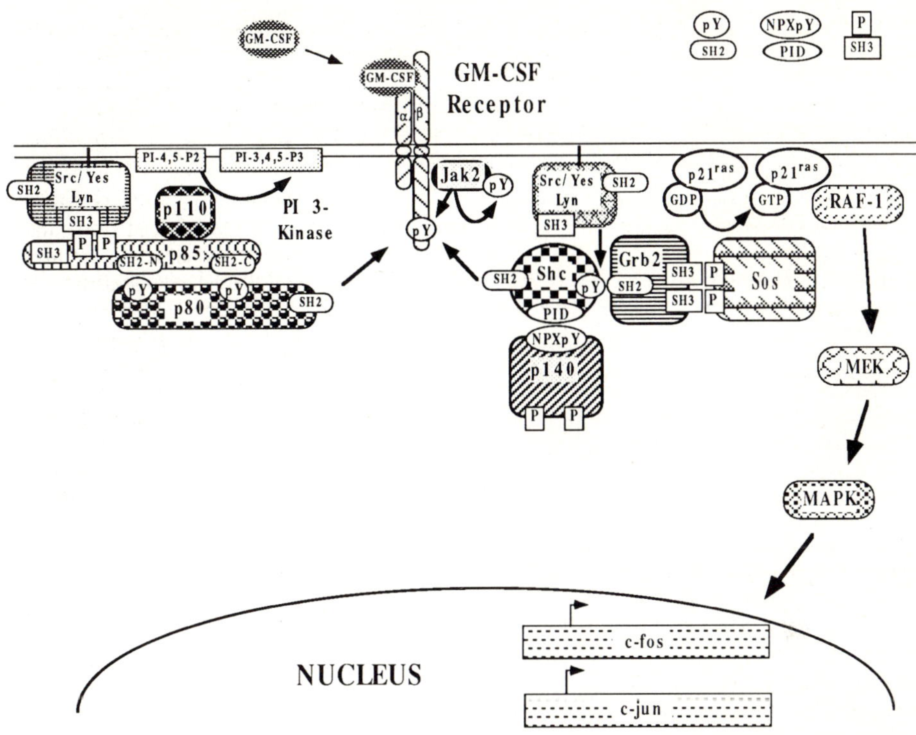

Fig. 3. Schematic diagram of the association of signaling proteins after stimulation of TF-1 cells with hGM-CSF. SH2, Src homology 2 domain; SH3, Src homology 3 domain; PID, phosphotyrosine-interacting domain; P, proline-rich region

p140, which binds to both Shc and Grb2. p140 is phosphorylated on tyrosine, presumably at an NPXY motif, which serves as a binding site for a novel phosphotyrosine-binding domain located at the N-terminus of Shc [3, 8, 11, 12]. p140 also binds directly to Grb2 through its SH3 domains, implying that p140 also has proline-rich sequences capable of binding SH3 domains [11, 18]. p140 is a good candidate for linking hGMR to the ras pathway, although direct binding of p140 to the receptor has not been demonstrated. The Shc/Grb2/p140 complex appears to be constitutively activated in some leukemias, providing evidence that deregulation of some of the signaling pathways described here may be a contributing factor in leukemogenesis. This and the direct demonstration that the normal subunits of hGMR have oncogenic potential [1] provide further experimental support to autocrine and paracrine models of leukemogenesis involving cytokines and their receptors.

References

1. Areces LB, Jücker M, San Miguel J, Mui A, Miyajima A Feldman RA (1993) Ligand-dependent transformation by the receptor for human granulocyte/macrophage colony-stimulating factor and tyrosine phosphorylation of the receptor β subunit. Proc Natl Acad Sci USA 90: 3963-3967
2. Auger KR, Serunian LA, Soltoff SP, Libby P, Cantley LC (1989) PDGF-dependent tyrosine phosphorylation stimulates production of novel polyphosphoinositosides in intact cells. Cell 57:167-175
3. Blaikie P, Immanuel D, Wu J, Li V, Margolis B (1994) A region of SHC distinct from the SH2 domain can bind tyrosine-phosphorylated growth factor receptors. J Biol Chem 269:32031-32034
4. Carpenter CL, Duckworth BC, Auger KR, Cohen B, Schaffhausen BS, Cantley LC (1990) Purification and characterization of phosphoinositide 3-kinase from rat liver. J Biol Chem 265:19704-19711
5. Corey S, Eguinoa A, Puyana-Theall K, Bolen JB, Cantley L, Molinedo F, Jackson TR, Hawkins PT, Stephens LR (1993) Granulocyte macrophage-colony stimulating factor stimulates both association and activation of phosphoinositide 3OH-kinase and src-related tyrosine kinase(s) in human myeloid derived cells. EMBO J 12:2681-2690
6. Egan SE, Giddlings BW, Brooks MW, Buday L, Sizeland AM, Weinberg RA (1993) Association of Sos ras exchange protein with Grb2 is implicated in tyrosine kinase signal transduction and transformation. Nature 363:45-51
7. Fantl WJ, Escobedo JA, Martin GA, Turck CW, del Rosario M, McCormick F, Williams, LT. (1992) Distinct phosphotyrosines on a growth factor receptor bind to specific molecules that mediate different signaling pathways. Cell 69:413-423
8. Gustafson TA, He W, Craparo A, Schaub CD, O'Neil TJ (1995) Phosphotyrosine-dependent interaction of SHC and Insulin Receptor Substrate 1 with the NPEY motif of the insulin receptor via a novel non-SH2 domain. Mol Cell Biol 15:2500-2508
9. Hayashida K, Kitamura T, Gorman DM, Arai KI, Yokota T, Miyajima A (1990) Molecular cloning of a second subunit of the receptor for human granulocyte-macrophage colony stimulating factor (GM-CSF): reconstitution of a high-affinity GM-CSF receptor. Proc Natl Acad Sci USA 87:9655-9659
10. Hu P, Margolis B, Skolnik EY, Lammers R, Ullrich A, Schlessinger J (1992) Interaction of phosphatidylinositol 3-kinase-associated p85 with epidermal growth factor and platelet-derived growth factor receptors. Mol Cell Biol 12:981-990
11. Jücker M, Feldman RA, Submitted
12. Kavanough WM, Williams LT (1994) An alternative to SH2 domains for binding to tyrosine phosphorylated receptors. Science 266:1862-1865
13. Kavanough WM, Klippel A, Escobedo JA, Williams LT (1992) Modification of the 85 kilodalton subunit of phosphatidylinositol-3 kinase in platelet-derived growth factor-stimulated cells. Mol Cell Biol 12:3415-3424
14. Keegan AD, Nelms K, White M,Wang L-M, Pierce JH, Paul,W (1994) An IL-4 receptor region containing an insulin receptor motif is important for IL-4-mediated IRS-1 phosphorylation and cell growth. Cell 76:811-820
15. Kitamura T, Tange T, Terasawa T, Chiba S, Kuwaki T, Miyagawa K, Piao Y-F, Miyazono K, Urabe A, Takaku F (1989) Establishment and characterization of a unique human cell line that proliferates dependently on GM-CSF, IL-3, or erythropoietin. J Cell Physiol 140:323- 334
16. Kitamura T, Sato N, Arai K-I, Miyajima A (1991) Expression cloning of the human IL-3 receptor cDNA reveals a shared β subunit for human IL-3 and GM-CSF receptors. Cell 66:1165-1174
17. Klippel A, Escobedo JA, Hu Q, Williams L (1993) A region of the 85-kilodalton (kDa) subunit of phosphatidylinositol 3-kinase binds the 110-kDa catalytic subunit in vivo. Mol Cell Biol 13:5560-5566
18. Lanfrancone L, Pelicci G, Brizzi MF, Arouica MG, Casciari C, Giuli S, Pegoraro L, Pawson T, Pelicci PG (1995) Overexpression of Shc proteins potentiates the proliferative response to the granulocyte-macrophage colony-stimulating factor and recruitment of Grb2/SoS and Grb2/p140 complexes to the receptor β subunit. Oncogene 10:907-917

19. Lowenstein EJ, Daly RJ, Batzer AG, Li W, Margolis B, LammersR, Ullrich A, Skolnik EY, Bar-Sagi D, Schlessinger J (1992) The SH2 and SH3 domain-containing protein GRB2 links receptor tyrosine kinases to ras signaling. Cell 70:431-442
20. McGlade J, Cheng A, Pelicci G, Pelicci P-G, Pawson T (1992) Shc proteins are phosphorylated and regulated by the v-src and v-fps protein-tyrosine kinases. Proc Natl Acad Sci (USA) 89:8869-8873
21. McGlade CJ, Ellis C, Reedijk M, Anderson D, Mbamalu , Reith AD, Panayotou G, End P, Bernstein A, Kazlauskas A, Waterfield MD, Pawson T (1992) SH2 domains of the p85α subunit of phosphatidylinositol 3-kinase regulate binding to growth factor receptors. Mol Cell Biol 12:991-997
22. Metcalf D (1989) The molecular control of cell division, differentiation commitment and maturation in haemopoietic cells. Nature 339:27-30
23. Miyajima A, Kitamura T, Harada N, Yokota, T, Arai K-I (1992) Cytokine receptors and signal transduction. Annu Rev Immunol 10:295-331.
24. Pleiman CM, Hertz W M, Cambier JC (1994) Activation of phosphatidylinositol-3' kinase by src-family SH3 binding to the p85 subunit. Science 263:1609-1612
25. Pellici G, Lanfrancone L, Grignani F, McGlade J, Cavallo F, Forni G, Nicoletti I, Grignani F, Pawson T, Pelicci PG (1992) A novel transforming protein (SHC) with an SH2 domain is implicated in mitogenic signal transduction. Cell 70: 93-104
26. Quelle FW, Shimoda K, Thierfelder W, Fischer, C, Kim A, Ruben SM, Cleveland JM, Pierce JH, Keegan AD, Nelms K, Paul WE, Ihle JN (1995) Cloning of murine Stat6 and human Stat6, Stat proteins that are tyrosine phosphorylated in responses to IL-4 and IL-3 but are not required for mitogenesis. Mol Cell Biol 15:3336-3343
27. Rozakis-Adcock M, McGlade J, Mbamalu G, Pelicci G, Daly R., Li W, Batzer A, Thomas S, Brugge J, Pelicci P G, Schlessinger J, Pawson T (1992) Association of the Shc and Grb2/Sem5 SH2-containing proteins is implicated in activation of the ras pathway by tyrosine kinases. Nature 360:689-692
28. Sachs L (1990) The control of growth and differentiation in normal and leukemic blood cells. Cancer 65:2196-206
29. Silvennoinen O, Witthuhn B, Quelle FW, Cleveland JL, Yi T, Ihle JN (1993) Structure of the JAK-2 protein tyrosine kinase and its role in IL-3 signal transduction. Proc Natl Acad Sci USA 90:8429-8433
30. Skolnik EY, Lee C-H, Batzer A, Vincentini LM, Zhou M, Daly R, Myers Jr MJ, Backer JM, Ullrich A, White MF, Schlessinger J (1993) The SH2/SH3 domain-containing protein GRB2 interacts with tyrosine-phosphorylated IRS1 and Shc: implications for insulin control of ras signalling. EMBO J 12:1929-1936
31. Sun XJ, Rothenberg P, Kahn CR, Backer JM, Araki E, Wilden PA, Cahill DA, Goldstein BJ, White M (1991) Structure of the Insulin Receptor Substrate IRS-1 defines a unique signal transduction protein. Nature 352:73-77
32. Torigoe T, O'Connor R, Santoli D, Reed JC (1992) Interleukin-3 regulates the activity of the LYN protein-tyrosine kinase in myeloid-committed leukemic cell lines. Blood 80:617-624

Myb: Proliferation and Maintenance of Stage-Specific Genes

Functional Analysis of the *c-myb* Proto-Oncogene

H. H. Lin[1], D. C. Sternfeld[2], S. G. Shinpock[3], R. A. Popp[1], and M. L. Mucenski[3]
[1]University of Tennessee Graduate School of Biomedical Sciences, Oak Ridge, TN 37831-8080, [2]DePauw University, Greencastle, IN 46135, and [3]Biology Division, Oak Ridge National Laboratory, Oak Ridge, TN 37831-8080

Introduction

The v-*myb* gene was initially identified as the oncogenic agent in the defective avian myeloblastosis virus (AMV), which induces acute myeloblastic leukemia *in vivo* and transforms myeloid cells *in vitro* (1). Klempnauer *et al* . (2,3) showed that the v-*myb* portion of AMV was transduced from seven internal exons of a cellular gene designated c-*myb*. A second replication-defective avian retrovirus, E26, was also found to contain v-*myb* as well as v-*ets* sequences (4,5). As with AMV, E26 contains a v-*myb* component which is an altered version of c-*myb*, with each containing both amino- and carboxy-terminal truncations of their gene products. The proteins encoded by AMV, E26, and c-*myb* have been localized to the nucleus (6,7).

Expression patterns of the c-*myb* gene indicate that the gene product has a primary role in hematopoiesis. High levels of c-*myb* expression are detected in immature cells of the myeloid, erythroid, and lymphoid lineages, with expression decreasing as these cells terminally differentiate (8-10). The integration of chronic retroviruses into the c-*myb* locus results in the synthesis of truncated gene products which induce myelogenous disease in mice (11) as well as B-cell lymphomas in chickens (12). The expression of the c-*myb* proto-oncogene is not restricted solely to hematopoietic cells, however, as the expression of this proto-oncogene has been detected in normal human mucosa (13), in developing neural tissue (14), and in murine embryonic stem (ES) cells (15). In addition, c-*myb* expression has been reported in a limited number of human malignancies of neuroectodermal and hematopoietic origin, as well as sporadically in colon, lung, and breast carcinomas (13, 16-20).

Aberrant c-*myb* expression has also given insight as to this proto-oncogene's possible function. Constitutive overexpression of c-*myb* cDNA has been shown to block the normal differentiation response of Friend erythroleukemia cells to dimethyl sulfoxide (21). However, the expression of a truncated message that encodes the DNA binding domain, but lacks the transcriptional activation domain of the c-*myb* gene product enhances differentiation of Friend murine erythroleukemia cells (22). Moveover, the v-*myb*-induced transformation of macrophages causes them to revert to a more immature phenotype (23). Lastly, antisense oligonucleotides to c-*myb* have been shown to inhibit the growth of T-cell cultures and cell lines (24) as well as most human myeloid cell lines tested (25).

The murine c-*myb* gene encodes a 636 amino acid product. The protein contains three functional domains: a DNA-binding domain, a transactivation domain, and a negative regulatory domain. The DNA binding domain consists of three imperfect 50-52 amino acid repeats designated R1, R2, and R3 located near the amino terminus of the gene product. A portion of R2 and all of R3 are required for DNA binding (26). Recent investigations into the similarity of repeated domains from various Myb proteins revealed the conservation of hydrophobic residues, suggesting the presence of three alpha-helical regions within each repeat (27). Further analysis of R2 and R3 revealed two consecutive helix-turn-helix motifs with unconventional turns, similar to those found in bacterial repressors and homeodomain proteins. The DNA binding domain has been shown to bind to a nucleotide consensus sequence. This portion of the gene product is evolutionarily well conserved in Myb family members. Through the use of c-*myb* deletion mutants, a transcriptional activation domain and a negative regulatory domain have also been identified in the Myb gene product (28). The *trans*-activation domain has been localized to a 50-residue region located carboxy-terminal to the DNA-binding domain. The domain appears to be similar to other *trans*-activation regions that have been identified, in that it is hydrophilic and slightly acidic (29). Deletion of the carboxy-terminal half of the Myb protein results in up to a 10-fold increase in Myb-specific *trans*-activation activity (30). This negative regulatory region contains a leucine zipper motif, similar to that found in other DNA binding proteins. This structure has been found to negatively regulate transactivation and transformation by the Myb gene product (31).

The identification of DNA-binding and transcriptional activation domains with its gene product, as well as consensus nucleotide binding site(s) located in the promoter region of hematopoietic and non-hematopoietic genes, indicated that Myb is a *trans*-acting transcription factor. Subsequently, several genes have been identified whose transcription is directly regulated by the Myb gene product. The *mim-1* gene was identified using differential hybridization screening to identify v-*myb* regulated genes in cells transformed by a temperature sensitive mutant of the oncogene (32). The promoter region of *mim-1* was shown to contain three Myb binding sites and these sites were found to confer v-*myb* dependent activation to a heterologous promoter. Other examples of Myb modulated genes include c-*myc* (33) and CD4 (34) which are *trans*-activated directly by Myb transcription factor binding. In addition, Myb appears to regulate the expression of several genes indirectly, indicating that as a transcription factor Myb is capable of using multiple mechanisms to regulate the expression of other genes. Considering the potential diversity ascribed to the functional domains of the Myb protein, this complex transcriptional regulation is not unexpected.

To define the biological function of the murine c-*myb* proto-oncogene *in vivo*, mice were generated using ES cells in which one allele of the gene was disrupted using standard gene targeting methodologies (35). While heterozygous mice are phenotypically indistinguishable from their wild-type littermates, homozygous mutant mice die at approximately 15.5 days of gestation, apparently due to anemia. At 12.5 and 13.5 days of gestation, hematocrit levels of all fetuses were approximately 35%, regardless of genotype. At 14.5 and 15.5 days of gestation however, there was a precipitous decline in hematocrit levels in mutant fetuses to approximately 5%, in comparison to approximately 40% for phenotypically normal littermates. During this time in embryogenesis, the site of erythropoiesis switches from the embyonic yolk sac to the fetal liver, where there is a rapid proliferation of hematopoietic progenitors. Erythrocytes from these two sites can be distinguished by their morphology; the yolk sac-derived red cells remain nucleated and larger than their fetal liver-derived enucleated counterparts. Analysis of peripheral blood from littermates between 12.5 and 15.5 days of gestation indicate that homozygous mutants are unable to switch from yolk sac to fetal liver erythropoiesis. The work presented in this paper further characterizes the hematopoietic potential of the c-*myb* mutant mice we have generated.

Materials and Methods

In vitro hematopoietic analysis

Fetal liver cells were plated in 0.8% methylcellulose-αMEM supplemented with 3u/ml erythropoietin and 1% PWMSCM acquired from Stem Cell Technologies (Vancouver, British Columbia) as a complete, pre-tested medium (HCC-3430). For the CFU-E analysis, cells were added to a final concentration of 1-5x10^5 cells/ml and plated in 0.1 ml aliquots. For BFU-E and CFU-GM analyses, cells were added to a final concentration of 4-5x10^4 cells/ml and were plated in 1 ml triplicate cultures. CFU-E-derived colonies, containing 8-32 hemoglobinized cells, were scored two days after plating; BFU-E and CFU-GM-derived colonies were scored 7-9 days after plating from the same 1 ml cultures. BFU-E are large bursts of 50 or more hemoglobinized cells. For these experiments, large colonies or bursts that contained greater than 50% hemoglobinized cells were counted as BFU-E although they may include mixed (CFU-GEMM or BFU-ME) colonies. DNA was extracted from yolk sacs or fetal tissues using standard methodologies and genotyped using PCR analysis (35). Primers used included 5'Ex6 (5'-GCAAGGTGGAACAGGAAGGCTACC) and 3'cEx6 (5'-GTGCTTCGGCGATGTGGTAATAGG).

RT-PCR analysis

Total RNA was isolated from frozen tissues (36). RNA preparations were subjected to DNase digestion prior to cDNA synthesis. Oligonucleotide primers were synthesized for a variety of genes, including cytokines and their receptors, hematopoietic markers, and genes containing the c-*myb* consensus binding sequence (37-42). Non-quantitative RT-PCR was carried out using standard methodologies (37).

Northern blot analysis

Northern blot analysis was performed on 10 ug of total liver RNA extracted from fetuses of heterozygous c-*myb* timed matings at 13.5 and 14.5 days of gestation, using a standard methodology (43). RT-PCR products of selected genes were used as hybridization probes in this analysis.

Differential display

Differential display was performed exactly as described in the RNAmap™ kit (GenHunter Corporation, Brookline, MA) using total RNA extracted from livers of 14.5 day of gestation littermates from a heterozygous c-*myb* timed mating. Differentially expressed cDNAs were reamplified using PCR and subcloned into a plasmid vector designed for the cloning of PCR products (TA Cloning Kit, Invitrogen, San Diego, CA). Sequencing was performed using the dideoxy chain termination methodology (44). DNA sequences were analyzed using software from the University of Wisconsin Genetics Computing Group (GCG).

Results

In vitro hematopoietic analysis

In vitro assays for hematopoietic colony-forming cells were used to determine the number of early (BFU-E) and late (CFU-E) erythroid progenitors and myeloid progenitors (CFU-GM) in 12.5 to 15.5 day of gestation fetuses from c-*myb* heterozygous matings. Results from this analysis are shown in Figure 1. Cells isolated from the fetal livers of wild-type and heterozygous animals at the same days of gestation give rise to similar numbers of hematopoietic colonies per 1×10^5 cells plated, as well as total number of hematopoietic progenitors per liver (data not shown). Homozygous mutant animals, however, show dramatically reduced numbers of hematopoietic progenitors in the fetal liver at all ages studied. A significant increase in liver cellularity due to the proliferation of hematopoietic progenitor cells between 12.5 days and 15.5 days of gestation was seen in wild-type and heterozygous mice ($1.5\text{-}30 \times 10^6$cells/liver). The homozygous mutant livers remain hypocellular, expanding from $0.4\text{-}4 \times 10^6$ cells per liver during the same days of gestation. This failure to proliferate leads to a critically low total number of hematopoietic progenitors in the mutant livers by 14.5 to 15.5 days of gestation. These results support what has been reported from the histological analysis of liver sections and the cytocentrifugation analysis of fetal liver cell suspensions of homozygous mutant c-*myb* animals and their phenotypically normal littermates (35).

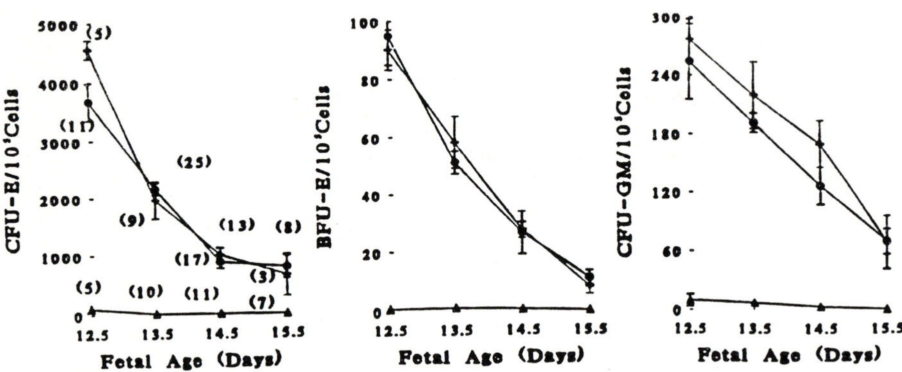

Fig 1. Hematopoietic progenitors in the fetal livers of wild-type (plus), heterozygous (closed circle), and homozygous mutant (closed triangle) animals from 12.5 to 15.5 days of gestation . The number of CFU-E/10^5 cells are presented in the graph on the left, the number of BFU-E/10^5 cells are presented in the middle graph, and the number of CFU-GM/10^5 cells are presented in the graph on the right. The numbers in parentheses are the number of fetuses analyzed at each time point. The same fetal liver suspensions were analyzed for all progenitor types. Error bars represent the standard error of the mean.

RT-PCR and Northern blot analyses

The Myb gene product is a sequence-specific DNA-binding protein that modulates expression of downstream genes. In an attempt to identify additional genes that may be regulated by the Myb protein, the expression pattern of a total of 20 genes was analyzed by RT-PCR analysis of total liver RNA isolated from littermates at 13.5 and 14.5 days of gestation. The genes chosen for analysis included genes known to be expressed early in hematopoiesis, selected cytokines and cytokine receptors, as well as genes known to contain putative Myb binding sites. Hypoxanthine phophoribosyltransferase (*HPRT*) expression served as a positive control. Results from this analysis are presented in Table 1. Of the 20 genes analyzed by non-quantitative RT-PCR analysis, only one gene, *GATA-1*, was found to be expressed at lower levels in mutant fetuses than in phenotypically normal littermates.

Table 1. RT-PCR analysis of fetal liver RNA

HPRT	*SCL*#	*IL-3*	*β-major globin*
c-*vav*	*NF-E2*#	*IL-3R*	*βH1 globin*
cdc-2	*GATA-1**#	*IL-1R*	*c-kit*
Epo#	*GATA-3*	*CSF-1R*	*Kit ligand (KL)*#
Epo-R#	*flk-2*	*G-CSFR*	*c-myc*

* : Differential expression detected by RT-PCR
\# : Expression also determined by Northern blot analysis

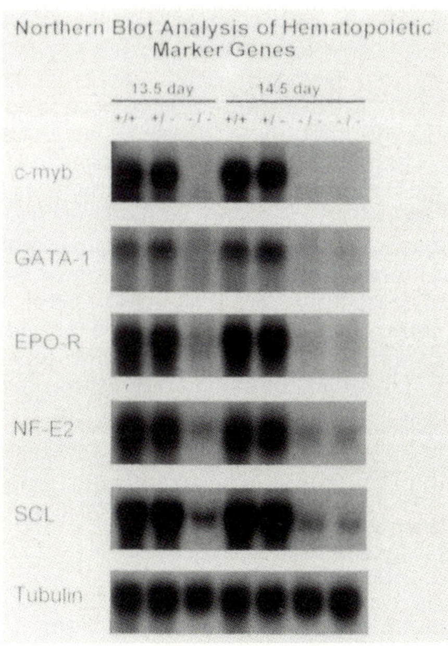

Fig 2. Northern blot analysis. Ten micrograms of total liver RNA from wild-type (+/+), heterozygous (+/-), and homozygous mutant (-/-) fetuses at 13.5 and 14.5 days of gestation were electrophoresed, blotted, and hybridized with probes for genes listed on the left. Tubulin served as a positive control.

Northern blot analysis was also used on a subset of genes analyzed by RT-PCR analysis to determine whether their expression was altered in total liver RNA isolated from wild-type, heterozygous, and homozygous mutant fetuses at 13.5 and 14.5 days of gestation. These results are shown in Figure 2. The expression of eight genes was analyzed. No full-length or truncated c-*myb* transcripts were detected in the mutant fetuses. As expected from the RT-PCR analysis, *GATA-1* expression was lower in the mutant fetuses than in phenotypically normal littermates. However, several genes which appeared to be expressed at similar levels by RT-PCR analysis were expressed at lower levels in mutant fetuses than in their phenotypically normal littermates. This included two transcription factors primarily involved in erythropoiesis, *NF-E2* and *SCL*, as well as the erythropoietin receptor (*EpoR*). Erythropoietin (*Epo*) expression was found to be similar in all fetuses regardless of phenotype while Kit ligand (*KL*) was found to be expressed at higher levels in the mutants than in phenotypically normal littermates (data not shown). Tubulin expression served as a loading control.

Differential Display

Differential display is a RT-PCR-based strategy that can be used to detect altered gene expression in eukaryotic cells and to isolate partial cDNAs of the affected genes. This methodology was used in an attempt to identify novel genes that are involved in hematopoiesis. Total RNA from the livers of 14.5 day of gestation littermates from a c-*myb* heterozygous mating was reversed transcribed and then used in this analysis. Preliminary results are shown in Table 2. Eight primer pairs were used in the PCR portion of this analysis, out of a total of eighty possible combinations. From this analysis, forty-five differentially expressed partial cDNAs were subcloned. To verify that the partial cDNAs were indeed differentially expressed, they were subcloned and used as probes in Northern blot analysis of total liver RNA from the three possible genotypes of fetuses at 13.5 and 14.5 days of gestation. Of the thirty-eight that have been used as probes, twenty-two have been shown to be differentially expressed (data not shown). Twelve of the differentially expressed partial cDNAs have been sequenced and eight represent putative novel genes based upon a lack of homology with any known sequences found in the Genbank database. The sequence of the four other clones were found to share identity with phospholipase c-alpha (two clones), mouse prothrombin or thrombin, and a cysteine protease inhibitor. Expression of these three genes has been detected in, but is not necessarily limited to, cells of hematopoietic lineage. Of the eight putative novel partial cDNAs that have been identified, four are expressed at higher levels in the livers of phenotypically normal fetuses, while four are expressed at higher levels in the livers of mutant fetuses (data not shown).

Table 2. Differential Display Analysis

Number of oligonucleotide primer pairs used: 8/80
Number of differentially expressed partial cDNAs subcloned: 45
Number of partial cDNAs used in Northern blot analysis: 38/45
Number of confirmed differentially expressed partial cDNAs: 22/38
Number of differentially expressed partial cDNAs sequenced: 12/22
Number of putative new genes bases upon sequence analysis: 8/12

Summary

Targeted mutagenesis studies were initiated to determine the normal biological function of the c-*myb* proto-oncogene. While heterozygous mice are phenotypically indistinguishable from their wild-type littermates, homozygous mutant fetuses die at approximately 15.5 days of gestation apparently due to anemia, which results from an inability to switch from embryonic yolk sac to fetal liver erythropoiesis. Studies are currently being done to determine the extent of hematopoietic abnormalities in the homozygous mutant fetuses.

In vitro assays for hematopoietic colony-forming cells have been used to determine the frequency of both erythroid and myeloid progenitors in the fetal livers of wild-type, heterozygous, and homozygous mutant c-*myb* fetuses. The reduced number of erythroid progenitors was not unexpected considering the mutant fetus's pale color and reduced hematocrit. The dramatically reduced number of colonies derived from myeloid progenitors in the mutant fetuses in comparison to the number detected in phenotypically normal littermates suggests that expression of the c-*myb* proto-oncogene is critical for the proliferation and/or differentiation of early hematopoietic progenitors and possibly hematopoietic stem cells. Other possible explanations would include a hematopoietic progenitor migration problem from the yolk sac to the fetal liver or a defect in the microenvironment of the liver. Whether the lymphoid lineage is also adversely affected by the lack of c-*myb* expression remains to be determined.

RT-PCR and Northern blot analyses were used in an attempt to identify downstream genes which may be directly or indirectly regulated by the Myb gene product. While the levels of expression of several genes involved in erythropoiesis (*GATA-1, NF-E2, SCL,* and *EpoR*) were reduced in the livers of homozygous mutant fetuses in comparison to phenotypically normal littermates and one gene, Kit ligand (*KL*), was expressed at higher levels in the mutant livers, these results must be viewed with caution. The livers of the mutant fetuses have been shown to be hypocellular in comparison to those of phenotypically normal littermates (35). It is possible that the Myb gene product is directly or indirectly modulating the expression of these genes. Conversely, the alteration in expression may be due to the reduced number or absence of specific hematopoietic lineages in the livers of the mutant fetuses.

Differential display has also been used to identify putative novel genes that are involved in hematopoiesis. Preliminary studies suggest that this may be a powerful methodology to compare the expression pattern of genes in the fetal liver of wild-type, heterozygous, and homozygous mutant littermates at 14.5 days of gestation. To date nearly 60% of the partial cDNAs subcloned analyzed have been shown to be differentially expressed. More importantly, 75% of the differentially expressed cDNAs that have been sequenced appear to encode novel genes. Whether any of these novel genes are involved in the c-*myb* transcriptional cascade remains to be determined.

Overall, analysis of the c-*myb* mutant fetuses have provided valuable insight into the biological function of this interesting proto-oncogene. The continued analysis of this resource will undoubtedly provide additional information concerning the role of the c-*myb* gene in hematopoiesis.

Acknowledgments

We thank Marilyn Kerley for excellent technical assistance. This work is supported by the USDOE under contract DE-AC0584OR21400 with Lockheed Martin Energy Systems, Inc.

References

1. Moscovici, C (1975) Leukemic transformation with avian myeloblastosis virus: present status. Curr Topics Microbiol Immunol 71:79-101
2. Klempnauer, KH, Gonda, TJ, Bishop, JM (1982) Nucleotide sequence of the retroviral leukemia gene v-myb and its cellular progenitor c-myb: the architecture of a transduced oncogene. Cell 31:453-463
3. Klempnauer, KH, Ramsay R, Bishop MJ, Moscovici MG, Moscovici C, McGrath JP, Levinson AD (1983) The product of the retroviral transforming gene v-myb is a truncated version of the protein encoded by the cellular oncogene c-myb. Cell 33:345-355
4. Roussel M, Saule S, Lagrou C, Rommens C, Beug H, Graf T, Stehelin D (1979) Three types of viral oncogenes of cellular origin specific for hematopoietic cell transfomation. Nature 281:452-455
5. Leprince D, Gegonne A, Coll J, de Taisne C, Schneeberger A, Lagrou C, Stehelin D (1983) A putative second cell-derived oncogene of the avian leukemia retrovirus E26. Nature 306:395-397
6. Klempnauer KH, Symonds G, Evan GI, Bishop JM (1984) Subcellular localization of proteins encoded by oncogenes of avian myeloblastosis virus and avian leukemia virus E26 and by the chicken c-myb gene. Cell 37:537-547
7. Boyle WJ, Lampert MA, Lipsick JS, Baluda MA (1984) Avian myeloblastosis virus and E26 virus oncogene products are nuclear proteins. Proc Natl Acad Sci USA 81:4265-4269
8. Westin EH, Gallo RC, Arya SK, Eva A, Souza LM, Baluda MA, Aaronson SA, Wong-Staal F (1982) Differential expression of the amv gene in human hematopoietic cells. Proc Natl Acad Sci USA 79:2194-2198
9. Gonda TJ, Metcalf D (1984) Expression of myb, myc, and fos proto-oncogenes during the differentiation of a murine myeloid leukemia. Nature 310:249-251
10. Sheiness D, Gardinier M (1984) Expression of a protooncogene (proto-myb) in hematopoietic tissues of mice. Mol Cell Biol 4:1206-1212
11. Shen-Ong GLC, Morse III HD, Potter M, Mushinski JF (1986) Two modes of c-myb activation in virus-induced mouse myeloid tumors. Mol Cell Biol 6:380-392
12. Kanter MR, Smith RE, Hayward WS (1988) Rapid induction of B-cell lymphomas: insertional activation of c-myb by avian leukosis virus. J Virol 62:1423-1432
13. Torelli G, Venturelli D, Colo A, Zanni C, Selleri L, Moretti L, Calabretta B, Torelli U (1987) Expression of c-myb protooncogene and other cell cycle-related genes in normal and neoplastic human colonic musosa. Cancer Res 47:5266-5269
14. Thiele CJ, Cohen PS, Israel MA (1988) Regulation of c-myb expression in human neuroblastoma cells during retinoic acid-induced differentiation. Mol Cell Biol 8:1677-1683
15. Dyson PJ, Poirier F, Watson RJ (1989) Expression of c-myb in embryonal carcinoma cells and embryonal stem cells. Differentiation 42:24-27.
16. Thiele CJ, McKeon C, Triche TJ, Ross RA, Reynolds CP, Israel MA (1987) Differential protooncogene expression characterizes histopathologically indistinguishable tumors of the peripheral nervous system. J Clin Invest 80:804-811
17. Alitalo K, Winquist R, Linn CC, de la Chapelle A, Schwab M, Bishop JM (1984) Aberrant expression of an amplified c-myb oncogene in two cell lines from a colon carcinoma. Proc Natl Acad Sci USA 81:4534-4538
18. Griffin CA, Baylin SB (1985) Expression of the c-myb oncogene in human small cell lung carcinoma. Cancer Res 45:272-275
19. Slamon DJ, Boone TC, Murdock DC, Keith DE, Press MF, Lavson RA, Souza LM (1986) Studies of the human c-myb gene and its product in human acute leukemias. Science 233:347-354
20. Slamon, DJ, deKernion JB, Verma IM, Cline MJ (1984) Expression of cellular oncogenes in human malignancies. Science 224:256-262
21. Clarke, MF, Kukowska-Latallo JF, Westin E, Smith M, Prochownik EU (1988) Constitutive expression of a c-myb cDNA blocks Friend murine erythroleukemia cell differentiation. Mol Cell Biol 8:884-8902

22. Weber BL, Westin EH Clarke MF (1990) Differentiation of mouse erythroleukemia cells enhanced by alternatively spliced c-myb cDNA. Science 249:1291-1293
23. Beug BL, Blundell P, Graf T (1987) Reversibility of differentiation and proliferative capacity in avian myelononocytic cells transformed by tsE26 leukemia virus. Genes Dev 1:277-286
24. Calabretta B, Sims RS, Valtieri M, Caraccolo D, Szylik C, Venturelli D, Ratayzak M, Beran M, Gewirtz AM (1991) Normal and leukemic hemapoietic cell manifest differntial sensitivity to inhibitory effects of c-myb antisense oligodeoxynucleotides in in vitro study relevant to bone marrow purging. Proc Natl Acad Sci USA 88:2351-2355
25. Venturelli D, Mariano MT, Szylik C, Valteri M, Lange B, Crist W, Link M, Calabretta B (1990) Down-regulated c-myb expression inhibits DNA synthesis of T-leukemia cells in most patients. Cancer Res 50:7371-7375
26. Luscher B, Eisenman RN (1990) New light on myc and myb. Part II Myb. Genes Dev 4:2235-2241
27. Frampton J, Leutz A, Gibson TJ, Graf T (1989) DNA binding domain ancestry. Nature 342:134
28. Grosser FW, LaMontagne K, Whittaker L, Stohr S, Lipsick JS (1992) A highly conserved cysteine in the v-Myb DNA-binding domain is essential for transformation and transcriptional trans-activation. Oncogene 7:1005-1009
29. Ptashne M (1988) How eukaryotic transcriptional activators work. Nature 335:683-689
30. Sakura H, Kanei-Ishii C, Nagase T, Nakagoshi H, Gonda T, Ishii S (1989) Delineation of three functional domains of the transcriptional activator encoded by the c-myb protooncogene. Proc Natl Acad Sci USA 86:5758-5762
31. Landschulz WH, Johnson PF, McKnight SL (1988) The leucine zipper: a hypothetical structure common to a new class of DNA binding proteins. Science 240:1759-1764
32. Ness SA, Marknell A, Graf T (1989) The v-myb oncogene product binds to and activates the promyelocyte-specific mim-1 gene. Cell 59:1115-1125
33. Evans JL, Moore TK, Keuhl WM, Bender T, Ting JPY (1990) Functional analysis of c-Myb protein in T-lymphocytic cell lines shows that it trans-activates the c-myc promoter. Mol Cell Biol 10:5747-5752
34. Siu G, Wuster AL, Lipsick JS, Hedrick SM (1992) Expression of the CD4 gene requires a MYB transcription factor. Mol Cell Biol 12:1592-1604
35. Mucenski ML, McLain K, Kier AB, Swerdlow SH, Schreiner CM, Miller TA, Pietryga DW, Scott Jr WJ, Potter SS (1991) A functional c-myb gene is required for normal murine fetal hepatic hematopoiesis. Cell 65:677-689
36. Chomczynski P, Sacchi N (1987) Single-step method of RNA isolation by acid guanidinum thiocyanate-phenol-chloroform extraction. Anal Biochem 162:156-159
37. Schmitt RM, Bruyns E, Snodgrass HR (1991) Hematopoietic development of embryonic stem cells in vitro: cytokine and receptor gene expression. Genes Dev 5:728-740
38. Keller G, Kennedy M, Papyannopoulou T, Wiles MV (1993) Hematopoietic commitment during embryonic stem cell differentiation in culture. Mol Cell Biol 13:473-486
39. McClanahan T, Dalrymple, S, Barkett M, Lee F (1993) Hematopoietic growth factor receptor genes as markers of lineage commitment during in vitro development of hematopoietic cells. Blood 81:2903-2915
40. Spurr NK, Gough AC, Lee MG (1990) Cloning of the mouse homologue of the yeast cell cycle control gene. DNA Seq 1:49-54
41. Andrews NC, Erdjument-Bromage H, Davidson MB, Tempst P, Orkin SH (1993) Erythroid transcription factor NF-E2 is a haematopoietic-specific basic-leucine zipper protein. Nature 362:722-728.
42. Stanton LW, Fahrlander PD, Tesser PM, Marcu KB (1984) Nucleotide sequence comparison of normal and translocated murine c-myb genes. Nature 310:423-425
43. Maniatis T, Fritsch EF, Sambrook J (1989) Molecular cloning: a laboratory manual. Cold Spring Harbor, New York.
44. Sanger F, Nicklen S, Coulson AR (1977) DNA sequencing with chain-terminating inhibitors. Proc Natl Acad Sci USA 74:5463-5467

Structure and Function of the Proteins Encoded by the *myb* Gene Family

C. Kanei-Ishii, T. Nomura, K. Ogata, A. Sarai, T. Yasukawa, S. Tashiro, T. Takahashi, Y. Tanaka, and S. Ishii
Tsukuba Life Science Center, The Institute of Physical & Chemical Research (RIKEN), 3-1-1 Koyadai, Tsukuba, Ibaraki 305, Japan

Introduction

The nuclear proto-oncogene c-*myb* is the cellular homologue of the v-*myb* gene carried by the chicken leukemia viruses avian myelobastosis virus (AMV) and E26, which transform avian myeloid cells *in vitro* and *in vivo* (for review, see ref. 1). c-*myb* expression is linked to the differentiation state of the cell, since expression is down-regulated during terminal differentiation of hemopoietic cells (2) and constitutive expression of introduced c-*myb* blocks the induced differentiation of erythroleukemia (3). In addition, antisense oligonucleotides to c-*myb* appear to impede *in vitro* hematopoiesis (4) and homozygous c-*myb* mutant mice displayed a specific failure of fetal hepatic hematopoiesis (5). These results all indicate a role for c-*myb* in maintaining the proliferative state of hematopoietic progenitor cells. Both c-Myb and v-Myb are transcriptional activators (6-9). The v-*myb* proteins (v-Myb) encoded by AMV and E26 are amino (N)- and carboxyl (C)-terminally truncated versions of c-Myb. In this report, we shall address the structure and function of each functional domain in c-Myb, and also the relationship between the retroviral-transforming v-*myb* genes and its cellular homologue, the c-*myb* gene, to ask whether changes in the control of transcription may account for the generation of the transformation phenotype.

Recognition of Specific DNA Sequences by Myb

Mouse c-Myb consists of 636 amino acids and has three functional domains responsible for DNA binding, transcriptional activation, and negative regulation (10; Fig. 1). For transcriptional activation by Myb, both the DNA-binding domain and the transcriptional activation domain, which contains a cluster of acidic amino acids, are necessary and sufficient. Both v-Myb and c-Myb recognize specific DNA sequences, AACNGN(A/T/C), of which the first A, third C, and fifth G are involved in very specific interactions (11, 12). The DNA-binding domain of c-Myb consists of three homologous tandem repeats of 51-52 amino acids (R1, R2, and R3 from the N terminus). Among the three repeats, R1 can be deleted without significant loss of DNA-binding activity, indicating that R1 is a minor player in sequence recognition (10, 13, 14). An interesting feature of the DNA-binding domain of Myb is the presence of three conserved tryptophans in each repeat (15). All of the Myb-related proteins, including the A-Myb and B-Myb (16), the C1

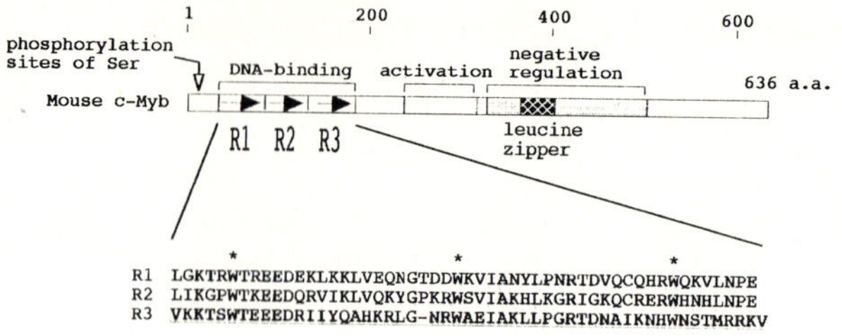

c-Myb-binding site (MBS-1)

1 2 3 4 5 6 7 8 9 10 11 12 13 14 15 16

\+ CCTAACTGACACACAT

\- GGATTGACTGTGTGTA

Fig. 1. Schematic representation of functional domains in mouse c-Myb. The three functional domains, which are responsible for DNA binding, transcriptional activation, and negative regulation, respectively, are shown. Arrows represent the three-fold tandem repeats of 51 or 52 amino acids. Amino acid sequence in each repeat and the DNA sequence recognized by c-Myb are shown below. The three tryptophans in each repeat are indicated by asterisks.

protein of *Zea mays*, and the yeast BAS1 protein, have three perfectly conserved tryptophans in each repeat with an interval of 18 or 19 amino acids. A conserved triplet of tryptophans with spacing similar to that of c-Myb is also found in the DNA-binding domains of the products of the *ets* gene family, indicating that these conserved tryptophans may represent a characteristic property of a group of DNA-binding proteins. Site-directed mutagenesis of these tryptophans showed that any single or multiple mutations of tryptophans to hydrophilic residues or alanine abolished or greatly reduced the sequence-specific DNA-binding activity, but mutations to hydrophobic amino acids retained considerable activity (17, 18). Raman spectroscopic study showed that these tryptophans were buried in the protein core. Based on these results, these three tryptophans were proposed to form a cluster in the hydrophobic core in each repeat.

Recently, the solution structures of R2R3-DNA complex, R1R2R3, and each of three repeats were determined by multi-dimensional NMR (Fig. 2) (19-21). The three repeats have similar overall architectures, each containing a helix-turn-helix variation motif. The three conserved tryptophans in each repeat are involved in a hydrophobic core. The third helix in each of R2 and R3 is a recognition helix.

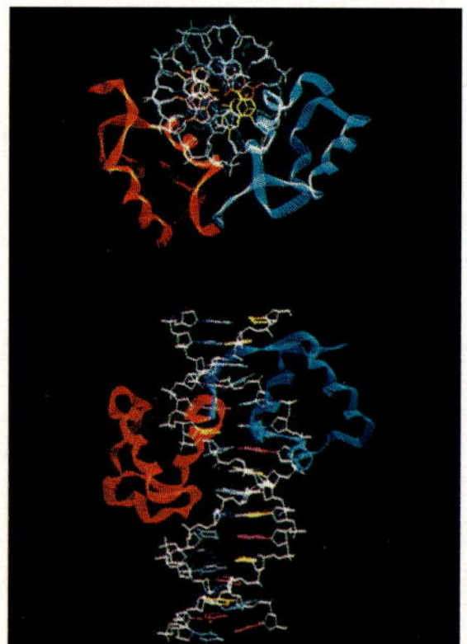

Fig. 2. Solution structure of R2R3-DNA complex determined by NMR. The backbone atoms in the R2 region of R2R3 are shown in red, and those in the R3 region are shown in blue. On the top, the position of two recognition helices in the major groove of DNA is shown in the view along the DNA double helix axis.

R2 and R3 are closely packed in the major groove, so that the two recognition helices contact each other directly to bind to a specific base sequence, AACNG, cooperatively. The three key base pairs in this sequence are specifically recognized by Asn-183 (R3), Lys-182 (R3), and Lys-128 (R2). In contrast, R1 does not bind tightly to DNA and R1 may fluctuate between free and bound states at a fast rate with only a minor population in the DNA-bound state. Thus, the homologous two repeats of Myb, which are directly connected in tandem, bind to the major groove of DNA continuously. In this sense, the binding of R2 and R3 is similar to that of transcription factor IIIA (TFIIIA)-type Zn fingers. Unlike these TFIIIA-type Zn fingers, however, the recognition helices of R2 and R3 are more closely packed together in the major groove. So far, this type of direct interaction between the recognition helices from different DNA-binding units appears to be unique among DNA-binding proteins.

Three hydrophobic amino acids in R2 is critical for activation of the *mim-1* gene, one of the *myb* target genes (8, 22). c-Myb and v-Myb encoded by E26 but not by AMV can activate the *mim-1* promoter. This difference can be attributed to the three point mutations in R2 of AMV v-Myb (Ile-91-Asn, Leu-106-His, Val-117-Asp). The *mim-1* promoter contains binding sites for both Myb and the myeloid-specific transcription factor NF-M (also called C/EBPβ or NF-IL6). Both the synergistic activation of the *mim-1* gene by these two factors and the close spacing of these binding sites (23, 24) suggest that these proteins are in direct contact. The solution structure of the R2R3-DNA complex indicates that these three hydrophobic residues are exposed to the solvent and are available for protein-protein contact with NF-M.

Modulation of Myb Activity by Negative Regulatory Domain

The negative regulatory domain (NRD) is important for modulation of c-Myb activity, because removal of this domain results in increase in both *trans*-activating and transforming capacities (10, 26). The potential leucine zipper structure is located in NRD (Fig. 3). This region was predicted to form an amphipathic α-helix and contains characteristic hydrophobic residues at every seventh position. Disruption of this leucine zipper by site-directed mutagenesis markedly increases c-

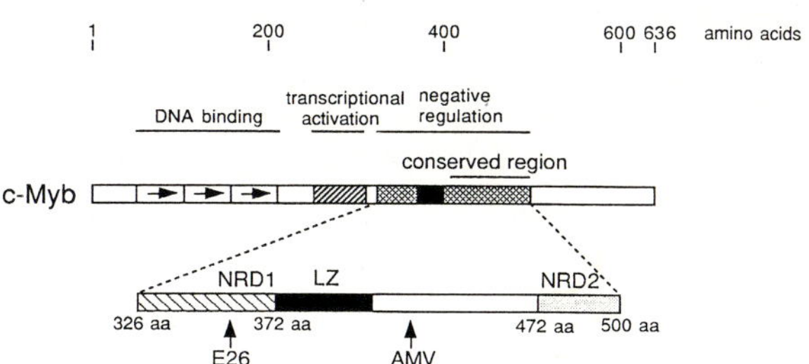

Fig. 3. Schematic representation of three subdomains in NRD. The N- and C-proximal subdomains, NRD1 and NRD2, are indicated by a hatched box and a stippled box, respectively. The leucine zipper structure is indicated by a closed box. The C-terminal end of v-Myb encoded by AMV or E26 is shown by an arrow.

Myb activity (27). The leucine zipper motif was originally identified as mediating dimerization of several DNA-binding proteins, such as the transcription factors C/EBP and Jun/Fos, but may also mediate interactions between other proteins including certain membrane proteins. Therefore, these results indicate that c-Myb activity is negatively regulated through the leucine zipper and imply the presence of an inhibitor(s) of c-Myb, which is likely to be important in the normal regulation of its activity. In fact, many cellular proteins were demonstrated to bind to the leucine zipper of c-Myb (27, 28). In addition to these proteins, c-Myb itself forms a dimer through the leucine zipper, and the c-Myb dimer cannot bind to DNA, indicating that c-Myb itself can also function as an inhibitor by binding to its own leucine zipper (29). As a consequence of the c-Myb dimer formation and the inability of the dimer to bind to DNA, in cotransfection assays, maximal Myb-induced *trans*-activation occurs with a low amount of wild type c-Myb, while higher levels of c-Myb reduce c-Myb-induced *trans*-activation. This apparent negative autoregulation is not observed with a c-Myb mutant containing an impaired leucine zipper.

Deletion of the region downstream from the leucine zipper also increases c-Myb activity (10). This suggest that besides the leucine zipper some subdomain that is also critical for negative regulation of c-Myb activity is localized in NRD. We have developed an assay system to precisely measure the DNA-binding activity of a series of C-truncated c-Myb mutants (Tanaka et al. submitted). Using this assay system, two subdomains (NRD1 and NRD2) that normally repress the DNA-binding activity have been identified in NRD (Fig. 3). The N-proximal subdomain NRD1, which is upstream of the leucine zipper, is the region between amino acids 326 and 372. The C-proximal subdomain NRD2, which is downstream of the leucine zipper, is the region between amino acids 472 and 500. Deletion of either of these subdomains increases the DNA-binding capacity, and additional effects of deletion of either of them were also observed. These results indicate that NRD contains three subdomains, NRD1, the leucine zipper, and NRD2, and suggest that c-Myb DNA-binding activity is independently regulated by multiple mechanisms through these subdomains.

Oncogenic Activation and Functional Domains

Analysis of various oncogenically activated *myb* genes suggested that truncation of the N or C-terminus of c-Myb can cause oncogenic activation (for review, see ref. 1). Both v-Myb proteins encoded by the chicken leukemia viruses AMV and E26 are N- and C-terminally truncated versions of c-Myb. Furthermore, integration of chronically transforming retroviruses into the c-*myb* locus can also result in truncation of the N or C terminus of c-Myb, inducing myelogenous diseases in mice and B-cell lymphomas in chickens. Removal of the NRD is responsible for oncogenic activation of C-truncated forms of Myb (26, 30, 31). v-Myb encoded by AMV lacks the C-proximal subdomain in NRD, NRD2, while v-Myb encoded by E26 lacks NRD2, the leucine zipper, and half of the N-proximal subdomain, NRD1 (Fig. 3). Deletion of any of these three subdomains increases DNA binding and *trans*-activation (32; Tanaka et al. submitted). Deletion of all of NRD by C-terminal truncation has the added effects of lack of each of them. Thus, these data support the view that C-truncated forms of *myb* transform by increasing the expression of target genes (Fig. 4).

On the other hand, the mechanism of oncogenic activation by N-terminal truncation is not clear. Both v-Myb proteins encoded by AMV and E26 lack the N-terminal 77 and 85 amino acids, which contain the N-terminal region upstream of the R1 and about half of R1. Based on the observations that phosphorylation of serines 11 and 12 of c-Myb by casein kinase II can inhibit the binding of c-Myb to DNA, it was speculated that loss of the casein kinase II (CK-II) phosphorylation site by N-terminal truncation could uncouple c-Myb activity from its normal physiological regulators (33). However, it was demonstrated that truncation of R1 is sufficient for transformation, but deletion of the CK-II phosphorylation site is not (34). R1 is not necessary to recognize the specific sequence, but stabilizes the Myb-DNA complex (12, 34). Truncation of R1 decreases the ability of Myb to bind DNA, indicating that the mechanism of oncogenic activation by N-terminal truncation is different from that by C-terminal truncation. Most of the target genes identified so far contain multiple c-Myb-binding sites including the high affinity sites, so the removal of R1 does not significantly decrease the level of transcription of these target genes. One hypothesis proposed by Dini and Lipsick (34) is that this decreased affinity for DNA results in the deregulation of a subset of c-Myb-regulated genes that control proliferation but not terminal differentiation. This hypothesis includes the possibility that the promoters of proliferation genes regulated by Myb may have Myb-binding sites with higher affinity or greater number than do differentiation genes. Another hypothesis is that N-terminal truncated c-Myb fails to repress transcription of some of the target genes due to the lack of a part of the DNA-binding domain (Fig. 4) (see below). This hypothesis includes the idea that the target genes with transcription repressed by c-Myb are critical for proliferation control of hematopoietic progenitor cells.

Target Genes

Several target genes of Myb including the *mim-1* and c-*myc* genes were identified. Ness et al. (8) used differential hybridization to screen for v-Myb-regulated genes in cells transformed by a ts mutant of v-*myb* and identified the *mim-1* gene. The *mim-1* gene encodes a secretable protein contained in the granules of promyelocytes. Several groups found that the promoter of the c-*myc* gene also contains multiple Myb-binding sites and its activity is stimulated by c-Myb (35-37). In addition, c-Myb also binds to the promoter regions of the *cdc2* and c-*myb* gene itself, and activates transcription (38, 39). However, the increase and/or deregulation of the expression of these identified target genes alone are thought not

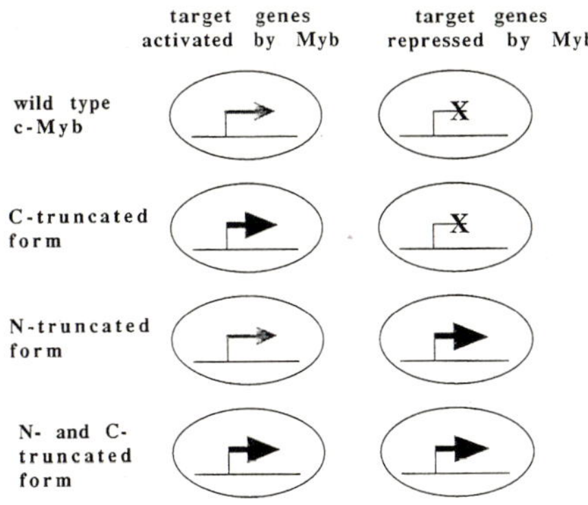

Fig. 4. Expression of two types of target genes by N-and/or C-truncated c-Myb. The levels of two types of target genes of which expression is activated or repressed by c-Myb are indicated. The C-terminal truncation of Myb increases expression of the c-Myb-induced target genes. The N-terminal truncation of Myb increases expression of the c-Myb-repressed target genes. Expression of the c-Myb-induced target genes is necessary for transformation, while the increased expression of the c-Myb-repressed target genes may stimulate transformation.

to be sufficient for cellular transformation. The results of extensive differential screenings suggest that c-Myb activates more of the genes than we expected.

Using an artificial promoter linked to the c-Myb-binding sites, we demonstrated that c-Myb can also repress transcription in some cases. We have found that the human c-*erbB*-2 promoter activity is repressed by c-Myb (40). The c-*erbB*-2 proto-oncogene (also called *neu* or HER2) encodes a 185-kDa transmembrane glycoprotein that has significant structural similarity to the EGF receptor (41). Among multiple Myb-binding sites in the c-*erbB*-2 promoter, two Myb-binding sites are critical for transcriptional repression by c-Myb. Myb represses the c-*erbB*-2 promoter activity by competing for DNA binding with positive regulators involving TFIID. The c-Myb mutant, which lacks the N-terminal 76 amino acids including about half of R1 in the DNA-binding domain, has much weaker activity than normal c-Myb to repress the c-*erbB*-2 promoter activity. If some target genes with transcription repressed by c-Myb are critical for proliferation control, the N-terminal truncated c-Myb lacking R1 would cause transformation by increasing the expression of these target genes (Fig. 4). Increase in the expression of this type of target gene is not sufficient for transformation of hematopoietic cells, because the transcriptional activation domain of c-Myb is necessary for transformation, but only the DNA-binding domain of c-Myb is required for repression of this type of target gene. Deregulation of this type of target gene may stimulate transformation, while expression of some target gene(s) of which expression is activated by c-Myb is essential.

Interestingly, c-Myb can activate the *hsp70* promoter without direct binding to its promoter region (42, 43). We have identified the heat shock element (HSE) as a responsible *cis*-element for *trans*-activation by c-Myb (43). c-Myb can activate the promoter linked to the tandem repeats of HSE, which contains no Myb-binding sites. For the HSE-dependent *trans*-activation by c-Myb, the leucine zipper in c-Myb is necessary, but not the transcriptional activation domain containing the acidic amino acid cluster. These results suggest that c-Myb can activate transcription of the gene containing HSEs by interacting with an unidentified *trans*-acting factor. The biological significance of this should be clarified to understand the physiological role of c-Myb.

Structure and Function of A-Myb and B-Myb

In addition to c-*myb*, two genes, A-*myb* and B-*myb*, are members of the *myb* gene family (16). Like c-Myb, A-Myb functions as a transcriptional activator in all types of cells examined (44). The functional domains of A-Myb are similar to those of c-Myb except for the leucine zipper (Fig. 5). The A-Myb activity is negatively regulated through an NRD downstream from the transcriptional activation domain. The NRD of A-Myb contains two subdomains at the positions corresponding to NRD1 and NRD2 of c-Myb. However, A-Myb does not have a functional leucine zipper, disruption of which would increase A-Myb activity.

We previously reported that human B-Myb is a transcriptional activator, and activates the artificial promoter containing multiple Myb-binding sites (MBS-I) (45) and the human c-*myc* promoter (37). Furthermore, the region downstream from the DNA-binding domain in the B-Myb molecule, which is rich in acidic amino acids, was found to be a transcriptional activation domain (Fig. 5) (46). In contrast, it was reported that chicken and murine B-Myb failed to *trans*-activate the promoter containing Myb-binding sites, and that B-Myb inhibited *trans*-activation

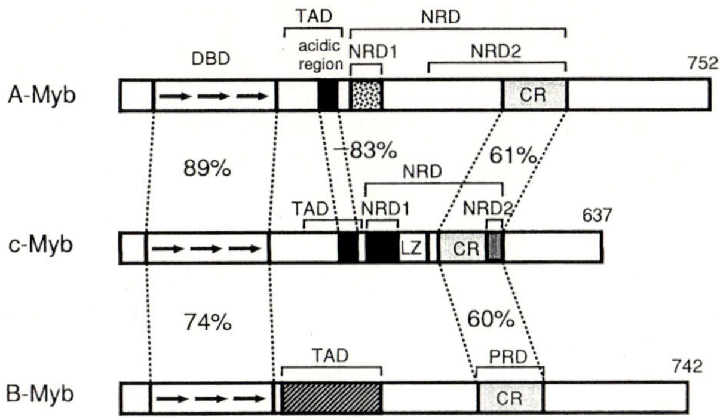

Fig. 5. Comparison of functional domains between three members of the *myb* gene family. The three tandem repeats in the DNA-binding domain are shown by arrows. The closed box indicates the region that is rich in acidic amino acids. The region conserved between the three members (CR) is indicated by a shaded box. DBD, DNA-binding domain; TAD, transcriptional activation domain; NRD, negative regulatory domain; PRD, positive regulatory domain; LZ, leucine zipper structure. Percentages indicate percent identity in each region encompassed by lines between human c-Myb and human A-Myb or between human c-Myb and human B-Myb.

of promoters by c-Myb, probably by competing for Myb-binding sites (47). Our recent analyses to resolve this discrepancy indicated that B-Myb is a cell type-specific transcriptional regulator (48). B-Myb functions as a transcriptional activator in CV-1 and HeLa cells, but not in NIH3T3 cells. Deletion analyses of B-Myb demonstrated that a region conserved between three members of the *myb* gene family (CR) is necessary for *trans*-activation by B-Myb. An *in vivo* competition assay suggest that regulatory factor(s) that bind to the CR of B-Myb are required for *trans*-activation (Fig. 5). Analyses using affinity resin show that multiple proteins bind to the CR of B-Myb and that the CR-binding proteins in CV-1 and HeLa cells are different from those in NIH3T3 cells. Thus, the CR-binding cofactor(s) appear to be critical for the cell type-specific *trans*-activation by B-Myb.

Although c-*myb* is expressed at high levels in immature hematopoietic cells, it is also expressed in non-hematopoietic tissues: colon and small intestine at moderated levels and testis, ovary, and lung at low levels. Tissue distribution of the B-*myb* mRNA is more broad than c-*myb*, and it is also expressed in spleen, placenta, and pancreas in addition to the tissues described above (48). Like the case of c-*myb*, expression of B-*myb* correlates with cellular proliferation. B-*myb* mRNA is not expressed in resting cells but is induced late in G1 to maximal levels that are maintained through S phase (49, 50). B-*myb* mRNA levels decrease when HL60 or U937 cells are induced to differentiate (49, 51). Inhibition of B-*myb* expression by introduction of a B-*myb* antisense construct diminished cell proliferation of hematopoietic cells and fibroblasts, but constitutive expression of B-*myb* induced a transformed phenotype (51). These results suggest that B-Myb is a positive regulator for proliferation like c-Myb. In contrast to c-*myb* and B-*myb*, A-*myb* is expressed only in testis and peripheral blood leukocytes (44). A-*myb* is expressed in resting T cells, and its levels gradually decrease after mitogenic stimulation (50). These facts suggest that physiological role of A-Myb is distinct from that of c-Myb and B-Myb.

Conclusions

Detailed analyses of the functional domain of c-Myb has led to the identification of the unique structure of its DNA-binding domain. We can now understand the molecular interaction between Myb and DNA at the atomic level. Similar strategies have also identified the specific motif in the c-Myb molecule, the leucine zipper, to which inhibitors can bind. These results have shown that the oncogenic activation of c-*myb* by C-terminal truncation results in increased *trans*-activating capacity due to the removal of both or either the inhibitor-binding site and/or the inhibitory region of DNA binding. This can be easily understood, since it is likely that overexpression of target genes induces cellular transformation. On the other hand, the mechanism of oncogenic activation by N-terminal truncation is still obscure, because it decreases the DNA-binding capacity of c-Myb by removal of R1 of the DNA-binding domain. Thus, the oncogenic activation of c-*myb* involves multiple mechanisms.

One of the obvious questions is how growth signal(s) can be transduced from growth factor(s) to the Myb protein in nuclei. A c-*myb* deficient homozygous mutant does not develop fetal blood cells, indicating a similarity between the disrupted c-*myb* phenotype and the dominant white-spotting and steel mutants which affect the c-kit receptor and its ligand SCF. These results imply that c-Myb may be the nuclear target of the c-kit signaling pathway. The inability of normal c-Myb to transform primary hematopoietic cells could therefore be due to its failure to escape normal posttranslational regulation by such cytokine receptor pathways. We have demonstrated that c-Myb activity is negatively regulated by inhibitors, and that the mutant to which inhibitors cannot bind has a transforming capacity. Therefore, the identification of inhibitor(s) by cDNA cloning may lead to understanding the mechanism of signal transduction from growth factor(s) to c-Myb.

Acknowledgments

We thank many collaborators for our work, especially, Tom Gonda, Rob Ramsay, Yoshifumi Nishimura, Haruki Nakamura, Saburo Aimoto, and Nobuo Nomura.

References

1. Graf T (1992) Myb: a transcriptional activator linking proliferation and differentiation in hematopoietic cells. Curr. Opin. Genet. Dev. 2:249-255
2. Gonda TJ, Metcalf D (1984) Expression of *myb*, *myc*, and *fos* proto-oncogenes during the differentiation of a murine myeloid leukemia. Nature 310:249-251
3. Todokoro K, Watson RJ, Higo H, Amanuma H, Kuramochi S, Yanagisawa H, Ikawa Y (1988) Down-regulation of c-*myb* gene expression is a prerequisite for erythropoietin-induced erythroid differentiation. Proc. Natl. Acad. Sci. USA 85:8900-8904
4. Gewirtz AM, Calabretta B (1988) A c-*myb* antisense oligodeoxynucleotide inhibits normal human hematopoiesis *in vitro*. Science 242:1303-1306
5. Mucenski ML, McLain K, Kier AB, Swerdlow SH, Schereiner CM, Miller TA, Pietryga DW, Scott WJ, Potter SS (1991) A functional c-*myb* gene is required for normal murine fetal hepatic hematopoiesis. Cell 65:677-689
6. Nishina Y, Nakagoshi H, Imamoto F, Gonda TJ, Ishii S (1989) Trans-activation by the c-*myb* proto-oncogene. Nucleic Acids Res. 17:107-117

7. Weston K, Bishop JM (1989) Transcriptional activation by the v-*myb* oncogene and its cellular progenitor, c-*myb*. Cell 58:85-93
8. Ness SA, Marknell A, Graf T (1989) The v-*myb* oncogene product binds to and activates the promyelocyte-specific *mim*-1 gene. Cell 59:1115-1125
9. Nakagoshi H, Nagase T, Kanei-Ishii C, Ueno Y, Ishii S (1990) Binding of the c-*myb* proto-oncogene product to the simian virus 40 enhancer stimulates transcription. J. Biol. Chem. 265:3479-3483
10. Sakura H, Kanei-Ishii C, Nagase T, Nakagoshi H, Gonda TJ, Ishii S (1989) Delineation of three functional domains of the transcriptional activator encoded by the c-*myb* proto-oncogene. Proc. Natl. Acad. Sci. USA 86:5758-5762
11. Biedenkapp H, Borgmeyer U, Sippel AE, Klempnauer KH (1988) Viral *myb* oncogene encodes a sequence-specific DNA binding activity. Nature 335: 835-837
12. Tanikawa J, Yasukawa T, Enari M, Ogata K, Nishimura Y, Ishii S and Sarai A (1993) Recognition of specific DNA sequences by the c-*myb* proto-oncogene product - role of three repeat units in the DNA-binding domain. Proc. Natl. Acad. Sci. USA 90:9320-9324
13. Klempnauer KH, Sippel AE (1987) The highly conserved amino-terminal region of the protein encoded by the v-*myb* oncogene functions as a DNA-binding domain. EMBO J. 6:2719-2725
14. Howe KM, Reakes CFL, Watson RJ (1990) Characterization of the sequence-specific interaction of mouse c-*myb* protein with DNA. EMBO J. 1:161-169
15. Anton IA, Frampton J (1988) Tryptophans in *myb* proteins. Nature 336: 719
16. Nomura N, Takahashi M, Matsui M, Ishii S, Date T, Sasamoto S, Ishizaki R (1988) Isolation of human cDNA clones of *myb*-related genes, A-*myb* and B-*myb*. Nucleic Acids Res. 16:11075-11089
17. Kanei-Ishii C, Sarai A, Sawazaki T, Nakagoshi H, He DN, Ogata K, Nishimura Y, Ishii S (1990) The tryptophan cluster: a hypothetical structure of the DNA-binding domain of the *myb* protooncogene product. J. Biol. Chem. 265:19990-19995
18. Saikumar P, Murali R, Reddy EP (1990) Role of tryptophan repeats and flanking amino acids in Myb-DNA interactions. Proc. Natl. Acad. Sci. USA 87:8452-8456
19. Ogata K, Hojo H, Aimoto S, Nakai T, Nakamura H, Sarai A, Ishii S Nishimura Y (1992) Solution structure of a DNA-binding unit of Myb: a helix-turn-helix-related motif with conserved tryptophans forming a hydrophobic core. Proc. Natl. Acad. Sci. USA 89:6428-6432
20. Ogata K, Morikawa S, Nakamura H, Sekikawa A, Inoue T, Kanai H, Sarai A, Ishii S, Nishimura Y (1994) Solution structure of a specific DNA complex of the Myb DNA-binding domain with cooperative recognition helices. Cell 79:639-648
21. Ogata K, Morikawa S, Nakamura H, Hojo H, Yoshimura S, Zhang R, Aimoto S, Ametani Y, Hirata Z, Sarai A, Ishii S, Nishimura Y (1995) Comparison of the free and DNA-complexed forms of the DNA-binding domain from c-Myb. Nature Structural Biology 2:1-12
22. Introna M, Golay J, Frampton J, Nakano T, Ness SA, Graf T (1990) Mutations in v-*myb* alter the differentiation of myelomonocytic cells transformed by the oncogene. Cell 63:1287-1297
23. Ness SA, Kownez-Leutz E, Casini T, Graf T Leutz A (1993) Myb and NF-M: combinatorial activators of myeloid genes in heterologous cell types. Genes Dev. 7:749-759
24. Burk O, Mink S, Ringwald M, Klempanuer KH (1993) Synergistic activation of the chicken *mim-1* gene by v-*myb* and C/EBP transcription factors. EMBO J. 12:2027-2038
25. Sarai A, Uedaira H, Morii H, Yasukawa T, Ogata K, Nishimura Y, Ishii S (1993) Thermal stability of the DNA-binding domain of the Myb oncoprotein. Biochemistry 32:7759-7764
26. Hu YL, Ramsay RG, Kanei-Ishii C, Ishii S, Gonda TJ (1991) Transformation by carboxyl-deleted Myb reflects increased transactivating capacity and disruption of a negative regulatory domain. Oncogene 6:1549-1553
27. Kanei-Ishii C, Macmillan EM, Nomura T, Sarai A, Ramsay RG, Aimoto S, Ishii S Gonda TJ (1992) Transactivation and transformation by Myb are negatively regulated by a leucine zipper structure. Proc. Natl. Acad. Sci. USA 89:3088-3092
28. Favier D, Gonda TJ (1994) Detection of proteins that bind to the leucine zipper motif of c-Myb. Oncogene 9: 305-311
29. Nomura T, Sakai N, Sarai A, Sudo T, Kanei-Ishii C, Ramsay RG, Favier D, Gonda TJ, Ishii S (1993) Negative autoregulation of c-Myb activity by homodimer formation through the leucine zipper. J. Biol. Chem. 268:21914-21923

30. Gonda TJ, Buckmaster C, Ramsay RG (1989) Activation of c-*myb* by carboxy-terminal truncation: Relationship to transformation of murine haemopoietic cells *in vitro*. EMBO J. 8:1777-1783
31. Grasser FA, Graf T, Lipsick JS (1991) Protein truncation is required for the activation of the c-*myb* proto-oncogene. Mol. Cell. Biol. 11:3987-3996
32. Ramsay RG, Ishii S, Gonda TJ (1991) Increase in specific DNA binding by carboxyl truncation suggests a mechanism for activation of Myb. Oncogene 6:1875-1879
33. Lüscher B, Christenson E, Litchfield DW, Krebs EG, Eisenman RN (1990) *Myb* DNA binding inhibited by phosphorylation at a site deleted during oncogenic activation. Nature 344:517-522
34. Dini PW, Lipsick JS (1993) Oncogenic truncation of the first repeat of c-Myb decreases DNA binding *in vitro* and *in vivo*. Mol. Cell. Biol. 13:7334-7348
35. Evans JT, Moore TL, Kuehl WM, Bender T, Ting JPY (1990) Functional analysis of c-Myb protein in T-lymphocytic cell lines shows that it *trans*-activates the c-*myc* promoter. Mol. Cell. Biol. 10:5747-5752
36. Zobel A, Kalkbrenner F, Guehmann S, Nawrath M, Vorbrueggen G, Moelling K (1991) Interaction of the v- and c-Myb proteins with regulatory sequences of the human c-*myc* gene. Oncogene 6:1397-1407
37. Nakagoshi H, Kanei-Ishii C, Sawazaki T, Mizuguchi G, Ishii S (1992) Transcriptional activation of the c-*myc* gene by the c-*myb* and B-*myb* gene products. Oncogene 7:1233-1240
38. Ku DH, Wen SC, Engelhard A, Nicolaides NC, Lipson KE, Marino TA, Calabretta B 91993) c-*myb* transactivates *cdc2* expression via Myb binding sites in the 5'-flanking region of the human *cdc2* gene. J. Biol. Chem. 268:2255-2259
39. Nicolaides NC, Gualdi R, Casadevall C, Manzella L, Calabretta B (1991) Positive autoregulation of c-*myb* expression via Myb binding sites in the 5' flanking region of the human c-*myb* gene. Mol. Cell. Biol. 11:6166-6176
40. Mizuguchi G, Kanei-Ishii C, Takahashi T, Yasukawa T, Nagase T, Horikoshi M, Yamamoto T, Ishii S (1995) c-Myb repression of c-*erbB*-2 transcription by direct binding to the c-*erbB*-2 promoter. J. Biol. Chem. 270: 9384-9389
41. Yamamoto T, Ikawa S, Akiyama T, Semba K, Nomura N, Miyajima N, Saito T, Toyoshima K (1986) Similarity of protein encoded by the human c-*erbB*-2 gene to epidermal growth factor receptor. Nature 319:230-234
42. Klempnauer KH, Arnold H, Biedenkapp H (1989) Activation of transcription by v-*myb*; evidence for two different mechanisms. Genes Dev. 3:1582-1589
43. Kanei-Ishii C, Yasukawa T, Morimoto RI, Ishii S (1994) c-Myb-induced *trans*-activation mediated by heat shock elements without sequence-specific DNA binding of c-Myb. J. Biol. Chem. 269:15768-15775
44. Takahashi T, Nakagoshi H, Sarai A, Nomura N, Yamamoto T, Ishii S (1995) Human A-*myb* gene encodes a transcriptional activator containing the negative regulatory domains. FEBS Lett. 358:89-96
45. Mizuguchi G, Nakagoshi H, Nagase T, Nomura N, Date T, Ueno Y, Ishii S (1990) DNA-binding activity and transcriptional activator function of the human B-*myb* protein compared with c-*myb*. J. Biol. Chem. 265:9280-9284
46. Nakagoshi H, Takemoto Y, Ishii S (1993) Functional domains of the human B-*myb* gene product. J. Biol. Chem. 268:14161-14167
47. Foos G, Grimn S, Klempnauer KH (1992) Functional antagonism between members of the *myb* family: B-*myb* inhibits v-*myb*-induced gene activation. EMBO J. 11:4619-4629
48. Tashiro S, Takemoto Y, Handa H, Ishii S (1995) Cell type-specific *trans*-activation by the B-*myb* gene product: requirement of the putative cofactor binding to the C-terminal conserved domain. Oncogene 10: 1699-1707
49. Reiss K, Travali S, Calabretta B, Baserga R (1991) Growth regulated expression of B-*myb* in fibroblasts and hematopoietic cells. J. Cell. Physiol. 148:338-343
50. Golay J, Capucci A, Arsura M, Casetellano M, Rizzo V, Introna M (1991) Expression of c-*myb* and B-*myb*, but not A-*myb*, correlates with proliferation in human hematopoietic cells. Blood 7:149-158
51. Arsura M, Introna M, Passerini F, Mantovani A, Golay, J (1992). B-*myb* antisense oligonucleotides inhibit proliferation of human hematopoietic cell lines. Blood 79:2708-2716

The c-Myb Negative Regulatory Domain

T. J. Gonda, D. Favier, P. Ferrao, E.M. Macmillan, R. Simpson[1] and F. Tavner

Hanson Centre for Cancer Research, Institute of Medical and Veterinary Science, Box 14, Rundle Mall PO, Adelaide SA 5000 Australia; [1]Joint Protein Structure Laboratory of the Ludwig Institute for Cancer Research and the Walter and Eliza Hall Institute of Medical Research, PO Royal Melbourne Hospital, Victoria 3050 Australia

Introduction

It is becoming increasingly clear that c-*myb* plays an essential role in controlling the proliferation and differentiation of haemopoietic cells. While this has been suggested for some time based on the preferential expression of *myb* in immature haemopoietic cells and the decrease in its expression on differentiation[1,2], more recent loss-of-function studies have provided additional confirmation. Disruption of c-*myb* by gene targeting results in an abrupt failure of the foetal haemopoietic system to develop in mouse embryos[3], while anti-sense oligonucleotides inhibit proliferation of both normal haemopoietic progenitor cells and transformed haemopoietic cell lines (eg ref. 4). Furthermore, the ability of naturally occurring[5] and recombinant retroviruses carrying activated forms of c-*myb*[6,7] to transform haemopoietic cells - but in general, not other cell types - *in vitro* argues for a dominant regulatory role for *myb* in haemopoiesis. Although its role has not been identified precisely, it is likely, on the basis of the above evidence and the ability of *myb* to inhibit the differentiation of certain inducible leukaemic cell lines[8-10], that a major function of c-*myb* is to maintain the proliferative state and immature characteristics of early haemopoietic cells.

The biochemical function of the proteins (Myb) encoded by *myb* genes is entirely consistent with such a role. Both normal and activated Myb proteins bind DNA in a sequence-specific manner and can activate transcription of genes bearing its cognate binding site[11-14]. Thus it is reasonable to assume that *myb* genes exert their effects by regulating the transcription of other cellular genes. The requirement for both the DNA-binding and transactivation functions of activated forms of Myb for transformation *in vitro*[15,16] fully supports this assumption.

This hypothesis, ie whereby *myb* genes maintain an immature proliferative state by activating the transcriptional expression of other cellular genes, raises several further questions:

(1) What are the identities of the genes which *myb* regulates?

(2) If *myb* acts to enforce an immature proliferative state, what mechanisms allow normal differentiation and the concomitant cessation of proliferation to occur?

(3) What is the basis for transformation by *myb* ie, how do activated forms differ from c-*myb*?

While some answers to these questions are available, they are at best incomplete and do not provide an adequate explanation of either the function of c-*myb* or the basis of transformation by *myb* oncogenes. For example, the best-documented example of a *myb* target gene is the avian *mim-1* gene[17]; however, it is very unlikely that *mim-1* plays a significant role in transformation by *myb* because its expression is activated by only one of the two transforming avian leukaemia viruses that carry v-*myb* genes. The ability of c-*myb* to transactivate c-*myc*[18,19], which is widely associated with and probably essential for cellular proliferation, is an example of a potentially significant target for *myb*; however, activation of *myc* expression cannot explain the effects of *myb* on the differentiation state of transformed cells, which are very different from those of *myc*[20]. Other proliferation-associated genes such as those that encode DNA polymerase α[21] and *cdc2* (ref. 22) are other possible targets. Thus it is clear that at least one, or more probably many, of the genes that mediate the effects of *myb* on haemopoietic cell proliferation and differentiation are yet to be identified. Another complicating factor is the apparent ability of *myb* to enhance expression of genes like *mim-1* that are more characteristic of differentiated cells than of immature progenitor cells eg those encoding lysozyme[23] and CD4 (ref. 24).

The observation that c-*myb* expression decreases at late stages of haemopoietic differentiation may appear to provide an answer to question (2) posed above. While this is almost certainly important in allowing terminal differentiation, there is strong evidence that Myb is also regulated post-translationally. For example, in some cases differentiation of myeloid[1,10] and erythroid[26] cell lines appears to precede the loss of c-*myb* expression. Since this would seem at odds with the ability of enforced *myb* expression to block differentiation and maintain an immature proliferative state, it may be that normal differentiation requires modulation of the activity as well as the level of expression of the c-Myb protein. Furthermore, although *myb* genes can clearly induce proliferation, c-*myb* expression in early haemopoietic cells does not appear to be modulated by mitogenic growth factors or vary during the cell cycle[27-29], in contrast to the situation for c-*myc*[27]; this again suggests the activity of the protein may be modulated.

Evidence that Myb activity can be regulated, as well as a partial answer to question (3) above, comes from a wide range of studies linking amino and/or carboxyl truncation of Myb proteins with oncogenic activation. This association is seen in the cases of v-Myb proteins and Myb proteins encoded by rearranged c-*myb* genes in tumour cell lines[30]. Furthermore, studies utilising c-*myb* constructs that encode truncated proteins have demonstrated a causal relationship between truncation and activation[6,7,17]. Using a series of carboxyl truncations of murine c-Myb, a region in the carboxyl portion of the molecule which appears to negatively modulate transactivation[31], DNA binding[32] and transformation[15] of haemopoietic cells, has been identified. The existence of this region, termed the negative regulatory domain (NRD), and the consequences of its disruption - ie enhanced activity - suggest that the activity of the c-Myb protein itself may normally be regulated *via* this domain.

In this paper, we will examine the function of the NRD both in terms of its role in regulating Myb activity and how this regulation is achieved at the molecular level. In doing so we will draw upon some recent studies, from our laboratories, on *in vitro* transformation by myb retroviruses and on identification of proteins that interact with the NRD.

Components of the Negative Regulatory Domain

The presence of a leucine zipper-like motif in Myb was pointed out by Biedenkapp *et al.*[12] while more recently, the leucine zipper has been identified as a critical element of the NRD. Specifically disrupting this structure by replacing the heptad repeat leucines with proline residues enhances both transactivation and transformation by Myb[33], so it is likely that the Myb leucine zipper is involved in an association between c-Myb and another protein which inhibits Myb function.

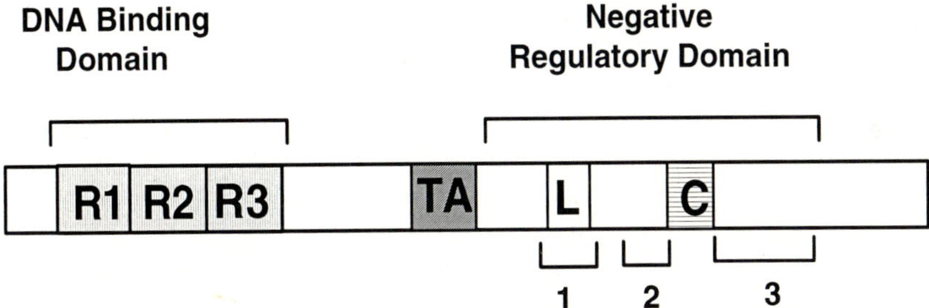

Fig. 1. Schematic of the c-Myb protein showing the major functional domains. **TA**, transactivation domain; **L** leucine zipper; and **C**, a highly conserved region shared with Drosophila Myb. The regions numbered 1, 2, and 3 delimit three apparently separate components of the NRD.

Other studies indicate that there may be components additional to the leucine zipper within the NRD. Evidence for this comes from observations that carboxyl truncations that delete part of the NRD without disrupting the leucine zipper can activate various functions of Myb. For example, truncation of avian c-Myb at a position equivalent to that of v-Myb^{AMV}, which contains an intact leucine zipper, results in activation as measured by *in vitro* transforming capacity[7]. Similar truncations also enhance the ability of c-Myb to transactivate reporter constructs in at least some[34,35] reports. It appears that studies from different laboratories that have attempted to define the elements of the NRD have sometimes produced different results. These discrepancies may be due to differences in transactivation assays, eg in the species and/or cell types, reporter constructs or Myb expression vectors employed in each case; in some studies a heterologous DNA binding domain has been substituted for that of Myb. Dubendorff *et al.*[35] have found that there are two C-terminal regions involved in negative regulation that are distinct from the leucine zipper motif, and observed no effect when the leucine zipper was deleted. There is little evidence to date that provides insight into how these other components of the NRD may function, except that it has been suggested that a C-terminal region of c-Myb may interact with the transactivation domain of Myb[35,36]. Fig. 1 represents an attempt to summarise, from several different studies[15,31-36], the locations of regions involved in negative regulation of Myb activity.

Proteins that interact with the c-Myb leucine zipper.

Identification of proteins that interact with the c-Myb leucine zipper.

One protein that may interact with the leucine zipper motif of the c-Myb NRD is c-Myb itself, since Nomura *et al.*[37] have provided evidence that c-Myb is capable of forming homodimers via the leucine zipper. These studies demonstrated physical association between the NRD and forms of Myb that contain the wild-type leucine zipper but not forms with a mutant leucine zipper. However, it has not been possible to demonstrate leucine zipper-mediated crosslinking of full-length c-Myb dimers in solution (R. Ramsay, personal communication), and it appears that quite high concentrations of Myb protein are needed to form dimers in solution. For these reasons, and because the degree to which mutating the leucine zipper enhances transactivation seems to vary between cell types[33], we considered it likely that other cellular proteins may interact with the leucine zipper *in vivo*.

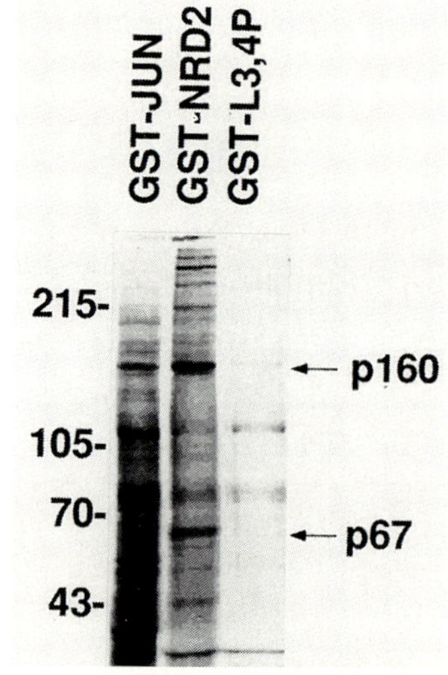

Fig. 2. Binding of 35S-Methionine-labelled nuclear proteins from FDC-P1 cells to fusion proteins containing wild-type Myb, mutated Myb (L3,4P) or Jun leucine zippers

The approach taken in our laboratory to identify such proteins was to use bacterially-expressed glutathione-S-transferase (GST)-NRD fusion proteins that include the leucine zipper as affinity probes. These studies[38] resulted in the identification of two proteins, p67 and p160, that bound to GST-NRD proteins containing the wild-type leucine zipper but not to those carrying a mutated version (see Fig. 2). Intriguingly, p160, but not p67, also bound to the c-Jun leucine zipper region. These proteins are both nuclear and in fact are closely related, as shown by peptide mapping with V8 protease. However, their distribution differs in that p160 is found in all the murine cell lines examined to date whilst p67 is found in a subset of immature myeloid cell lines.

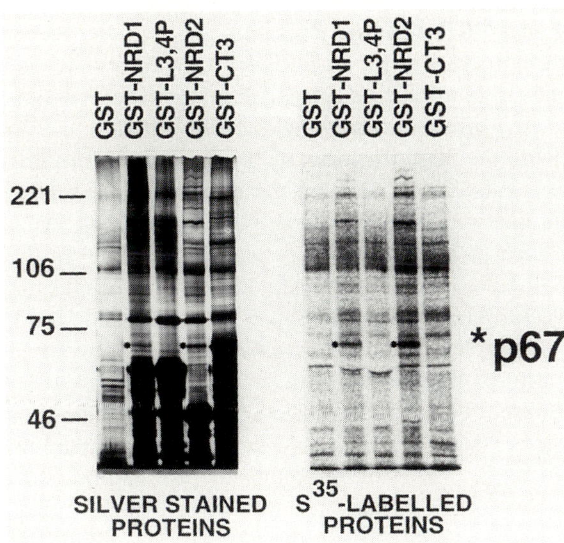

Fig. 3. Detection of p67 by metabolic labelling and silver staining after binding of cell extracts to GST fusion proteins. The bands corresponding to p67 are indicated by asterisks.

In order to further characterise these proteins, we undertook the molecular cloning of the corresponding cDNAs. That this could be approached by purification of p67 from FDC-P1 cells was suggested by the observation that p67 could be detected by silver staining as well as metabolic labelling from relatively small cell numbers (~10^7). Fig. 3 illustrates that protein bands corresponding to p67 could be detected in FDC-P1 nuclear extracts after binding to fusion proteins containing the Myb NRD, but not to fusion proteins containing a mutated leucine zipper or other regions of c-Myb. Subsequent scaling-up allowed the purification of sufficient protein (see Fig. 4) to obtain amino acid sequence data on five tryptic peptides. These sequences did not match any proteins found in the major protein sequence databases and lead us to conclude that p67 is indeed a novel protein.

The sequences were then used to design degenerate PCR primers with which we attempted to amplify a corresponding cDNA fragment from reverse-transcribed FDC-P1 RNA. A short fragment was amplified using some of the sequences corresponding to two of these peptides; the sequence of this fragment contained an open reading frame as well as additional sequences from each peptide. Using a combination of further PCR steps, cDNA library screening and 5'RACE PCR, we isolated cDNA clones that comprise the entire coding sequence of a novel protein. While a full description of these clones will be published elsewhere (FT, DF, RS and TJG, manuscript in preparation), the major feature of the sequence is that it is predicted to encode a protein of molecular weight 152kD. Thus we believe that the clones correspond to p160 rather than p67; while somewhat surprising, this was at least consistent with the V8 protease mapping data which indicated that p67 and p160 are closely related[38]. This in turn suggests that p67 may be derived from p160 by proteolytic cleavage, a notion supported by the presence of all five tryptic peptide sequences in the amino-terminal portion of the predicted p160 sequence. If this is the case, cleavage must be tightly regulated since p67 has been found in only a limited number of cell lines to date.

Possible Functions of the NRD and NRD-Binding Proteins.

Currently, we are in the process of characterising the ability of the translation products of these clones to bind to the Myb and Jun leucine zippers, as well as attempting to determine what effect they may have on Myb function. As discussed above, we embarked on the identification of leucine zipper-binding proteins because of the negative regulatory function of the leucine zipper. Thus we imagined that proteins that interact with the leucine zipper might inhibit DNA binding and/or transactivation. Since p67 and p160 were identified solely on the basis of their ability to bind to the Myb leucine zipper, however, we have no evidence that they are responsible for the inhibitory effects of the NRD and could, in principle, play other roles. For example, one or both of p67 and p160 may act to prevent Myb homodimer formation and thus enhance the activity of full-length c-Myb, since c-Myb homodimers are likely to be inactive in DNA binding or transactivation[37]. The level of active Myb protein could be regulated in part by the amounts of p67 and/or p160 present in the cell. An additional level of regulation could be provided by post-translational modification of c-Myb or p67/p160, eg by phosphorylation, which in turn could modulate the degree of association between these proteins.

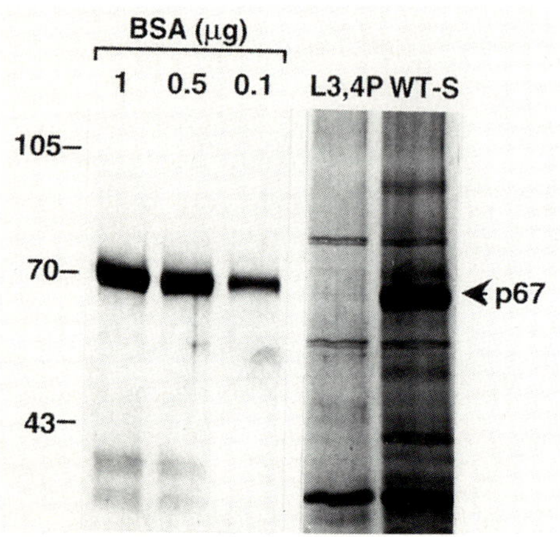

Fig. 4. Partial purification of p67. The left-hand lanes contain BSA standards, while the two right-hand lanes contain 5% of each preparation of nuclear proteins affinity-purified from FDC-P1 cells by binding to the indicated mutant and wild-type GST-NRD fusion proteins. The indicted p67 band was excised from the remainder of the sample and used for tryptic digestion and microsequencing. Proteins were detected by Coomassie Blue staining.

There are still other conceivable functions for p67 and p160. For example one or both might function as transcriptional "co-activators", although such an activity could not be essential for all Myb functions since deletion or mutation of the leucine zipper actually enhances transactivation of reporter constructs as well as transformation[15,31,33]. However, as pointed out above, Myb can enhance transcription of genes that are associated more with mature cell function than with proliferation or an immature state (eg *mim*-1(ref. 17), lysozyme[23], and CD4 (ref. 24). It is conceivable that p67 and/or p160 may be required specifically for transcription of such genes, either by activating transcription or by involvement in interactions between Myb and other factors, such as Ets[39] and NF-M[40,41], with which Myb synergises. Prevention of Myb's ability to activate some "differentiation genes" would not interfere with, and may actually augment, its ability to transform haemopoietic cells.

Transformation by Myb and the Role of the NRD

Discussion of the possible effects of NRD-binding proteins on Myb function leads back to the more general question of the role of the NRD. While it is clear that c-Myb can transactivate and bind to DNA, albeit less efficiently than carboxyl-truncated forms (ie with a disrupted NRD), studies in the avian system have found that c-Myb has no detectable transforming activity[7]. On the other hand, published studies from our laboratory indicate that murine Myb constructs with a complete carboxyl terminus have very weak - but detectable - transforming activity[6,33]. Moreover, constitutive expression of c-Myb is, as mentioned previously, capable of blocking the induced differentiation of myeloid and erythroid cell lines. It is hard to imagine that this activity of Myb is unrelated to its ability to transform primary haemopoietic cells, since transformation of primary haemopoietic cells by Myb similarly results in the maintenance of a relatively immature proliferative state. In view of these observations, and recent experiments (EM, PF and TJG, unpublished) in which we have observed higher levels of transformation by c-Myb than previously reported[6,15,33], we have reinvestigated the ability of c-Myb to transform murine haemopoietic cells.

Transformation by c-Myb in the Murine System

Because the retroviral vectors used in our previous studies of transformation all expressed forms of Myb that (in addition to any other alteration) lacked the amino-terminal 18 residues of c-Myb[6], it was necessary for the present studies to use a vector that expressed completely normal c-Myb. For this purpose, we used the RUFNeo retroviral vector[42] or a derivative thereof that lacks the Neo^R gene (J.Rayner, EM, PF and TJG, unpublished); in each case insertion of appropriate Myb constructs into the multiple cloning site allowed expression of proteins with a complete amino terminus.

In considering factors that may account for variability, both within our own experiments and possibly between murine and avian systems, in the degree of transformation observed with c-Myb, we focussed on the conditions under which *myb* retrovirus-infected primary cells were cultured. This was in part because we had previously observed[43] that suspension culture of *myb*-transformed haemopoietic cells (MTHCs) required an as-yet-unidentified autocrine growth factor in addition to GM-CSF. Our results support this notion, as we have found that the degree of transformation (measured using the colony assay described in ref. 6) obtained depends on the density at which the infected cells are cultured (PF, L. Ashman and TJG, submitted). Importantly, we also found that the relative transforming abilities of normal and truncated c-Myb varies with cell density. This is illustrated in Table 1, which shows the result of an experiment in which foetal liver cells were infected, as described previously[6,15,33], with the RUFNeo(FL) and RUFNeo(CT3) viruses which express full-length c-Myb and CT3Myb[31] (which lacks the entire NRD), respectively.

Table 1. Effect of cell density on transformation by *myb* retroviruses

Virus	Colonies[a] formed following suspension culture at	
	2×10^4/ml	1.6×10^5/ml
none	0	0
RUFNeo	0	0
RUFNeo(FL)	2.5	24.5
RUFNeo(CT3)	27.5	38

[a]Colonies formed per 5,000 cells plated in methylcellulose-containing medium; each value is the mean of duplicate assays.

Following suspension culture at two different densities for 7 days in the presence of GM-CSF, cells were plated (at the same density in all cases) in semi-solid medium. (Note that uninfected cells, or cells infected with the RUFNeo vector, did not generate colonies under these conditions.) Culturing cells infected with RUFNeo(FL) at the higher density generated 10-fold more colonies than culturing the same cells at low density, whereas the difference seen with RUFNeo(CT3)-infected cells was only 1.4 fold. Correspondingly, it can be seen that at the higher density, the difference in transformation between CT3 Myb and normal c-Myb was only 1.6-fold, while at the lower density the difference was much greater (11-fold). This effect has been confirmed in similar experiments in which the infected cells were cultured at a wider ranger of densities (PF, L. Ashman and TJG, submitted).

The most probable explanation for the effects of cell density involves the production of and/or response to an autocrine (or paracrine) factor that is required for the proliferation or survival of MTHCs, as reported previously[43]. Since it appears that CT3Myb-transformed cells are capable of survival and proliferation in suspension culture at lower cell density than are c-Myb-transformed cells, it is likely that CT3Myb-expressing cells either produce more of the autocrine factor or respond to lower concentrations. This in turn implies that the production of or responsiveness to this factor depends on Myb activity, suggesting either that transcription of the gene encoding the autocrine factor is regulated by Myb itself, or that Myb is involved in the signalling pathways triggered by the factor. Burk *et al.*[41] have also raised the possibility that Myb may be involved in growth factor-activated signalling pathways on the basis of their studies showing cooperation between Myb and C/EBP transcription factors.

To summarise these findings, we can conclude that enforced expression of c-Myb can indeed result in transformation of primary haemopoietic cells under appropriate conditions. The difference in transforming activity between c-Myb and truncated Myb (lacking the NRD) may be more a reflection of their effects on growth factor production or responsiveness than of their abilities to maintain or enforce an immature phenotype. Whether this is due to different thresholds of Myb activity for these two effects or to a more qualitative difference caused by deletion of the entire NRD, including components additional to the leucine zipper (see above), remains to be determined.

Conclusions

Observations from our laboratory and from several other groups have indicated that the regulation of c-Myb activity by elements in the carboxyl region of the protein - the NRD - is complex and probably results from the actions of several distinct elements. Since one such element is a leucine zipper motif, it is likely that interactions between c-Myb and other cellular proteins contribute to the modulation of Myb activity. Two such proteins, p67 and p160, have recently been identified and partially characterised.

Disruption of the NRD clearly enhances transformation of haemopoietic cells by Myb but, contrary to prevalent beliefs, enforced expression of normal c-Myb is also capable of transformation at least in the murine system. Differences in the properties of cells transformed by normal c-Myb and forms lacking the NRD suggest that production of or response to an autocrine growth factor may play a role in transformation by Myb.

Note: Figure 2 is reproduced with permission from Stockton Press Ltd, Hampshire, UK

References

1. Gonda TJ, Metcalf D (1984) Expression of *myb*, *myc* and *fos* proto-oncogenes during the differentiation of a murine myeloid leukaemia. Nature 310:249-251
2. Westin EH, Gallo RC, Arya SK, Eva A, Souza LM, Baluda MA, Aaronson SA, Wong-Staal F (1982) Differential expression of the *amv* gene in human hematopoietic cells. Proc Natl Acad Sci USA 79:2194-2198
3. Mucenski ML, McLain K, Kier AB, Swerdlow SH, Schreiner CM, Miller TA, Peitryga DW, Scott Jr WJ, Potter SS (1991) A functional c-*myb* gene is required for normal murine fetal hepatic hematopoiesis. Cell 65:677-689
4. Anfossi G, Gewirtz AM, Calabretta B (1989) An oligomer complementary to c-*myb*-encoded mRNA inhibits proliferation of human myeloid leukemia cell lines. Proc Natl Acad Sci USA 86:3379-3383

5. Moscovici C, Gazzolo L (1982) Transformation of hemopoietic cells with avian leukaemia viruses. In: Klein G, (ed) Advances in Viral Oncology. Raven Press, New York, Vol 1, pp 83-106
6. Gonda TJ, Buckmaster C, Ramsay RG (1989) Activation of c- *myb* by carboxy-terminal truncation: relationship to transformation of murine haemopoietic cells *in vitro*. EMBO J 8:1777-1783
7. Grasser FA, Graf T, Lipsick JS (1991) Protein truncation is required for the activation of the c-*myb* proto- oncogene. Mol Cell Biol 11:3987-3996
8. Clarke MF, Kukowska-Latallo JF, Westin E, Smith M and Prochownik EV. (1988) Constitutive expression of a c-myb cDNA blocks Friend murine erythroleukemia cell differentiation. Mol Cell Biol 8:884-892.
9. Yanagisawa H, Nagasawa T, Kuramochi S, Abe T, Ikawa Y, Todokoro K (1991) Constitutive expression of exogenous c- *myb* gene causes maturation block in monocyte-macrophage differentiation. Biochimica et Biophysica Acta 1088:380- 384.
10. Patel G, Kreider B, Rovera G and Reddy EP. (1993) v-*myb* blocks granulocyte colony stimulating factor-inuced myeloid cell differentiation but not proliferation. Mol Cell Biol 13:2269-2276
11. Biedenkapp, H., Borgmeyer, U., Sippel, A.E. & Klempnauer, K.-H. (1988) Viral *myb* oncogene encodes a sequence-specific DNA binding activity. Nature 335:835-837.
12. Nishina, Y., Nakagoshi, H., Imamoto, F., Gonda, T.J. & Ishii, S. (1989) Trans-activation by the *c-myb* proto-oncogene. Nuc. Acids. Res. 17:107-117.
13. Weston K, Bishop JM (1989) Transcriptional activation by the v-*myb* oncogene and its cellular progenitor, c-*myb*. Cell 58:85-93
14. Nakagoshi H, Nagase T, Kanei-Ishii C, Ueno Y, Ishii S (1990) Binding of the c-*myb* proto-oncogene product to the simian virus 40 enhancer stimulates transcription. J Biol Chem 265:3479-3483
15. Hu Y, Ramsay RG, Kanei-Ishii C, Ishii S, Gonda TJ (1991) Transformation by carboxyl-deleted Myb reflects increased transactivating capacity and disruption of a negative regulatory domain. Oncogene 6:1549-1553
16. Lane, T., Ibanez, C., Garcia, A., Graf, T. & Lipsick, J. (1990) Transformation by v-*myb* correlates with transactivation of gene exoression. *Mol.Cell.Biol* 10:2591-2598.
17. Ness SA, Marknell A, Graf T (1989) The v-*myb* oncogene product binds to and activates the promyelocyte-specific *mim-1* gene. Cell 59:1115-1125
18. Evans JL, Moore TL, Kuehl WM, Bender T, Ting JP-Y (1990) Functional analysis of c-Myb protein in T-lymphocytic cell lines shows that it *trans*-activates the c-*myc* promoter. Mol Cell Biol 10:5747-5752
19. Zobel A, Kalkbrenner F, Guehmann S, Nawrath M, Vorbrueggen G and Moelling K. (1991) Interaction of the v- and c-Myb proteins with regulatory sequences of the human c-myc gene Oncogene 6:1397-1407
20. Durban EM, Boettiger D (1981) Differential effects of transforming avian RNA tumor viruses on avian macrophages. Proc Natl Acad Sci USA 78:3600-3604
21. Venturelli D, Travali S, Calabretta B (1990) Inhibition of T-cell proliferation by a MYB antisense oligomer is accompanied by selective down-regulation of DNA polymerase α expression. Proc Natl Acad Sci USA 87:5963- 5967
22. Ku D-H, Wen S-c, Engelhard A, Nicolaides NC, Lipson KE Marino TA and Calabretta B (1993) c-*myb* transactivates *cdc2* expression via Myb binding sites in the 5'-flanking region of the human *cdc2* gene. J Biol Chem 268:2255-2259
23. Introna M, Golay J, Frampton J, Nakano T, Ness SA and Graf T (1990) Mutations in v-*myb* alter the differentiation of myelomonocytic cells transformed by the oncogene. Cell 63:1287-1297
24. Siu G, Wurster AL, Lipsick J and Hendrick SM (1992) Expression of the CD4 gene requires a Myb transcription factor. Mol Cell Biol 12:1592-1604

25. Selvakumaran M, Lieberman DA and Hoffman-Liebermann B. (1992) Deregulated c-*myb* disrupts Interleukin-6- or Leukemia Inhibitory Factor-induced myeloid differentiation prior to c-*myc*: role in leukemogenesis. Mol. Cell. Biol. 12:2493-2500.
26. Ramsay RG, Ikeda K, Rifkind RA and Marks PA (1986) Changes in gene expression associated with induced differentiation of erythroleukemia: Proto-oncogenes, globin genes and cell division. Proc Natl Acad Sci USA 83:6849-6853.
27. Dautry F, Weil D, Yu J, Dautry-Varsat A (1988) Regulation of pim and myb mRNA accumulation by interleukin 2 and interleukin 3 in murine hematopoietic cell lines. J Biol Chem 263:17615-17620.
28. Asura M, Luchetti MM, Erba E, Golay J, Rambaldi A & Introna, M. (1994) Dissociation between ^{p93}B-myb and ^{p75}c-myb expression during the proliferation and differentiation of human myeloid cell lines. Blood 83:1778-1790.
29. Catron KM, Purkerson JM, Isakson PC & Bender TM. (1992) Constitutive versus cell cycle regulation of c-*myb* mRNA expression correlates with developmental stages in murine B lymphoid tumours. J Immunol 148:934-942.
30. Shen-Ong GLC, Potter M, Mushinski JF, Lavu S, Reddy EP (1984) Activation of the c-*myb* locus by viral insertional mutagenesis in plasmacytoid lymphosarcomas. Science 226:1077-1080
31. Sakura H, Kanei-Ishii C, Nagase T, Nakagoshi H, Gonda TJ, Ishii S (1989) Delineation of three functional domains of the transcriptional activator encoded by the c-*myb* protooncogene. Proc Natl Acad Sci USA 86:5758-5762
32. Ramsay RG, Ishii S, Gonda TJ (1991) Increase in specific DNA binding by carboxyl truncation suggests a mechanism for activation of Myb. Oncogene 6:1875-1879
33. Kanei-Ishii C, Macmillan EM, Nomura T, Sarai A, Ramsay RG, Aimoto S, Ishii S, Gonda TJ (1992) Transactivation and transformation by Myb are negatively regulated by a leucine zipper structure. Proc Natl Acad Sci USA 89:3088-3092.
34. Kalkbrenner F, Guehmann S and Moelling K. (1990) Transcriptional activation by human c-*myb* and v-*myb* genes. Oncogene 5:657-661.
35. Dubendorff JW, Whittaker LJ and Lipsick, JL. (1992) Carboxy-terminal elemnts of c-Myb negatively regulate transcriptional activation in *cis* and *trans*. Genes Dev. 6:2424-2535.
36. Vorbrueggen G, Kalkbrenner F, Guehmann S and Moelling K. (1994) The caroboxyterminus of human c-myb protein stimulates activated transcription in *trans*. Nuceic Acids Res. 22:2466-2475.
37. Nomura T, Sakai N, Sarai A, Sudo T, Kanei-Ishii C, Ramsay RG, Favier D, Gonda TJ and Ishii S. (1993) Negative autoregulation of c-Myb activity by homodimer formation through the leucine zipper. J Biol. Chem 268:21914-21923
38. Favier D and Gonda TJ. (1994) Detection of proteins that bind to the leucine zipper motif of c-Myb. Oncogene 9:305-311
39. Dudek H, Tantravahi RV, Rao VN, Reddy ESP and Reddy EP (1992) Myb and Ets proteinbs cooperate in transcriptional activation of the *mim-1* promoter. Proc Natl Acad Sci USA 89:1291-1295
40. Ness SA, Kowenz-Leutz E, Casini T, Graf T and Leutz A. (1993) Myb and NF-M: combinatorial activators of myeloid genes in heterologous cell types. Genes Dev. 7:749-759
41. Burk O, Mink S, Ringwald M and Klempnauer K-H. (1993) Synergistic activation of the chicken *mim*-1 gene by v-*myb* and C/EBP transcription factors. EMBO J 12:2027-2938
42. Rayner JR and Gonda TJ. (1994) A simple and efficient procedure for generating stable expression libraries by cDNA cloning in a retroviral vector. Mol. Cell. Biol. 14:880-887.
43. Macmillan EM and Gonda TJ. (1994) Murine myeloid cells transformed by *myb* require fibroblast-derived or autocrine growth factors in addition to granulocyte-macrophage colony-stimulating factor for proliferation. Blood 83:209-216.

Myeloid-Specific Transcription

Terminal Myeloid Gene Expression and Differentiation Requires the Transcription Factor PU.1

M.C. Simon[1,2,3,4], M. Olson[1], E. Scott[1], A. Hack[1], G. Su[4], and H. Singh[1,3,4]
[1]Howard Hughes Medical Institute, Departments of [2]Medicine and [3]Molecular Genetics and Cell Biology, [4]Committee on Immunology, University of Chicago, Chicago, IL 60637

Introduction

Hematopoiesis is a multi-stage developmental process which yields at least eight distinct lineages including monocytes, granulocytes, lymphocytes, megakaryocytes, and erythrocytes. During myelopoiesis, pluripotent hematopoietic stem cells become committed as myeloid precursor cells which differentiate into morphologically and functionally distinct end-stage macrophages and neutrophils. Multipotent progenitors differentiate into macrophages through successive intermediates involving monoblasts, promonocytes, monocytes, and then macrophages. Granulocytic development involves the differentiation of progenitors into myeloblasts, promyelocytes, myelocytes, and then neutrophils. Hematopoiesis in the developing mouse embryo is initiated in the yolk sac on the seventh day of gestation. Primitive macrophages first appear in yolk sac blood islands on day 9, bypassing the differentiation pathway of the monocytic series, to become fetal macrophages in various tissues [13, 22]. Yolk sac monocytes and macrophages differ in enzyme cytochemistry, structural characteristics, and functional properties when compared to monocytic cells generated at later stages of hematopoetic development. For example, primitive macrophages lack myeloperoxidase, an enzyme that is prominent in fetal and adult myeloid cells. Granulocytes are not produced in large numbers in the yolk sac. Beyond day 12 of gestation, yolk sac hematopoiesis declines and the fetal liver becomes the predominant hematopoetic organ until birth. During the fetal liver period, myeloid cells are a minor population initially, increase in numbers in the late fetal stage, and include both monocytes and granulocytes [13]. In adult animals, bone marrow stem cells give rise to neutrophils and monocyte-derived macrophages. Myeloid development appears to be regulated by a combination of hematopoietic growth factors, growth factor receptors, and lineage restricted transcription factors.

The transcription factor PU.1 had previously been implicated as an important regulator of hematopoiesis. PU.1 is expressed exclusively in the hematopoietic system; it is highly expressed in B lymphocytic, granulocytic, and monocytic cells and expressed to a lesser extent in immature erythroid cells [5, 6, 8]. PU.1 has many presumptive targets in the B and monocytic lineages. In the monocytic lineage these include the genes for the macrophage colony-stimulating factor receptor (M-CSFR), the granulocyte colony-stimulating factor receptor (G-CSFR), the granulocyte-monocyte colony-stimulating factor receptor (GM-CSFR), the Fc-receptor, the scavenger receptor, and the myeloid integrin CD-11b [12, 14, 15, 21, 26]. We have recently analyzed hematopoietic defects caused by a mutation in the murine PU.1 locus generated using gene targeting in embryonic stem (ES) cells [19]. Disruption of the PU.1 gene results in prenatal lethality as homozygous mutant embryos die between days 16 and 18 of gestation. Mutant embryos produce normal numbers of megakaryocytes and erythroid precursors, however the numbers of enucleated erythrocytes are reduced in some of the day 16.5 embryos. An invariant consequence of the mutation is a multilineage defect in the development of B and T lymphocytes, monocytes, and granulocytes. Histological staining for lysozyme and myeloperoxidase as well as FACs analysis (Mac -1, GR-1) suggested an inhibition of myelopoiesis at the level of differentiating progenitors. Consistent with these findings, colony-forming assays using yolk sac and fetal liver cells from mutant embryos yield normal numbers of CFU-E and CFU-Meg, but no CFU-GM, CFU-G, CFU-M or CFU-GEMM (Olson, M., E. W. Scott, A. Hack, G. Su, H. Singh, and M. C. Simon, manuscript in preparation).

Many of the putative target genes of PU.1 are thought to be critical for early myeloid cell proliferation and differentiation. In this report, we summarize the effect of the PU.1 mutation on the expression of a number of hematopoietic and myeloid cell-specific genes. We have analyzed gene expression in the yolk sac and fetal liver of mutant mice and also during the *in vitro* differentiation of PU.1 mutant ES cells into hematopoietic cells. We conclude that several early myeloid target genes are transcribed normally in the absence of PU.1, reflecting their probable expression in multipotent progenitor cells. Therefore, altered expression of these genes such as GM-CSFR and G-CSFR cannot account for the lack of myeloid cells in mutant mice. In contrast, later stage myeloid target genes are expressed at lower levels or not at all. We conclude that terminal myeloid gene expression and/or differentiation requires an active PU.1 protein.

Myeloid Gene Expression in PU.1 Mutant Embryos

To determine if early myeloid maturation or gene expression is taking place in PU.1$^{-/-}$ embryos, sensitive RT-PCR analyses were performed with myeloid cell-specific genes and the results are summarized in Table 1.

Table 1. Summary of gene expression in yolk sac, fetal liver, and differentiated ES cells

Gene	Yolk Sac			Fetal Liver			Differentiated ES cells[a]		
	WT[b]	Het	Hom	WT	Het	Hom	WT[b]	Het	Hom
HPRT	+[c]	+	+	+	+	+	+	+	+
GATA-2	+	+	+	+	+	+	+	+	+
GM-CSFR	+	+	+	+	+	+	+	+	+
G-CSFR	+/-	+/-	+/-	+	+	+/-	+	+	+/+
MPO	+	+	+/-	+	+	+/-	+	+	+/-
c-fes	nd[d]	nd	nd	nd	nd	nd	+	+	-
CD-18	+	+	+	+	+	+	+	+	+
CD-11b	+/-	+/-	-	+	+	+/-	+	+	-
M-CSFR	+	+	+	+	+	+	+	+	-
CD-64	nd	nd	nd	nd	nd	nd	+	+	-
ß-globin	nd	nd	nd	nd	nd	nd	+	+	+

[a] ES cells were differentiated for 11 days

b RNA was isolated from wild-type (WT), PU.1 Heterozygous Mutant (Het), and PU.1 Homozygous Mutant (Hom) samples

c The PCR reactions were electrophoresed and scored as detectable signal (+), trace signal (+/-), and no signal (-)

d Not determined

To compensate for variable RNA yields, the amount of cDNA synthesized was calibrated by using the relative expression of hypoxanthine phosphoribosyltransferase (HPRT).

GM-CSFR binds granulocyte-macrophage colony stimulating factor which stimulates the proliferation and differentiation of granulocyte and macrophage precursors as well as their more mature counterparts [24]. G-CSFR is another receptor that, upon ligand binding, stimulates both multipotent precursors and cells restricted to the neutrophilic granulocyte lineage [4]. Myeloperoxidase (MPO) is an enzyme exclusively found in myeloid cells and synthesized in the azurophilic granules of promyelocytes and promonocytes [10]. The CD-11b/CD-18 complex forms the Mac-1 surface receptor on monocytes, macrophages, and granulocytes. Mac-1 is first expressed during the myelocytic and monoblastic developmental stages and is upregulated during granulocytic and monocytic differentiation [17]. The receptor for M-CSF is expressed at high levels in monocytes and macrophages but is poorly expressed in immature myeloid cells and mature granulocytes [20]. RT-PCR analysis of myeloid gene expression in the day 15.5 fetal livers of wild type, heterozygous, and homozygous embryos indicates that the MPO and CD-11b genes are expressed at lower but detectable levels in PU.1 mutants (see Table 1). In contrast, GM-CSFR, G-CSFR, CD-18, and M-CSFR genes are expressed at approximately the same levels in the mutant as the wild type and heterozygous fetal livers. Myeloid gene expression was also analyzed in pooled samples of day 10.5 wild type, heterozygous, and homozygous mutant yolk sacs. CD-11b is not expressed in PU.1 mutant yolk sac cells. However, mRNAs for many of the other myeloid genes including GM-CSFR, G-CSFR, MPO, CD-18, and M-CSFR are detected. In both the fetal liver and yolk sac preparations, the failure to see pronounced effects on G-CSFR and M-CSFR expression may be attributed to the presence of non-myeloid cells which are capable of expressing these genes. G-CSFR is predominantly found on precursor and mature neutrophils, but is also expressed on endothelial cells, placenta, and trophoblastic cells. M-CSFR is expressed during mouse embryogenesis in trophoblastic cells and in unidentified cells in the visceral yolk sac [16]. However, GM-CSFR is not expressed in non-hematopoietic tissues. Therefore expression of GM-CSFR in hematopoietic cells is not dependent on PU.1 and cannot account for the myeloid defects.

PU.1 is Required for the Development of Myeloid Cells during ES Cell Differentiation

To further explore the role of PU.1 in myelopoiesis, we have developed a simplified *in vitro* system based on differentiation of ES cells into hematopoietic cells. This differentiation system recapitulates mouse hematopoietic development at 6.5-7.5 days of gestation [7]. We generated ES cells that lack intact PU.1 alleles by sequential gene targeting and differentiated them into embryoid bodies (EBs) that contain multiple hematopoietic cell lineages. In order to determine if granulocytes and monocytes/macrophages were being produced, cytospin preparations were performed on wild type, heterozygous and PU.1 homozygous mutant EBs after 11 days of *in vitro* differentiation. When stained with May-Grunwald-Geimsa or an antibody to the macrophage-specific marker F4/80, no granulocytes and monocytes/macrophages are detected in the PU.1 homozygous mutant embryoid bodies EBs. In contrast, wild type and heterozygous EBs produce large numbers of myeloid cells.

The expression of a number of myeloid-specific genes was analyzed including GM-CSFR, G-CSFR, MPO, *c-fes,* CD-11b, CD-18, M-CSFR, and CD-64. The high affinity receptor of IgG, CD-64 (FCγRI), is expressed on monocytes, macrophages, and activated neutrophils [1]. *c-fes* is a tyrosine kinase that is expressed in monocytes, macrophages, neutrophils, and their precursors [11, 18]. It has also been detected in purified CD-34$^+$ hematopoietic stem cells and progenitors [3]. The genes expressed in early stages of myelopoiesis include GM-CSFR, G-CSFR, MPO, and *c-fes*. We also analyzed the expression of two additional genes associated with early hematopoiesis: CD-34 and GATA-2. CD-34 is detected in both hematopoietic progenitors and stem cells. The human CD-34 protein is only expressed on progenitors, and is lacking on terminally differentiated cells [2]. Its presence on murine stem cells (CFU-S) has been confirmed [9]. GATA-2 has been shown to be critical in early progenitor cell development [23]. From our *in vitro* cultures it is clear that CD-34 is expressed in differentiated wild type, heterozygous and homozygous mutant EBs. We obtained similar results for GATA-2 and also observe wild type levels of GATA-2 expression in the yolk sacs and fetal livers of mutant embryos. GM-CSFR, G-CSFR, and MPO are also expressed in the differentiated wild type, heterozygous, and PU.1 homozygous mutant EBs. Unlike other early myeloid genes, *c-fes* expression is significantly reduced by the PU.1 mutation. We have recently demonstrated that a functional binding site for PU.1 is located within the *c-fes* promoter (Juang, G., A. Heydemann, K. Hennessy, M. Parmacek, and M. C. Simon, manuscript submitted). To date, *c-fes* is the only early myeloid gene whose expression is significantly diminished in the PU.1 mutant EBs.

When RT-PCR analysis was carried out on genes expressed later in myeloid development, the expression of several genes was found to be significantly reduced by the PU.1 mutation. In the PU.1 mutant EBs, CD-18 expression is notably lower whereas CD-11b, M-CSFR and CD-64 are completely absent. ß-major globin, which is expressed exclusively in the erythroid lineage, is not altered by the PU.1 mutation indicating that erythroid production proceeds normally in the PU.1 null EBs. Taking the cytological and RT-PCR data together, it appears that PU.1 is not essential for much of the early program of myeloid gene expression but becomes critical at later stages.

Discussion

We have previously demonstrated using gene targeting that there is a multi-lineage defect in the generation of lymphocytic, monocytic, and granulocytic cells in day 16.5 PU.1$^{-/-}$ embryos. In the current studies, we have focused on how the PU.1 mutation affects myeloid development and gene expression. By analyzing embryonic and fetal hematopoietic organs from PU.1$^{-/-}$ mice and by *in vitro* differentiation of PU.1 null ES cells, we have observed a notable effect on late myeloid gene expression. Importantly, the expression of many early myeloid targets of PU.1 is unimpaired by the PU.1 mutation.

It is possible that myelopoesis is initiated but arrested at an early stage of myeloid development. In our studies, the CD34, GATA-2, GM-CSFR , and G-CSFR genes are expressed at wild type levels in PU.1$^{-/-}$ EBs. We also detect MPO mRNA in PU.1$^{-/-}$ EBs and yolk sacs. This enzyme is exclusively found in myeloid cells; specifically in the azurophilic granules. MPO mRNA can be detected in late myeloblastic, promyelocytic and promonocytic stages of differentiation but it decreases upon further myeloid maturation [17]. In our previous study, the MPO protein could not be detected in the PU.1$^{-/-}$ fetal livers by histochemical staining. We do detect low levels of MPO mRNA in the fetal livers of mutant embryos. It is possible that early progenitors transcribe low levels of MPO RNA and further differentiation is required for efficient expression of MPO protein. The results of the RT-PCR analysis of early hematopoietic markers and MPO suggest that hematopoietic progenitors including early myeloid precursors are not altered by the PU.1 mutation.

In the *in vitro* system, a number of late myeloid genes are not expressed in the PU.1$^{-/-}$ differentiated ES cells including CD-11b, CD-64, and M-CSFR. Low levels of CD-18 are detectable. The CD-11b/CD-18 (Mac-1) complex is present on myelocytes, monoblasts, granulocytes, monocytes, and macrophages. However, CD-18 is also part of the LFA-1 surface complex found at the late myeloblastic stage of development. If myeloid differentiation is initiated but does not progress further, the low levels of CD-18 may be due to the presence of LFA-1 on these early myeloid precursors. Previous studies have shown that M-CSFR is expressed during mouse embryogenesis in trophoblastic cells and in unidentified cells in the visceral yolk sac. This may explain why we observe this late myeloid marker in the yolk sac but not in the differentiated ES cells.

It is significant that several presumptive PU.1 target genes are unaffected by the PU.1 mutation: GM-CSFR and G-CSFR. GM-CSFR and G-CSFR could be regulated by the related ETS-family member, Spi-B, which is also present in early progenitor cells. One early myeloid target gene for PU.1, *c-fes*, is expressed at substantially lower levels in PU.1 mutant EBs. *c-fes* is expressed in the same cells that are positive for CD-34, GATA-2, GM-CSFR and G-CSFR. Given that these cells are present in PU.1$^{-/-}$ EBs, we conclude that PU.1 is essential for *c-fes* expression. *c-fes* has also been implicated in myeloid maturation [25] and therefore could account for the arrest in myeloid development that we observe in PU.1 mutants.

In conclusion, early myelopoiesis is relatively unaffected by a mutation in the PU.1 gene. We suggest that multipotent progenitors (expressing GATA-2, GM-CSFR, G-CSFR, and CD34) are formed at normal numbers. Two early markers of myelopoeisis are also transcribed: MPO and CD-18. However, later developmental events are blocked and later markers are absent: CD-11b, M-CSFR and CD-64. We have established an *in vitro* model of early myelopoiesis that is dependent on an intact PU.1 gene. These cells will be extremely useful for further studies on the role of PU.1 in the commitment of multipotent progenitors to myeloid development.

References

1. Allen J, Seed B (1989) Isolation and expression of functional high affinity Fc receptor complementary DNAs. Science 243:378-381
2. Berenson RJ, Andrews RG, Bensinger WI, Kalamasz D, Knitter G, Buchner CD, D Bernstein (1988) Antigen CD34+ marrow cells engraft lethally irradiated baboons. Journal of Clinical Investigation 81:951
3. Care A, Mattia G, Montesoro E, Parolini I, Russo G, Colombo M, Peschle C (1994) c-fes expression in ontogenetic development and hematopoietic differentiation. Oncogene 9:739-747
4. Demitri GD, Griffin JD (1991) Granulocyte colony-stimulating factor and its receptor. Blood 78:2791-2808

5. Galson D, Hensold J, Bishop T, Schalling M, D'Andrea A, Jones C, Auron P, Houseman D (1993) Mouse b-globin DNA-binding protein B1 is identical to a proto-oncogene, the transcription factor Spi-1/Pu.1, and is restricted in expression to hematopoietic cells. Mol. Cell. Biol. 13:2929-2941
6. Hromas R, Orazi A, Neiman R, Maki R, Van Beveran C, Moore J, Klemsz M (1993) Hematopoietic lineage- and stage-restricted expression of the ETS oncogene family member PU.1. Blood 82:2998-3004
7. Keller G, Kennedy M, Papayannopoulou T, Wiles M (1993) Hematopoietic commitment during embryonic stem cell differentiation in culture. Mol. Cell. Biol. 13:473-486
8. Klemsz M, McKercher S, Celada A, Van Beveren C, Maki R (1990) The macrophage and B cell-specific transcription factor PU.1 is related to the ets oncogene. Cell 61:113-124
9. Krause DS, Ito T, Fackler MJ, Smith OM, Collector MI, Sharkis SJ, WS May (1994) Characterization of murine CD34, a marker for hematopoietic progenitor and stem cells. Blood 84:691-701
10. Lubbert M, Miller C, Koeffler H (1991) Changes of DNA methylation and chromatin structure in the human myeloperoxidase gene during myeloid differentiation. Blood 78:345-356
11. MacDonald I, Levy J, Pawson T (1985) Expression of the mammalian c-fes protein in hematopoietic cells and identification of a fes-related protein. MCB 5:2543-2551
12. Moulton K, Semple K, Wu H, Glass C (1994) Cell-Specific Expression of the Macrophase Scavenger Receptor Gene Is Dependent on PU.1 and a Composite AP-1/ets Motif. Molecular and Cellular Biology 14:4408-4418
13. Naito M (1993) Macrophage heterogeneity in development and differentiation. Archives of Histology and Cytology 56:331-351
14. Pahl H, Scheibe R, Zang D, Chen H-M, Galson D, Maki R, Tenen D (1993) The proto-oncognene PU.1 regulates expression of the myeloid-specific CD11b promoter. J. Biol. Chem. 7:5014-5020
15. Perez C, Wietzerbin J, Benech P (1993) Two cis-DNA elements involved in myeloid cell-specific expression and gamma interferon (IFN-γ) activation of the human high-affinity Fcγ receptor gene: a novel IFN regulatory mechanism. Mol. Cell. Biol. 13:2182-2192
16. Regenstreif LJ, Rossant J (1989) Expression of the *c-fms* proto-oncogene and of the cytokine, CSF-1, during mouse embryogenesis. Developmental Biology 133:284-294
17. Rosmarin A, Weil S, Rosner G, Griffin J, Arnaout M, Tenen D (1989) Differential expression of CD11b/CD18 (Mol) and myeloperoxidase genes during myeloid differentiation. Blood 73:131
18. Samarut J, Mathey-Prevot B, Hanafusa H (1985) Preferential expression of the c-Fps protein in chicken macrophages and granulocytic cells. Mol. Cell. Biol. 5:1067-1073
19. Scott E, Simon M, Anastasi J, Singh H (1994) Requirement of transcription factor PU.1 in the development of multiple hematopoietic lineages. Science 265:1573-1577
20. Sherr CJ (1990) Colony-stimulating factor-1 receptor. Blood 75:1
21. Smith LA, Gonzalez DA, Hohaus S, Tenen D (1994) The myeloid specific granulocyte colony stimulating factor (G-CSF) receptor promoter contains a functional site for the myeloid transcription factor PU.1 (Spi-1). Blood 84:372a
22. Takahashi K, Yamamura F, Naito M (1989) Differentiation, maturation, and proliferation of macrophages in the mouse yolk sac. Journal of Leukocyte Biology 45:87-96

23. Tsai F -Y, Keller G, Kuo FC, Weiss M, Chen J, Rosenblatt M, Alt FW, Orkin SH (1994) An early heamatopoietic defect in mice lacking the transcription factor GATA-2. Nature 371:221-226
24. Wognum AW, Westerman Y, Visser TP, Wagemaker G (1994) Distribution of receptors for granulocyte-macrophage colony-stimulating factor on immature CD34+ bone marrow cells, differentiating monomyeloid progenitors, and mature blood cell subsets. Blood 84:764-774
25. Yu G, Grant S, Glazer R (1988) Association of p93$^{c\text{-}fes}$ tyrosine protein kinase with granulocytic/monocytic differentiation and resistance to differentiating agents in HL-60 leukemia cells. Mol. Pharmacol. 33:384-388.
26. Zhang D, Hetherington C, Chen H, Tenen D (1994) The macrophase transcription factor PU.1 directs tissue-specific expression of the macrophase colony-stimulating factor receptor. Mol. Cell. Biol. 14:373-381

A Role for STAT Family Transcription Factors in Myeloid Differentiation

F. Barahmand-pour, A. Meinke, Matthias Kieslinger, A. Eilers and T. Decker
Vienna Biocenter, Institute of Microbiology and Genetics, Dr. Bohr-Gasse 9, A-1030 Vienna, Austria

Abstract

STAT family transcription factors regulate gene expression in response to a wide variety of cytokines. A transcription factor designated differentiation-induced factor (DIF), activated by treatment of myeloid cells with the differentiating agents interferon-gamma (IFN–γ), granulocyte-macrophage colony-stimulating factor (GM-CSF), colony-stimulating factor-1 (CSF-1) or during phorbol ester-induced differentiation, was characterized as a 112kDa protein related to, but not identical with known isoforms of STAT 5. Taken together with previously published results, our data suggest an important function for members of the STAT 5 subfamily in regulating gene expression during the process of myeloid differentiation.

Introduction

Hematopoietic cells committed to the myeloid lineage undergo a complex series of differentiation steps before reaching the stage of mature granulocytes or monocytes. While differentiating, the cells are continuously reprogrammed for the expression of stage-specific genes. Each differentiation step thus requires the activation of transcription factors that cause progression within the lineage and untimely transcription factor activity may result in failure to differentiate and thus promote leukemic growth.

Intracellular signals originating from cell surface receptors are crucial in directing hematopoiesis. Ligand binding of the colony-stimulating factor receptors for G-CSF, GM-CSF, the macrophage lineage-specific CSF-1, or by the IFN–γ receptor, induces partial or complete differentiation of granulocyte-macrophage precursors, promyelocytes or promonocytes into mature granulocytes and macrophages [1, 2]. Intracellular signals originating from these receptors must therefore target the transcription factors required for an altered pattern of gene expression. One way of rapidly directing transcription from the cell surface is the activation of the JAK-STAT signaling path [3, 4]. The cytoplasmic domains of most cytokine receptors, including those for G-CSF, GM-CSF or IFN–γ have been shown to associate with one or more protein tyrosine kinases (PTKs) of the Janus Kinase (JAK) family. Ligand binding induces activation of these PTKs and subsequently the PTKs themselves as well as the cytoplasmic receptor domains are phosphorylated on tyrosine. Phosphotyrosine residues then serve as docking sites for STAT family transcription factors that bind to these sites using their *src* homology 2 (SH2) domains. The bound transcription factors are phosphorylated on tyrosine and released from the receptors. They form either homo- or heterodimers and translocate to the cell nucleus where they bind to DNA designated gamma-interferon activation site (GAS)- like elements after the prototype sequence [5]. In many cases, DNA-binding of STATs directly correlates with the onset of transcription from a neighbouring gene.

During recent years we have studied how the differentiation stage of a hematopoietic cell affects immediate early gene activation by cytokines. Using the promonocytic cell line U937 which can be differentiated into a monocytic stage in cell culture, we were able to demonstrate that activation of STAT transcription factors by IFN–γ changes upon monocytic differentiation, indicating that STAT target genes may be important in determining the stage-specific biological effect of cytokines [6, 7]. In this report we summarize some of our most recent findings, indicating that STAT family transcription factors may not only determine stage-specific cytokine responses, but in addition also be involved in the differentiation processes per se.

Materials and Methods

All reagents and experimental procedures have previously been described [6-8].

Results

Activation of DIF by Myeloid Differentiation Factors.

The original detection of DIF resulted from studies addressing transcription factor activation by IFN–γ in U937 promonocytes and monocytes. IFN–γ promotes differentiation in promonocytes while triggering cellular activation in monocytes. Transcription of genes containing a GAS element within their promoters changes upon monocytic differentiation. These genes might therefore contribute to the altered biological response to IFN–γ in promonocytes versus monocytes [6, 7]. Using a GAS sequence in electrophoretic mobility shift assays (EMSA) we noted that IFN–γ activated binding of two distinct transcription factors to this element (Fig. 1).

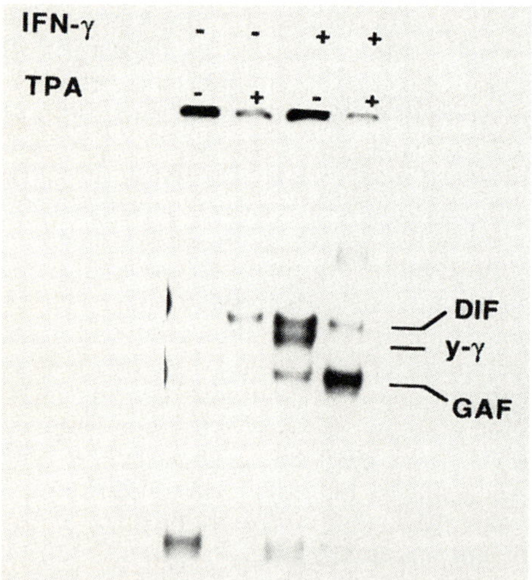

Fig. 1. Transcription factor activation by IFN-γ in differentiated (TPA) or undifferentiated U937 cells. Nuclear extracts were reacted with a GAS oligonucleotide from the IFP53 gene promoter and analyzed in an electrophoretic mobility shift assay (EMSA). GAF, gamma-interferon activation factor; DIF, differentiation-induced factor. Reproduced with permission of ASM publications.

The higher mobility factor was predominantly activated in monocytes and identified as the IFN–γ activation factor (GAF), a dimer of STAT1 [9, 10]. The second IFN–γ-activated factor, DIF, was observed almost exclusively in promonocytes and might therefore be part of the differentiation response to the cytokine. In accordance with this assumption, DIF activation was observed during phorbol ester-induced monocytic differentiation (Fig. 2a) and after stimulation of U937 promonocytes with GM-CSF but not with G-CSF (Fig. 2b). U937 cells do not express CSF-1 receptors, however, in BAC1.2F5 cells that represent an immature macrophage stage, activation occurred in response to the the lineage-specific CSF-1 (Fig. 2c).

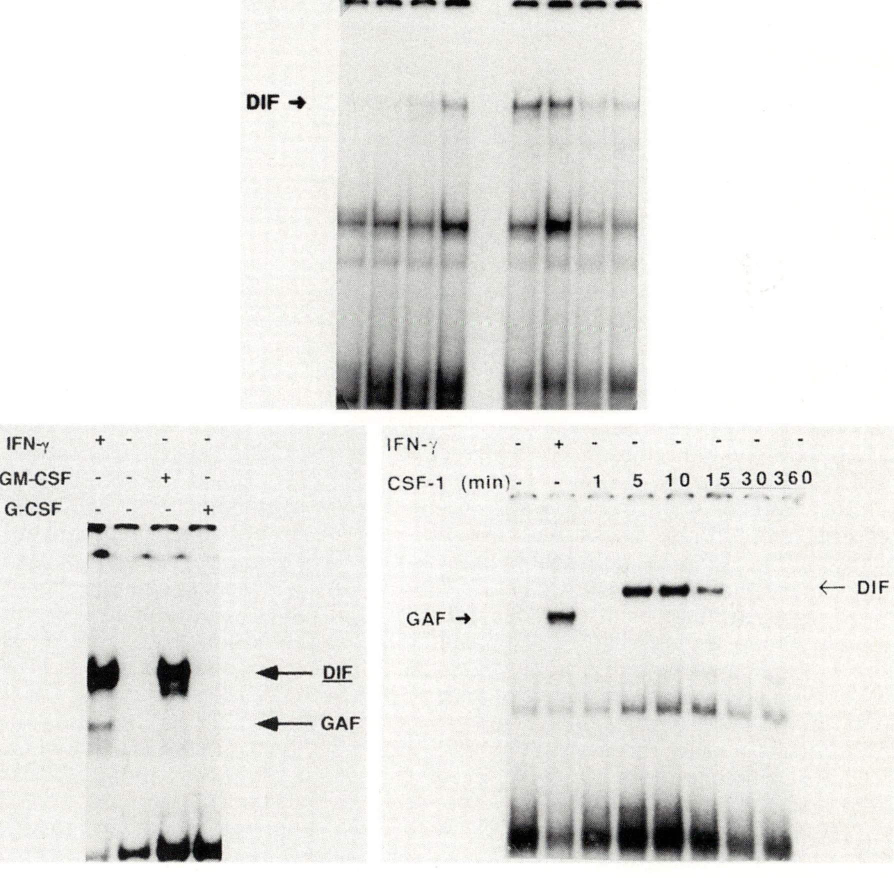

Fig. 2. Activation of the differentiation-induced factor (DIF) during TPA-mediated differentiation (**a**) or by the colony-stimulating factors GM-CSF and CSF-1 (**b, c**). U937 cells were treated with cytokines for 30 min or with TPA for the indicated periods. Control extracts were obtained from IFN-γ-treated cells to mark the position of the gamma-interferon activation factor (GAF, the STAT1 dimer). DIF activity in nuclear extracts was determined in the EMSA with the IFP53 GAS as a probe. Reproduced with permission of AMS publications and Elsevier Science publishers.

The common denominator of all tested agents is to promote differentiation of myeloid- or macrophage lineage precursors. The activated, GAS-binding factor was therefore designated differentiation-induced factor (DIF). The inability of G-CSF to activate DIF, despite the presence of the G-CSF receptor on U937 cells, suggested a role for DIF in monocytic, but not in granulocytic differentiation. This assumption was only partially verified using HL60 cells that represent a granulocyte-macrophage precursor stage. In HL60, DIF activation occured in response to stimuli of granulocyte differentiation (DMSO, G-CSF), albeit at much lower levels than in response to those of monocytic differentiation (TPA, GM-CSF; unpublished results). In primary chicken Ets26 monoblasts [11], DIF activation occurred specifically in response to cMGF, a myeloid differentiation factor in chicken, but not in response to IGF, a factor promoting survival but not differentiation of these cells (M. Kieslinger, H. Beug and T. Decker, unpublished results). Further consistent with a role of DIF in a myeloid cell differentiation response, TPA- or IFN–γ treatment of fibroblasts or epithelial cells did not produce activated factor (unpublished results).

Biochemical Characterization of DIF

Several features of DIF suggested it might be a STAT family transcription factor: i) its rapid activation by cell surface receptors known to employ the JAK-STAT path, ii) its association with GAS sequences, iii) the ability of anti-phosphotyrosine antibodies to prevent formation of the DIF-GAS complex [6], indicating that the activated factor contains phosphotyrosine. We therefore reacted antibodies to known STAT proteins with DIF-containing extracts and tested the resulting complexes in the EMSA. Antibodies against STAT 1-4 were without effect, however, antibodies to the N-terminus of STAT 5 caused an altered mobility of the DIF-GAS complex (a DIF supershift [8]), indicating an immunological relationship between DIF and STAT 5. STAT 5 has previously been characterized from several species and proteins ranging in molecular weight between 77 and 96kDa have been reported [12-18]. To assess whether DIF might be STAT 5, we isolated the factor using GAS oligonucleotide-mediated precipitation [8] or adsorption to GAS affinity matrices followed by elution of bound protein, SDS PAGE and Western blotting. The membranes were stained with specific STAT 5 antibodies (Fig. 3a) or with anti-phosphotyrosine antibodies (Fig. 3b). STAT 5 antibodies predominantly stained a 112kDa protein present in extracts from IFN–γ or GM-CSF-treated U937 cells (Fig. 3a) or in extracts from CSF-1-treated BAC1.2F5 cells (data not shown). In addition, a minor band of about 97kDa was observed. Both the 112kDa and 97kDa bands stained with anti-phosphotyrosine antibodies (Fig. 3b). These findings indicate that the IFP53 GAS sequence [19] preferentially binds a 112kDa, STAT5-related protein in extracts of either GM-CSF or IFN-γ-treated cells. The 97 kDa minor protein most likely represents STAT5. The following additional experiments confirm this assumption: i) an antibody directed to the SH3 domain of STAT 1 which crossreacts with STAT 5 (97kDa [20]) does not recognize DIF/p112, but precipitates the 97kDa species. ii) three distinct antisera to the STAT 5 C-terminus crossreact only weakly with DIF, indicating related but distinct carboxy termini of STAT 5 and DIF/p112 (data not shown).

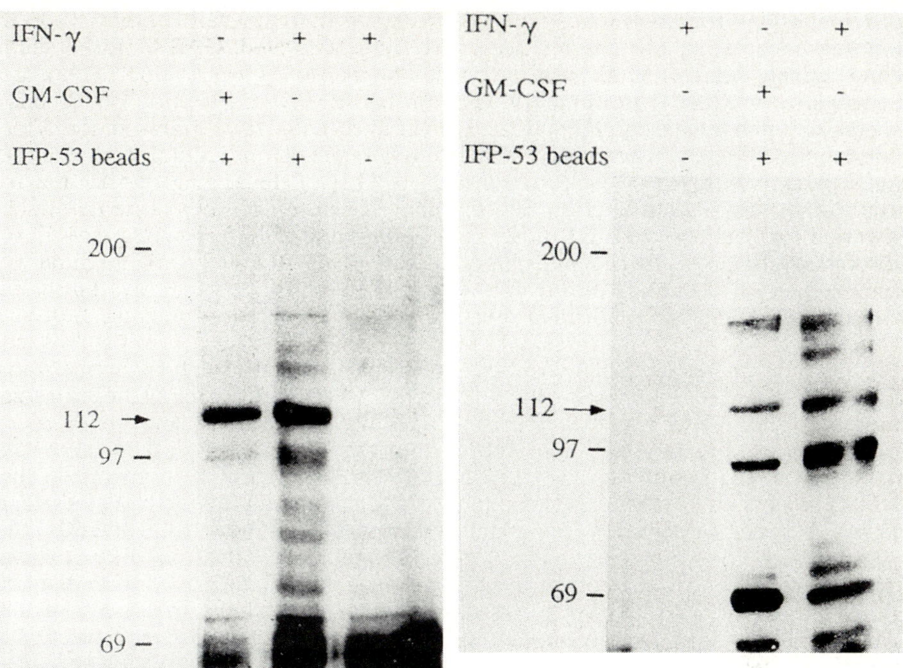

Fig. 3. DIF is, or includes, a 112kDa, STAT5-related, tyrosine-phosphorylated protein. Nuclear extracts from GM-CSF or IFN-γ-treated U937 cells were subjected to GAS-oligonucleotide-mediated precipitation. Precipitated protein was analyzed by SDS-PAGE and Western blotting, followed by staining with either STAT5 N-terminal antibodies (**a**) or anti-phosphotyrosine antibodies (**b**). Reproduced with permission of Elsevier Science publishers.

Discussion

We have investigated transcription factor activation in response to differentiation factors of myeloid hematopoietic cells. Our results provide evidence for an involvement of DIF, a STAT family protein, in the differentiation response to all tested stimuli. DIF activation in response to such stimuli was observed in hematopoietic precursor cells, but not fibroblasts or epithelial cells. In the case of IFN–γ, DIF activation is lost once monocytic differentiation of U937 promonocytes has occurred. Both IFN-γ and TPA induce a G1 arrest of U937 cells clearly linking DIF to a differentiating, not a proliferative response. Taken together, these results support the conclusion that DIF may be a necessary component for the alteration of differentiation stage-specific gene expression during maturation of myeloid cells. It remains to be tested whether DIF activation also occurs in other lineages and what other differentiation factors might cause its activation.

While being potentially necessary for differentiation, DIF activation is apparently not sufficient for this process because U937 or HL60 cells represent leukemic cells that have lost the ability to differentiate in response to e.g. GM-CSF, but DIF activation nevertheless occurs after treatment with GM-CSF. Receptors for differentiation factors activate several signaling paths targeting transcription factors and it is not surprising that more than one of these is required for the complete differentiation program. Interestingly, DIF may itself be targeted not only by JAK kinases but also by a signaling path involving MAP kinase. BAC1.2F5 cells constitutively expressing the v-raf oncogene lose the ability to activate MAP kinase in response to CSF-1, presumably due to a constitutive expression of the dual specificity phosphatase MKP-1 [21, 22]. In such cells JAK kinase signals are normal, but DIF activation by CSF-1 is strongly reduced (our unpublished results). Full DIF activation may therefore require the activity of converging signaling paths originating from the same cell surface receptor. We have recently reported a similar situation for GAF, the STAT 1 dimer in differentiating U937 cells and others have shown an influence of serine phosphorylation on the activity of APRF, a dimer of activated STAT 3 [23-25]. STAT factor activity may thus be generally controlled through an input of several distinct signaling paths.

Biochemical studies have demonstrated DIF to be related to, but not identical with STAT 5. Is there a general role for this STAT subfamily in differentiation processes? STAT 5 was first purified and cloned from rat and sheep as a transcription factor, MGF, involved in the transcription of milk proteins in mammary gland epithelium [12, 13]. MGF/STAT5 is therefore involved in the prolactin-induced maturation of mammary gland epithelium into actively lactating tissue. In mice, transcription factors induced in myeloid cells by IL-3 and GM-CSF were shown to be murine STAT 5 homologues. Purification and cloning demonstrated the existence of several STAT 5 isoforms, both derived from distinct genes (STAT 5a and 5b) and, most likely, from proteolytical processing of the larger isoforms [14, 15]. In human cells, STAT 5 was recently reported to be activated by receptors of the IL-2 receptor family during T cell activation [17, 18], and also by growth hormone, erythropoietin and GM-CSF [16]. STAT 5 activation might thus be linked to either differentiation or functional maturation of different cell lineages. However, most receptors, like the GM-CSF receptor, that mediate the differentiation of hematopoietic cells are also capable of generating proliferative signals. Thus, activation of STAT 5 by these receptors might also involve the transcription factor in a proliferative response.
It is tempting to speculate that different STAT 5 isoforms might perform distinct functions, either linked to proliferation or differentiation/maturation. One possible support for this stems from the finding that the smaller versions of muSTAT 5 were predominantly found in more immature myeloid stages whereas the larger forms occured in more mature precursor cells [15]. DIF, as a STAT 5 related transcription factor, displays a binding preference for different GAS sequences that is distinct from STAT 5 and it may therefore be expected to activate a set of target genes distinct from those of STAT 5. Investigating the function of individual STAT 5-related transcription factors may be an important step towards understanding the regulation of stage-specific gene expression during myeloid differentiation.

Acknowledgement

Funding for our research was through grants P09721-MED and P10486-MED to TD from the austrian Fonds zur Förderung der wissenschaftlichen Forschung. We thank Fabrice Goilleux, Bernd Groner and Chris Schindler for providing STAT5-reactive antibodies.

References

1. Metcalf D (1989) The molecular control of cell division, differentiation commitment and maturation in haematopietic cells. Nature 339:27-30.

2. Romeo G, Fiorucci G Rossi GB (1989) Interferons in cell growth and development. Trends in Genetics 5:19-24

3. Ihle JN, Kerr IM (1995) Jaks and Stats in signaling by the cytokine receptor superfamily. Trends in Genetics 11:69-74

4. Darnell jr JE, Kerr IM, Stark GR (1994) Jak-STAT pathways and transcriptional activation in response to IFNs and other extracellular signaling proteins. Science 264:1415-1421

5. Lew DJ, Decker T, Strehlow I, Darnell jr JE (1991) Overlapping elements in the guanylate-binding protein gene promoter mediate transcriptional induction by alpha- and gamma interferons. Mol Cell Biol 11:182-191

6. Eilers A, Baccarini M, Horn F, Hipskind RA, Schindler C, Decker T (1994) A factor induced by differentiating signals in cells of the macrophage lineage binds to the gamma interferon activation site. Mol Cell Biol 14:1364-1373

7. Eilers A, Seegert D, Schindler C, Baccarini M, Decker T (1993) The response of the gamma interferon activation factor (GAF) is under developmental control in cells of the macrophage lineage. Mol Cell Biol 13:3245- 3254

8. Barahmand-pour F, Meinke A, Eilers A, Gouilleux F, Groner B, Decker T (1995) Colony-stimulating factors and interferon-γ activate a protein related to MGF-Stat 5 to cause formation of the differentiation-induced factor in myeloid cells. FEBS Letters 360:29-33

9. Decker T, Lew D J, Mirkovitch J, Darnell jr JE (1991) Cytoplasmic activation of GAF, an IFN-γ–regulated DNA-binding factor. EMBO J 10:927-932

10. Shuai K, Horvath C, Tsai-Huang LH, Qureshi SA, Cowburn D, Darnell jr JE (1994) Interferon activation of the transcription factor Stat91 involves dimerization through SH2-phosphotyrosyl peptide interactions. Cell 76:821-828

11. Beug H, Blundell PA, Graf T (1987) Reversibility of differentiation and proliferative capacity in avian myelomonocytic cells transformed by tsE26 leukemia virus. Genes and Development 1:277-286

12. Wakao H, Schmitt-Ney M, Groner B (1992) Mammary gland-specific nuclear factor is present in lactating rodent and bovine mammary tissue and composed of a single polypeptide of 89kDa. J Biol Chem 267:16365-16370

13. Wakao H, Gouilleux F, Groner B (1994) Mammary gland factor (MGF) is a novel member of the cytokine regulated transcription factor gene family and confers the prolactin response. EMBO J 13:2182-2191

14. Mui AL-F, Wakao H, O'Farrell A-M, Harada N, Miyajima A (1995) Interleukin-3, granulocyte-macrophage colony stimulating factor and interleukin-5 transduce signals through two STAT5 homologs. EMBO J 14:166-1175

15. Azam M, Erdjument-Bromage HE, Kreider BL, Xia M, Quelle F, Basu R, Saris C, Tempst P, Ihle JN, Schindler, C (1995) Interleukin-3 signals through multiple isoforms of Stat5. EMBO J 14:1402-1411

16. Goilleux F, Pallard C, Dusanter-Fourt I, Wakao H, Haldosen L-A, Norstedt G, Levy D, Groner B (1995) Prolactin, growth hormone, erythropoietin and granulocyte-macrophage colony stimulating factor induce MGF-Stat5 DNA binding activity. EMBO J 14:2005-2013

17. Hou J, Schindler U, Henzel WJ, Wong SC, McKnight S.L (1995) Identification and purification of human Stat proteins activated in response to interleukin-2. Immunity 2:321-329

18. Lin, J-X, Migone T-S, Tsang M, Friedmann M, Weatherbee JA, Zhou L, Yamauchi A, Bloom ET, Mietz J, John S, Leonard WJ (1995) The role of shared receptor motifs and common Stat proteins in the generation of cytokine pleiotropy and redundancy by IL-2, IL-4, IL-7, IL-13 and IL-15. Immunity 2:331-339

19. Strehlow I, Seegert D, Frick C, Bange F-C, Schindler C, Boettger EC, Decker T (1993) The gene encoding IFP 53/tryptophanyl-tRNA synthetase is regulated by the gamma-interferon activation factor. J Biol Chem 268:16590-16595

20. Rothman P, Kreider B, Azam M, Levy D, Wegenka U, Eilers A, Decker T, Horn F, Kashleva H, Ihle JN, Schindler C (1994) Cytokines signal through tyrosine phophorylation of a family of related transcription factors. Immunity 1:457-468

21. Büscher D, DelloSbarba PD, Hipskind RA, Rapp UR, Stanley ER, Baccarini M (1993) *v-raf* confers CSF-1-independent growth and leads to immediate early gene expression without MAP-Kinase activation. Oncogene 8:3323-3332

22. Krautwald S, Büscher D, Dent T, Ruthenberg K, Baccarini M (1995) Suppression of growth factor-mediated MAP kinase activation by v-raf in macrophages: a putative role for the MKP-1 phosphatase. Oncogene 10:1187-1192

23. Eilers A, Georgellis D, Klose B, Schindler C, Ziemiecki A, Harpur AG, Wilks AF and Decker T (1995) Differentiation-regulated serine phosphoporylation of STAT 1 promotes GAF activation in macrophages. Mol Cell Biol 15:000

24. Lütticken C, Coffer P, Yuan J, Schwartz C, Schindler C., Kruijer W., Heinrich PC, Horn F (1995) Interleukin-6-induced serine phosphorylation of transcription factor APRF: evidence for a role in interleukin-6 target gene induction. FEBS Letters 360:137-143

25. Zhang X, Blenis J, Li H-C, Schindler C. Chen-Kiang, S (1995) Requirement of Serine phosphorylation for differential formation of Stat-promoter complexes in gp130 signaling. Science 267:1990-1994

Regulation of C/EBPβ/NF-M Activity by Kinase Oncogenes

G. Twamley-Stein, E. Kowenz-Leutz, S. Ansieau, and A. Leutz

Max Delbrück Centrum for Molecular Biology (MDC), Robert Rössle Str. 10, 13122 Berlin, Germany.

Abstract

CAAT Enhancer Binding proteins (C/EBP) belong to a family of transcription factors which are implicated in a number of developmental and growth regulatory processes. One member of this family known as C/EBPβ (called NF-M in the chicken system) is particularly important in myelomonocytic cells because it is targeted by kinases and collaborates with the Myb oncoprotein to induce the expression of myeloid specific genes. Experiments dissecting the structure of NF-M suggest that it is a repressed transcription factor. Using the yeast two-hybrid system we showed that a negative regulatory domain masks the transactivation domain. Examination of NF-M mutants suggests that kinase (proto-) oncogenes uncover the concealed transcriptional activity by phosphorylation of the negative regulatory domain.

Introduction

In several cell types C/EBP proteins are of major importance in a number of basic functions including proliferation, growth arrest, differentiation and the immune as well as inflammatory responses (Akira and Kishimoto, 1992; Juan et al., 1993; Poli et al., 1990). In the hematopoietic system, C/EBPα, β and δ appear to play decisive roles in myelopoiesis (Scott et al., 1992) and collaborate with the Myb oncoprotein to induce the expression of myeloid specific target genes even in heterologous cell types such as erythroblasts or fibroblasts (Ness et al., 1993). Recently, it was shown that mice containing a homozygous null mutation in the C/EBPβ gene were defective in macrophage functions and resulted in pup death shortly after birth (Tanaka et al., 1995). The activity of C/EBPβ protein appears to be regulated by a number of factors that induce macrophage activation including, bacterial lipopolysaccharide (LPS), cytokines (Akira et al., 1990; Nakajima et al., 1993; Katz et al., 1993; Sterneck et al., 1992) growth factor receptors and kinase oncogenes (Katz et al., 1993; Sterneck et al., 1992). The targeting of NF-M by kinase proto-oncogenes as well as its collaboration with Myb suggests involvement in leukemogenesis since two oncogene pathways converge on C/EBPβ to regulate its activity.

Here we focus on how kinase oncogenes alter the transcriptional efficacy of NF-M. Like its mammalian counterparts, NF-M is a phosphoprotein which immediately implicates phosphorylation in its regulation. Several phosphorylation sites have been described to date in mammalian C/EBPβ. In P19 embryonal carcinoma cells, NF-Il6 (human C/EBPβ) can be activated by a Ras regulated MAP kinase through phosphorylation of threonine 235 (Nakajima et al., 1993). In glial cells the Calcium/Calmodulin dependent Kinase II (CaMKII) also activates C/EBPβ (Wegner et al., 1992). Unlike the MAP kinase site, CaMKII-induced phosphorylation occurred in the C-terminal dimerization domain (serine 276). In Hepatoma cells it was shown that phorbol-ester treatment could activate rat C/EBPβ and enhance its transcriptional efficacy. In this case activation was believed to be brought about by phosphorylation of a PKC-site situated in the N-terminal transactivation domain (serine 105). The PKC target site however was not a substrate for purified PKC (Trautwein et al., 1993) instead, purified PKA could phosphorylate the site *in vitro*. However, PKA did not appear to be the responsible kinase *in vivo*, since stimulation of the cAMP pathway did not augment this phosphorylation. Further, this phosphorylated residue is neither conserved in human nor in chicken C/EBPβ. Contrary to these results, in Pheochromacytoma (PC12) cells C/EBPβ is phosphorylated after induction of the cAMP pathway, although the targeted sites have not been defined (Metz and Ziff, 1991). Surprisingly, an additional report suggested that both PKA and PKC phosphorylated another site within the DNA binding domain of C/EBPβ (serine 240 in the rat protein) (Trautwein et al., 1994). However phosphorylation of serine 240 resulted in a marked reduction of DNA binding, indicating loss rather than gain of C/EBPβ mediated transactivation. Therefore to date, accumulated data on mammalian C/EBPβ phosphorylation are difficult to reconcile with the activation of NF-M by oncogenic kinases.

The cloning of the chicken C/EBPβ homologue (NF-M) revealed a striking feature of its structure, allowing the relationship between C/EBPβ phosphorylation, structure and function to be examined by an alternative approach. Sequence comparison of NF-M with its mammalian counterparts revealed in addition to the homologous C-terminal DNA binding and dimerization domains, seven other conserved regions (CR´s). These CR´s lie between divergent ala/gly/pro rich stretches within the N-terminus of the protein and suggest that C/EBPβ consists of discreet domains that are linked by flexible hinges (i.e. "beads on a string") (Katz et al., 1993). In the light of this consideration, such a structure/function relationship could integrate functionally independant domains into the same protein thus allowing the convergence of several signaling pathways on C/EBPβ. This possiblity was examined by deletion analysis of NF-M (Kowenz-Leutz et al., 1994). Here we describe the involvment of kinase oncogenes in regulating the activity of a transcription supressor domain in NF-M/ C/EBPβ.

Materials and Methods

Plasmid constructions and tissue culture: NF-M mutants were constructed as described in Kowenz-Leutz et al., 1994. The ts-verbB transformed HD3 erythroblast cell line (as described by Beug et al.,1979) were grown in Dulbecco's modified Eagle medium (DMEM) supplemented with 8% fetal calf serum (FCS), 2% heat inactivated chicken serum, 10mM HEPES (pH7.2), penicillin and streptomycin at 37°C, 5%CO_2.

Transfections: 2x10^7 Cells were transfected by DEAE-Dextran as described (Ausubel et al., 1987; Katz et al., 1993; Kowenz-Leutz et al., 1994) and seeded in 10 ml of tissue culture medium. Cells were grown at 36°C and harvested 24h later.

Metabolic labeling, immune-precipitations and tryptic mapping: For *in vivo* labeling experiments, cells were transfected and harvested as above, washed twice in phosphate-free DMEM plus dialysed 10%FCS and incubated with 1mCi/ml of [^{32}]P-orthophosphate for 4h. They were then washed, lysed and immune-precipitated as described in Kowenz-Leutz et al., 1994. Proteins were separated by SDS-polyacrylamide gel electrophoresis and transferred to immobilon. The labeled proteins were localized by exposure to X-ray film and cut out. Membrane slices were blocked in 0.5% polyvinylpyrrolidone, 100mM acetic acid for 30 minutes at 37°C, after which samples were washed 5X in H_20 and digested in 50mM Ammonium-bicarbonate, with 2x 2μg trypsin per sample. The membrane was removed and the samples were lyophilized. Specimens were loaded on Kodak Cellulose thin-layer-chromatograpy plates and electrophoresed at pH1.9 at 1KV for 30 minutes followed by ascending chromatography in phosphochromo buffer (Boyle et al., 1991).

Results

In order to examine how oncogenic kinases and signal transduction are involved in C/EBPβ activation, we used an erythroblast cell line (HD3) which contains a temperature sensitive mutant of the v-erbB kinase (the viral, constitutively active oncogenic form of the EGF-receptor). Thus C/EBPβ induced the expression of the resident myeloid specific gene #126 (encoding a potential calcium binding protein) at the permissive temperature (36°C) when the kinase was active, but not at the non-permissive temperature (42°C) when the kinase was inactive (Kowenz-Leutz et al., 1994). Cotransfection of NF-M with either Ras, Raf or polyoma middle T antigen complimented the inactivated tsv-erbB kinase and induced expression of #126 and other NF-M target genes even at the non-permissive temperature. Thus NF-M which is normally present in the inactive state can induce myeloid-specific gene expression after activation of the signaling cascade.

Chimeric mutants of NF-M that lack either CR5, CR7, or CR57 (noted as ΔCR) were found to be more active than the wild-type protein in reporter assays. Since all chimeras were equivalently expressed, we assumed that CR5 and CR7 might be implicated in the regulation of NF-M activity. Significantly, in temperature shift experiments using tsv-erbB expressing HD3 cells, it was found that ΔCR5, ΔCR7 and ΔCR57 were all active at the permissive as well as the non-permissive temperature (Kowenz-Leutz et al., 1994). This implied that removal of these regulatory domains constitutively activated the NF-M protein allowing it to by-pass the requirement for receptor-mediated signal transduction.

To examine whether NF-M phosphorylation correlated with gene activation we determined the relative ratios of specific phosphate incorporation into the constitutively active ΔCR7 and ΔCR57 mutants. Removal of CR7 or both 5 and 7 reduced phosphate incorporation by 90% and 95% respectively compared to the wild-type protein (Kowenz-Leutz et al., 1994). These results suggest that most of the kinase oncogene-induced phosphorylation occurs in the negative regulatory regions of NF-M.

To compare activated states of NF-M with their respective phosphorylation patterns, we examined the ΔCR7 and ΔCR57 mutants by *in vivo* phosphorylation followed by phosphopeptide analysis (Fig. 1). Removal of CR7 results in the loss of a number of major phosphopeptides. The additional removal of CR5 did not detectably alter any other phosphopeptides. These data therefore demonstrate that the major target of phosphorylation is CR7.

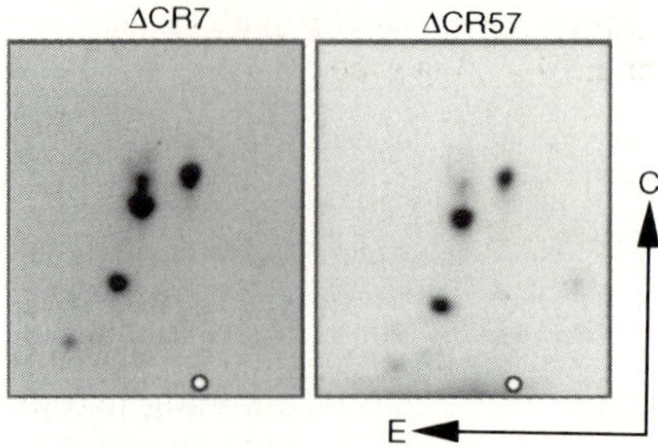

Fig. 1. Phosphorylation of the regulatory part of NF-M. HD3 cells were transfected with expression vectors encoding ΔCR7 and ΔCR57 and harvested after 24h at the permissive temperature. Cells were labeled with orthophosphate, immune-precipitated, separated by PAGE and immunoblotted. Labeled proteins were digested with trypsin and subjected to two-dimensional phosphotryptic analysis.

CR7 contains many conserved residues which could be phosphorylated, one of which is a MAP kinase consensus sequence (Katz et al., 1992; Nakajima et al., 1993). To address the functional role of the MAP kinase site, a serine to aspartic acid mutation was introduced at this position (NF-M S220D) and temperature-shift experiments were performed as described above. Like the ΔCR5, ΔCR7 and ΔCR57 deletion mutants but unlike the wild-type protein, NF-M S220D also induced expression of the myeloid-specific genes at the non-permissive temperature (Kowenz-Leutz et al., 1994). The main implication of this finding is that phosphorylation of NF-M at its MAP kinase site is sufficient to remove repression and induce activation of target genes.

In order to examine whether the MAP kinase target (Ser 220) in NF-M is phosphorylated in tsv-erbB transformed erythroblasts *in vivo*, two dimensional tryptic maps of the wild type NF-M and the MAP kinase mutant (NF-M S220D) were compared (Fig. 2). It is apparent from these maps that there are no major differences between the wild-type and mutant protein. However there is one minor difference, namely that one phosphopeptide is reduced in intensity (indicated by an arrow in Fig. 2). Ser 220 of NF-M is thus phosphorylated *in vivo* but the minor difference observed suggests that even if the MAP kinase site is of major regulatory importance, it is only a minor phosphorylation site in the protein.

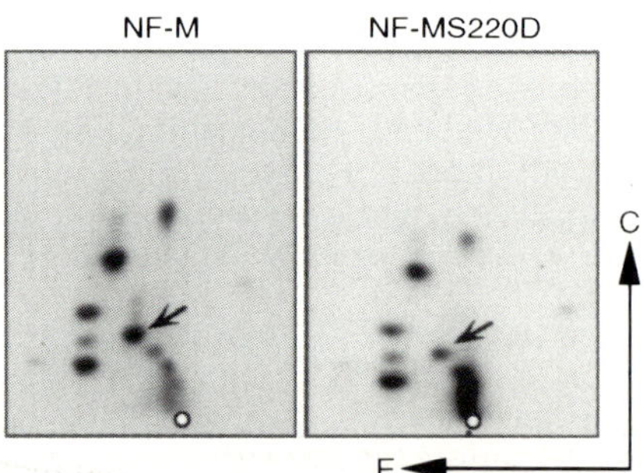

Fig. 2. The MAP kinase mutant is a minor phosphorylation site by two-dimensional phosphotryptic analysis. HD3 cells were transfected with expression vectors encoding NF-M or NF-M mutated at its MAP kinase site (NF-M S220D) and harvested after 24h at the permissive temperature. Cells were labeled with orthophosphate, immune-precipitated, separated by PAGE and immunoblotted. Labeled proteins were digested with trypsin and subjected to two-dimensional phosphotryptic analysis.

Besides the regulatory region which covers CR´s 5 to 7 as recently described (Kowenz-Leutz et al., 1994), a second region of NF-M, the leucine zipper, was described as having a crucial role in regulation of C/EBPβ activity (Wegner et al., 1992). We asked whether the leucine zipper played a role in kinase-oncogene induced activation of NF-M. Therefore we exchanged its leucine zipper for that of the CREB protein (cAMP Responsive Element Binding protein). Like the wild-type protein,the leucine zipper mutant (NF-MCREBLZ) forms homodimers and is active at the permissive but not at the non-permissive temperature (i.e. it requires tsv-erbB kinase activity). This observation suggests that the leucine zipper is not implicated in NF-M activation by kinase oncogenes. Examining two-dimensional phosphotryptic maps of the mutant NF-MCREBLZ compared with the wild type protein indicates that the leucine zipper is phosphorylated (Fig.3). However these phosphorylations are not required for NF-M activation and appear to have no effect on CR7 phosphorylation (Fig. 1). Thus we conclude that CR7 phosphorylation can occur independantly of leucine zipper phosphorylation.

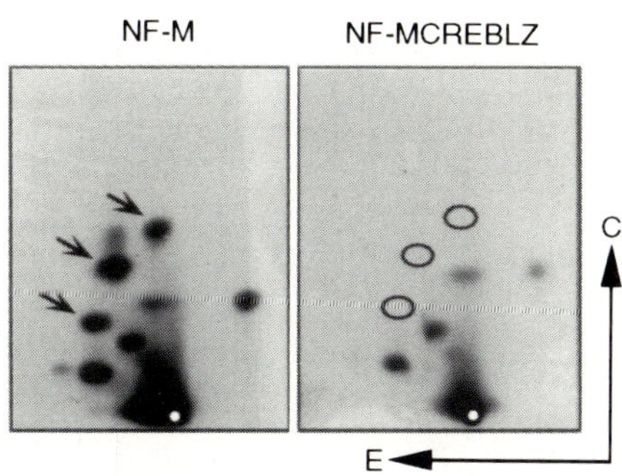

Fig. 3. Phosphorylation of the NF-M leucine-zipper. HD3 cells were transfected with expression vectors encoding NF-M or NF-MCREBLZ and harvested after 24h at the permissive temperature. Cells were labeled with orthophosphate, immune-precipitated, separated by PAGE and immunoblotted. Labeled proteins were digested with trypsin and subjected to two-dimensional phosphotryptic analysis.

In discovering that CR7 and CR5 were crucial for the regulation of NF-M by signaling pathways, we asked how such a mechanism could work. Thus we tested the possibility that these domains could repress NF-M activity by an intramolecular mechanism. To do this we used a genetic complimentation test known as the yeast two-hybrid system (Fields and Song 1989). In this system, both CR5 and CR7 were found to interact with the amino-terminal transactivation domain of NF-M. In the case of CR7 this is particularly striking as mutation of the MAP kinase site abrogates the interaction (Table. 1). C/EBPβ may thus adopt an inactive conformation in which its transactivation domain is normally masked by CR5 or CR7 (as shown in Fig. 4). Modification of either domain (for example by phosphorylation of the MAP kinase site in CR7) may induce a conformational change that releases the transactivation domain of NF-M/ C/EBPβ and therefore allows activation of target genes.

Table 1. NF-M is regulated by an intra-molecular mechanism

DNA binding hybrid (in pGBT9)	Transactivation hybrid (in pGAD)	Color of colonies
Gal DBD-	Gal AD	white
Gal DBD-	Gal AD- CR 1234	white
Gal DBD- CR 7	Gal AD-	white
Gal DBD- CR 5	Gal AD-	white
Gal DBD- CR 7D	Gal AD-	white
Gal DBD- CR 5	Gal AD- CR 1234	blue
Gal DBD- CR 7	Gal AD- CR 1234	blue
Gal DBD- CR 7D	Gal AD-CR1234	white

Discussion

Our results suggest that a negative regulatory domain (CR7) within NF-M/ C/EBPβ is itself negatively regulated or neutralized by the downstream action of oncogenic kinases. Since a number of other sites in the same domain can be phosphorylated, it appears that several pathways may converge on this region to derepress its activity.

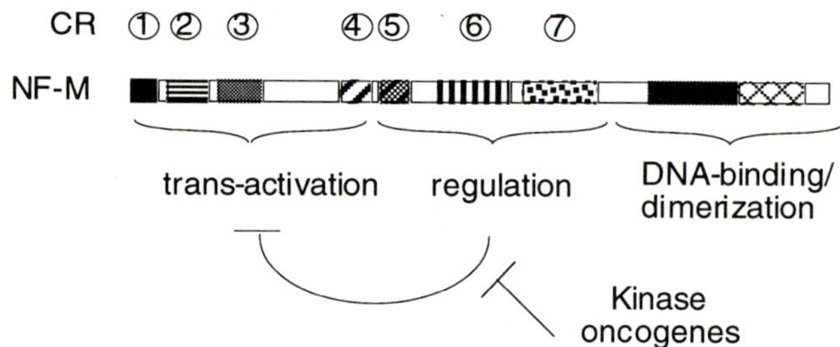

Fig. 4. Functional analysis of NF-M indicates a regulated modular structure. Sequence comparison suggests nine discreetly conserved regions (CR´s) in C/EBPβ proteins (patterned boxes) interrupted by non-conserverd segments (white boxes). The transactivation domain consists of four individual modules, CR´s 1-4. CR´s 5, 6 and 7 comprise the regulatory region in the middle of the protein. It is this region that conceals the transactivation potential of the molecule and that is regulated by kinase oncogenes.

A number of kinases belonging to different signal transduction pathways have been found to target distinct regions in the C/EBPβ protein, including the transactivatory, regulatory and basic DNA binding/ dimerization domains to bring about gene activation. Why are so many kinases involved in C/EBPβ regulation? In many of the above-mentioned cases, the kinases which phosphorylate C/EBPβ cause it to induce expression from certain cell-specific promoters. For example it is clear that the cAMP pathway activates C/EBPβ in PC12 cells but not in HepG2 cells (Metz and Ziff, 1991; Trautwein et al., 1994). Thus the plethora of kinases known to phosphorylate C/EBPβ do not appear to be redundant but may function in a tissue restricted fashion to activate cell-type specific genes .

We are currently using two-dimensional phosphopeptide analysis to establish the presence and identity of various minor phosphorylation sites within NF-M.

Here we present data that strongly correlate activation of NF-M by kinase oncogenes with concurrent phosphorylation of a negative regulatory domain. All constitutively activated mutants of NF-M examined (ΔCR7, ΔCR57 and NFM220D) had in common at least the removal of a phosphopeptide consisting of the MAP kinase site. In addition the wild-type protein and NFMCREBLZ mutant contained a phosphorylated residue at this site when activated by kinase oncogenes. Strong evidence from the yeast two-hybrid system indicates how this site may be important in NF-M activation, since mutation of the MAP kinase site abrogates interaction between CR7 and the transactivation domain of the protein. This fits with the model that phosphorylation of this amino acid derepresses the transcription factor and can therefore be mimicked either by exchange of the MAP kinase site or by the complete removal of the domain in which it is located. Though removal of CR5 also enhanced NF-M transactivation, it did not appear to contain any major phosphorylation sites indicating it may be targeted by another mechanism (e.g. by protein-protein interaction). Although both CR5 and CR7 interact with the transactivation domain, removal of either one is sufficient for constitutive activation. It is thus tempting to speculate that CR5 and CR7 communicate with each other during activation of NF-M. Alternatively they could represent mutually exclusive ways of activating the protein in response to different signaling pathways.

Ras induces a massive activation of NF-M in P19 cells (data not shown), but is a poor activator compared to middle T antigen in HD3 cells at the non-permissive temperature (Kowenz-Leutz et al., 1994). This implicates other signaling pathways in NF-M activation. Analysis of two-dimensional tryptic maps of NF-M show that CR7 is targeted through phosphorylation in response to oncogenic kinases. Though the MAP kinase site is clearly important for this response it is not the only target in CR7. In fact, several other phosphorylated residues are present within the same region but their functions have not been defined individually. Our data do not exclude that these other phosphorylation sites may be required for phosphorylation of the MAP kinase site. In fact this is believed to be the case for one phosphorylation site in CR7 (Nakajima et al., 1993). Such phosphorylations would thus participate in activation of NF-M either through direct structure modification or by regulating the ability of NF-M to interact with other factors.

In addition to CR7, the leucine zipper was also found to be phosphorylated in the kinase-activated wild-type NF-M as well as in the activated mutants. Furthermore, one phosphorylation site within this region has already been identified as activating C/EBPβ in response to CaMKII in glial cells (Wegner et al., 1992). However the significance of these phosphorylations are not known at the present time since the leucine zipper can be replaced without affecting kinase-oncogene induced activation of NF-M in HD3 erythroblasts.

It is evident from these and other data that the modular structure of NF-M together with secondary modifications such as phosphorylation, assist it in the coordination of multiple cellular signaling pathways and thus the tissue-specific expression of target genes. The results of our experiments show that CR7 (and potentially CR5 also) is a target for cellular signals that regulate proliferation and differentiation.

References

Akira, S., Isshiki, H., Sugita, T., Tanabe, O., Kinoshita, S., Nishio, Y., Nakajima, T., Hirano, T., and Kishimoto, T. (1990). A nuclear factor for IL-6 expression (NF-IL6) is a member of a C/EBP family. EMBO J. *9*, 1897-1906.

Akira, S., and Kishimoto, T. (1992). IL-6 and NF-IL6 in acute-phase response and viral infection. Immunol Rev *127*, 25-50.

Ausubel, F. M., Brent, R., Kingston, R. E., Moore, D. D., Seidman, J. G., Smith, J. A., and Struhl, K. (1987). Current protocols in molecular biology. (New York-Chichester-Brisbane-Toronto-Singapore: Greene Publishing Assoc. and Wiley-Interscience).

Boyle, W. J., van der Geer, P. and Hunter., T. Meth. Enzym. 201, 110-149 (1991)

Fields, S. and Song, O. (1989). A novel genetic system to detect protein-protein interactions. Nature, *340*, 245-247

Juan, T. S., Wilson, D. R., Wilde, M. D., and Darlington, G. J. (1993). Participation of the transcription factor C/EBP delta in the acute-phase regulation of the human gene for complement component C3. Proc Natl Acad Sci U S A *90*, 2584-8.

Katz, S., Kowenz, L. E., Muller, C., Meese, K., Ness, S. A., and Leutz, A. (1993). The NF-M transcription factor is related to C/EBP β and plays a role in signal transduction, differentiation and leukemogenesis of avian myelomonocytic cells. Embo J *12*, 1321-32.

Kowenz-Leutz, E., Twamley, G., Ansieau. S., and Leutz, A. (1994). Novel mechanism of C/EBPβ (NF-M) transcriptional control: Activation through derepression. Genes & Development *8*, 2781-2791.

Metz, R., and Ziff, E. (1991b). cAMP stimulates the C/EBP-related transcription factor rNFIL-6 to *trans*-locate to the nucleus and induce c-*fos* transcription. Genes Dev. *5*, 1754-1766.

Nakajima, T., Kinoshita, S., Sasagawa, T., Sasaki, K., Naruto, M., Kishimoto, T., and Akira, S. (1993). Phosphorylation at threonine-235 by ras-dependent mitogen-activated protein kinase cascade is essential for transcription factor NF-IL6. Proc. Natl. Acad. Sci. USA *90*, 2207-2211.

Ness, S. A., Kowenz, L. E., Casini, T., Graf, T., and Leutz, A. (1993). Myb and NF-M: combinatorial activators of myeloid genes in heterologous cell types. Genes Dev *7*, 749-59.

Poli, V., Mancini, F. P., and Cortese, R. (1990). IL-6DBP, a nuclear protein involved in interleukin-6 signal transduction, defines a new family of leucine zipper proteins related to C/EBP. Cell *63*, 643-653.

Scott, L. M., Civin, C. I., Rorth, P., and Friedman, A. D. (1992). A novel temporal expression pattern of three C/EBP family members in differentiating myelomonocytic cells. Blood *80*, 1725-35.

Sterneck, E., Müller, C., Katz, S., and Leutz, A. (1992b). Autocrine growth induced by kinase type oncogenes in myeloid cells requires AP-1 and NF-M, a myeloid specific, C/EBP-like factor. EMBO J. *11*, 115-126.

Tanaka, T., Akira, S., Yoshida, K., Umemoto, M., Yoneda, Y., Shirafuji, N., Fujiwara, H., Suematsu, S., Yoshida, N., and Kishimoto, T. (1995) Targeted disruption of the NF-IL6 gene discloses its essential role in bacterial killing and tumor cytotoxicity by macrophages. Cell. *80*, 353-361.

Trautwein, C., Caelles, C., van, d. G. P., Hunter, T., Karin, M., and Chojkier, M. (1993). Transactivation by NF-IL6/LAP is enhanced by phosphorylation of its activation domain. Nature *364*, 544-7.

Trautwein, C., van der Geer, P., Karin, M., Hunter, T., and Choijkier, M.
Protein kinase A and C site-specific phosphorylations of LAP (NF-IL6) modulate its binding affinity to DNA recognition elements. J. Cli. Inv. *93*, 2554-2561.

Wegner, M., Cao, Z., and Rosenfeld, M. G. (1992). Calcium-regulated phosphorylation within the leucine zipper of C/EBP beta. Science *256*, 370-3.

Function of PU.1 (Spi-1), C/EBP, and AML1 in Early Myelopoiesis: Regulation of Multiple Myeloid CSF Receptor Promoters

D.-E. Zhang, S. Hohaus, M.T. Voso, H.-M. Chen, L.T. Smith, C.J. Hetherington, and D.G. Tenen. Hematology/Oncology Division, Beth Israel Hospital, Harvard Medical School, Boston, MA 02215, USA

Introduction

Recent studies of the regulation of normal myeloid genes, as well as the study of leukemias, has suggested that transcription factors play a major role in hematopoietic differentiation as well as leukemogenesis.[1-3] In order to understand the process of normal myeloid differentiation, it is important to identify and characterize the transcription factors which specifically activate important genes in the myeloid lineage. In addition, these factors may play a role in acute myelogenous leukemia (AML), in which this normal differentiation program is blocked. Receptors for growth factors play an important role in myelopoiesis.[4-7] In order to further understand the mechanisms directing the expression of these key regulators of hematopoiesis, we have initiated studies investigating the transcription factors activating the expression of several important myeloid colony-stimulating factor (CSF) receptors. Our previously published studies have demonstrated that the protooncogene PU.1 regulates the M-CSF receptor (CSF-1 receptor, c-fms).[8] We have also demonstrated that PU.1 is expressed in early human multipotential CD34+ cells, and specifically upregulated during commitment of these progenitors to the myeloid lineage.[9,10] Competitor double stranded oligonucleotides which inhibit PU.1 and Spi-B function can specifically block myeloid colony formation, indicating that expression and upregulation of PU.1 may play an important role in early myeloid development.[9]

We have now gone on to characterize a second family of factors which regulate the M-CSF receptor promoter, C/EBP.[11] We have also identified two additional important PU.1 and C/EBP target CSF receptors in addition to the M-CSF receptor: the GM-CSF receptor α [12] and the G-CSF receptor.[13] In all three of these myeloid CSF receptors, a relatively small (less than 100 base pair) upstream region directs specific expression in myeloid cell lines, and within these regions lie functionally important C/EBP sites and PU.1 sites. These results suggest that PU.1 and C/EBP direct the cell-type specific expression of multiple CSF receptor promoters, further establish the role of PU.1 as a key regulator of myelopoiesis, and point to C/EBP as an additional important factor in this process. These findings, indicating that at least three important myeloid CSF receptors are regulated by PU.1 and C/EBP, suggest why inactivation of PU.1 leads to a loss of myeloid cell development, and point to a major role of C/EBP in myelopoiesis.

We have recently described that a third factor, AML1, regulates the M-CSF receptor promoter.[11] This factor also regulates the GM-CSF and IL3 promoters,[14,15] suggesting that it, like PU.1 and C/EBPα, may play a major role in myeloid development by activating multiple myeloid CSF signalling pathways.

How do myeloid promoters work?

Myeloid promoters in general appear to differ from what has been described previously for tissue specific promoters.[16] In general, a relatively small upstream region (usually a few hundred base pairs) is capable of directing cell type specific expression in tissue culture studies. (See Fig. 1, references in Table 1, as well as [17]).

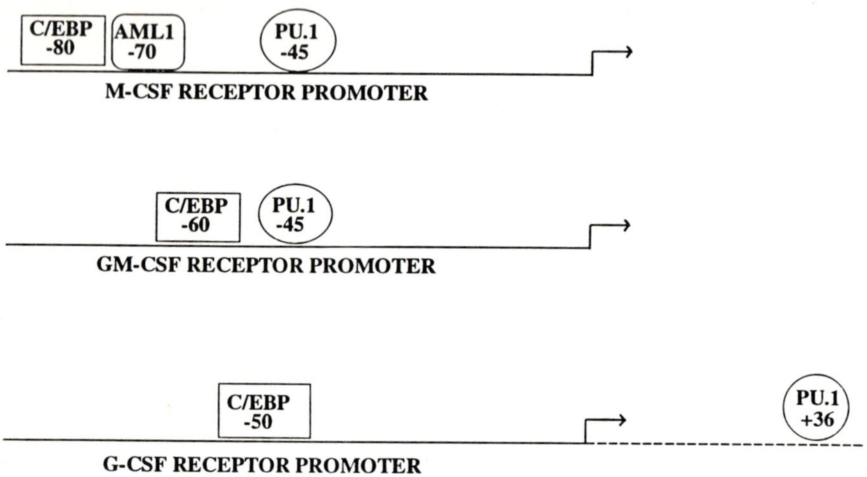

Fig. 1. Structure of the myeloid CSF receptor promoters. As with most myeloid promoters, there is no well defined TATA box, and the major transcription start site is designated with the arrow. Shown are the locations of binding sites for PU.1, C/EBP, and AML1. In myeloid cells, the major C/EBP gene product binding to these sites is C/EBPα. The PU.1 site in the G-CSF receptor promoter is located in the 5' untranslated region, at bp +36.

For example, almost all of the specificity and activity detected in transfection of tissue culture cells by the M-CSF receptor, GM-CSF receptor α, and G-CSF receptor promoters are contained within 80, 70, and 74 bases of the major transcription start sites.[11-13] Although most of these promoters direct lineage and developmental specific expression, in general they lack a TATA box or defined initiator sequence. Consistent with their lack of a TATA box, these promoters often are dependent on a functional Sp1 site, which not only mediates activity, but also specificity and inducibility with differentiation.[18-20]

Interestingly, many myeloid promoters have a functional PU.1 binding site upstream of the transcription start site at a location corresponding to one similar to that of a TATA box. Since PU.1 can bind to the TATA binding protein (TBP) *in vitro*,[21] one mechanism by which PU.1 could activate a whole class of myeloid promoters is by recruiting TBP, the primary component of the basal transcription factor TFIID, and the rest of the transcriptional apparatus. Consistent with this mechanism are studies showing that mutation of a PU.1 site in the FcγR1 promoter can abolish myeloid specific expression, and replacement of this mutant site with a TATA box can restore myeloid expression.[22]

Regulation of PU.1 Expression and Function

In the past two years, a number of investigators have identified the transcription factor PU.1 as a possible regulator of myelopoietic development. PU.1 was first isolated as the Spi-1 oncogene,[23,24] subsequently shown to be a member of the ets family of transcription factors.[25] PU.1 is expressed at highest levels in mature B cells and myeloid cells. Recently, it has been shown that PU.1 is expressed at low levels in murine ES cells and human CD34+ stem cells, and specifically upregulated with myeloid differentiation.[9] Studies using competitor oligonucleotides in human CD34+ cells, as well as targeted inactivation in mice, point to a major functional role for PU.1 in myeloid development.[9,26] Consistent with these findings has been the identification in the past two years of multiple myeloid genes regulated by PU.1 (See Table 1). PU.1 also plays a major role in B cell gene expression, and is important for lymphocyte development.[26]

Table 1. Myeloid Gene Targets Regulated by PU.1/Spi-1

Gene Target	Reference
CD11b (MAC-1α)	Pahl et al, JBC 268:5014, 1993 [27]
M-CSF (CSF-1) receptor	Zhang et al, MCB 14:373, 1994 [8]
FcγR1	Eichbaum et al, J Exp Med 179, 1985, 1994 [22]
	Perez et al, MCB 14:5023, 1994 [28]
FcγRIIIA	Feinman et al, EMBO J 13:3852, 1994 [29]
Chicken lysozyme	Ahne et al, JBC 269:17794, 1994 [30]
CD18 (MAC-1 β)	Rosmarin et al, PNAS 92:801, 1995 [31]
Macrophage scavenger receptor	Moulton et al, MCB 14:4408, 1994 [32]
IL-1β	Kominato et al, MCB 15:58, 1995 [33]
Neutrophil elastase	Srikanth et al, JBC 269:32626, 1994 [34]
	Nuchprayoon et al, MCB 14:5558, 1994 [35]
Lentivirus	Carvalho et al, J Virol 67:3885, 1993 [36]
	Maury, J Virol 68:6280, 1994 [37]
G-CSF Receptor	Smith et al, Blood 84:372a, 1994 [13]
GM-CSF Receptor	Hohaus et al, accepted to MCB pending revision, 1995 [12]

Previous studies have shown that PU.1 regulates the M-CSF receptor by binding to a site 45 nucleotides upstream of the major transcription start site.[8] More recently, we have identified functional PU.1 binding sites in the promoters for the GM-CSF receptor α at almost the identical relative position, and a site at +36 of the G-CSF receptor promoter.[12,13] In all three of these sites, PU.1 is the only factor in myeloid cells binding to this site, and specific mutations which abrogate PU.1 binding decrease promoter activity significantly in PU.1 expressing cells only.[8,12,13] A prediction of these studies is that murine hematopoietic cells with targeted disruption of both alleles of the PU.1 gene will have decreased or absent expression of these three myeloid CSF receptors, and, if so, their role in myelopoiesis can be tested by restoring their expression singly or in combination in these targeted cells.

Role of C/EBP factors in Regulation of Myeloid CSF Receptor Promoters

A second major regulator of these three myeloid CSF receptor promoters is the family of C/EBP factors. C/EBP was first isolated as a protein with a basic region/leucine zipper structure which binds a CAAT site (hence the name CAAT enhancing binding protein), and subsequently shown to be a member of a family of differentially expressed genes.[38,39] A number of studies have shown the importance of C/EBP proteins in adipogenesis, as well as regulating several avian myeloid genes.[40] These and other studies suggested that in the hematopoietic system, C/EBP factors may be specifically expressed in myeloid cells.[41] Alan Friedman's group demonstrated their regulated expression in a murine myeloid cell line, further suggesting that they might play a role in myeloid development.[42] It was recently shown that C/EBPβ (NF-IL6) regulates the IL1β promoter, consistent with the role of this member of the family in activation by cytokines like IL6 and LPS.[43] Targeted disruption of C/EBPβ leads to defects in killing of bacteria and tumor cells by macrophages.[44]

Our own studies have now shown that important C/EBP binding sites are located in the proximal promoter regions of the same three myeloid CSF receptors regulated by PU.1, and that the C/EBP and PU.1 sites are separated by as little as 15 (GM-CSF receptor α) to 80 (G-CSF receptor) bases (Fig. 1). Although all three C/EBP proteins (C/EBPα, C/EBPβ, and C/EBPδ) can bind to the M-CSF receptor and GM-CSF receptor α sites and trans-activate the promoter in cells which do not normally express significant amounts of C/EBP proteins, to date we have identified C/EBPα as the major form that binds to and activates these promoters in myeloid cells lines (D.E.Z., C.J.H., S. Hiebert, and D.G.T, manuscript in preparation [11,12]). These studies suggest that the relative abundance of the different C/EBP proteins during myeloid differentiation may be critical to CSF receptor expression, and we are currently measuring the expression of the different C/EBP genes during myeloid differentiation of human CD34+ cells. Our prediction is that targeted disruption of C/EBPα may produce a phenotype in myeloid cells similar to that of the PU.1 knockout mice, in which the PU.1 related factor Spi-B is unable to compensate for loss of PU.1, perhaps due to a lack of significant levels of Spi-B expression in myeloid cells.[10]

Role of the AML1 Factor in Regulation of the M-CSF Receptor Promoter

In addition to PU.1 and C/EBPα, we have identified a functionally critically binding site in the M-CSF receptor promoter for the AML1 transcription factor, which has been described in detail in other chapters of this volume.[11] AML1 is a member of the core binding factor family,[45-47] and is expressed in a variety of hematopoietic cells, with significant levels in myeloid and T cells.[11,48-51] AML1 forms a fusion protein, AML1/ETO, in t(8:21) leukemias, one of the most common translocations found in AML.[52-54] Therefore, identification of AML1 targets will be important to understand the pathogenesis of this form of leukemia. To date, AML1 sites have been identified in the promoters for the IL3 and GM-CSF growth factor genes,[14,15] but not in other myeloid receptor promoters, such as G-CSF receptor and GM-CSF receptor α. Therefore, alteration of the AML1 protein could affect the expression of three myeloid growth factor pathways (IL3, GM-CSF, and M-CSF through its receptor). The AML1 site in the M-CSF receptor lies just downstream of the C/EBP site, separated by only 10 bases (Fig. 1), and although C/EBP and AML1 bind to these two adjacent sites independently, mutation of either site dramatically decreases promoter function to low levels similar to that of mutating both sites.[11] These results suggest that although we do not observe evidence for physical interactions between AML1 and C/EBP in gelshift experiments, it is likely that these two factors interact positively in their functional activation of M-CSF receptor expression. We are currently testing the effect of the AML1/ETO fusion protein on M-CSF receptor promoter function.

The M-CSF receptor promoter provides a striking example of how the combinatorial action of transcription factors can lead to lineage specific expression. To date, we have identified three major factors activating the promoter (Fig. 1). When one takes into account the lineage specificity of the three factors (PU.1 in myeloid and B cells, C/EBPα in myeloid cells, and AML1 predominantly in T and myeloid cells), it is clear why this promoter is primarily active in the myeloid lineage, the only one in which all three are present in significant amounts.

Summary and Conclusions

Our studies of the promoters of the myeloid CSF receptors (M, GM, and G) in cell lines have led to the findings that the promoters are small, and are all activated by the PU.1 and C/EBP proteins. To date, we have only found evidence for involvement of C/EBPα, although further experiments will be needed to exclude the role of C/EBPβ and C/EBPδ in receptor gene expression. These studies suggest a model of hematopoiesis (Fig. 2) in which the lineage commitment decisions of multipotential cells are made by the alternative patterns of expression of certain transcription factors, which then activate growth factor receptors which allow those cells to respond to the appropriate growth factor to proliferate and survive. For example, expression of GATA-1 activates its own expression,[55] as well as that of the erythropoietin receptor,[56] inducing these cells to be capable of responding to erythropoietin. Similarly, expression of PU.1 activates its own promoter,[57] and turns on the three myeloid CSF receptors (M, GM, and G), pushing these cells along the pathway of myeloid differentiation. C/EBP proteins, particularly C/EBPα, are also critical for myeloid receptor promoter function, and may also act via autoregulatory mechanisms. Murine C/EBPα has a C/EBP binding site in its own promoter.[58] Human C/EBPα autoregulates its own expression in adipocytes by activating the USF transcription factor.[59] Myeloid genes expressed later during differentiation, such as CD11b, are also activated by PU.1, which is expressed at highest levels in mature myeloid cells,[10] but not by C/EBPα, which is downregulated in a differentiated murine myeloid cell line.[42] Consistent with this model are the findings that overexpression of PU.1 in erythroid cells blocks erythroid differentiation, leading to erythroleukemia,[23,24,60] and overexpression of GATA-1 in a myeloid line blocks myeloid differentiation.[61]

While these findings have provided some framework for understanding myeloid gene regulation, there are a number of critical questions to be addressed in the near future:

- What is the pattern of expression of the C/EBP proteins during the course of myeloid differentiation and activation of human CD34+ cells?
- What is the effect of targeted disruption and other mutations of the C/EBP and AML1 proteins on myeloid development and receptor expression?
- What are the interactions among these three different types of factors (ets, basic region-zipper, and *Runt* domain proteins) to activate the promoters?
- What is the effect of translocations, mutations, and alterations in expression of these factors, particularly in different forms of AML?

Acknowledgments

We gratefully acknowledge the assistance of Steven Ackerman, Scott Hiebert, Giuseppina Nucifora, Alan Friedman, Richard Maki, Linda Shapiro, Tom Look, and the many other past and current members of our laboratory who contributed to this work.

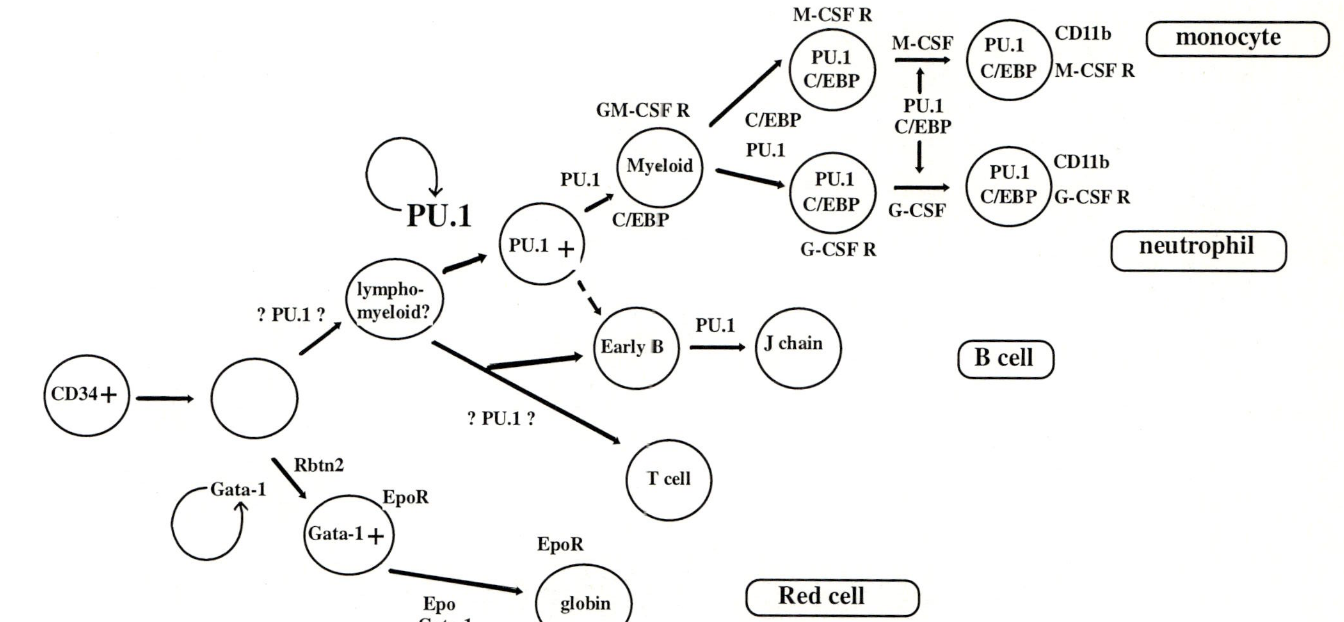

Fig. 2. Transcription factor model of hematopoietic lineage commitment. As discussed in the text, transcription factor activation in multipotential progenitors leads to autoregulation, induction of growth factor receptor expression and lineage commitment. For example, induction of GATA-1 induces expression of GATA-1 itself as well as the Epo receptor, allowing these progenitors to respond to Epo and further proliferate and differentiate into red cells. Alternatively, induction of PU.1 induces expression of PU.1 and the three myeloid CSF receptors. Recent data indicates that C/EBP proteins also induce expression of these same three receptors, leading to responsiveness of these myeloid progenitors to GM, G, and/or M-CSF. PU.1, expressed at high levels in mature monocytes and neutrophils, can also activate late genes such as CD11b, an adhesion molecule critical for phagocytic function. PU.1 may act on an early lymphomyeloid precursor, or on myeloid and B cell progenitors; its role in T cell development is not certain at this time.

References

1. Lubbert M, Herrmann F, Koeffler HP (1991) Expression and regulation of myeloid-specific genes in normal and leukemic myeloid cells. Blood 77:909-924
2. Nichols J, Nimer SD (1992) Transcription factors, translocations, and leukemia. Blood 80:2953-2963
3. Rabbitts TH (1994) Chromosomal translocations in human cancer. Nature 372:143-149
4. Rapoport AP, Abboud CN, DiPersio JF (1992) Granulocyte-macrophage colony-stimulating factor (GM-CSF) and granulocyte colony-stimulating factor (G-CSF): receptor biology, signal transduction, and neutrophil activation. Blood Rev 6:43-57
5. Nagata S, Fukunaga R (1991) Granulocyte colony-stimulating factor and its receptor. Prog Growth Factor Res 3:131-141
6. Sherr CJ (1990) Colony-stimulating factor-1 receptor. Blood 75:1-12
7. Metcalf D (1993) Hematopoietic Regulators - Redundancy or Subtlety. Blood 82:3515-3523
8. Zhang DE, Hetherington CJ, Chen HM, Tenen DG (1994) The macrophage transcription factor PU.1 directs tissue specific expression of the macrophage colony stimulating factor receptor. Mol Cell Biol 14:373-381
9. Voso MT, Burn TC, Wulf G, Lim B, Leone G, Tenen DG (1994) Inhibition of hematopoiesis by competitive binding of the transcription factor PU.1. Proc Natl Acad Sci U S A 91:7932-7936
10. Chen HM, Zhang P, Voso MT, Hohaus S, Gonzalez DA, Glass CK, Zhang DE, Tenen DG (1995) Neutrophils and monocytes express high levels of PU.1 (Spi-1) but not Spi-B. Blood 85:2918-2928
11. Zhang DE, Fujioka KI, Hetherington CJ, Shapiro LH, Chen HM, Look AT, Tenen DG (1994) Identification of a region which directs monocytic activity of the colony-stimulating factor 1 (macrophage colony-stimulating factor) receptor promoter and binds PEBP2/CBF (AML1). Mol Cell Biol 14:8085-8095
12. Hohaus S, Petrovick MS, Voso MT, Sun Z, Tenen DG (1995) PU.1 (Spi-1) and C/EBPα regulate the expression of the granulocyte-macrophage colony stimuating factor receptor α. Mol Cell Biol, accepted for publication pending revision.
13. Smith LA, Gonzalez DA, Hohaus S, Tenen DG (1994) The myeloid specific granulocyte colony stimulating factor (G-CSF) receptor promoter contains a functional site for the myeloid transcription factor PU.1 (Spi-1). Blood 84:372a
14. Frank R, Zhang J, Hiebert S, Meyers S, Nimer S (1994) AML1B but not the AML1/ETO fusion protein can transactivate the GM-CSF promoter. Blood 84:229a
15. Cameron S, Taylor DS, Tepas EC, Speck NA, Mathey-Prevot B (1994) Identification of a critical regulatory site in the human interleukin-3 promoter by in vivo footprinting. Blood 83:2851-2859
16. Goodbourn S, Maniatis T (1988) Overlapping positive and negative regulatory domains of the human beta-interferon gene. Proc Natl Acad Sci U S A 85:1447-1451
17. Pahl HL, Rosmarin AG, Tenen DG (1992) Characterization of the myeloid-specific CD11b promoter. Blood 79:865-870
18. Chen HM, Pahl HL, Scheibe RJ, Zhang DE, Tenen DG (1993) The Sp1 transcription factor binds the CD11b promoter specifically in myeloid cells in vivo and is essential for myeloid-specific promoter activity. J Biol Chem 268:8230-8239
19. Zhang DE, Hetherington CJ, Tan S, Dziennis SE, Gonzalez DA, Chen HM, Tenen DG (1994) Sp1 is a critical factor for the monocytic specific expression of human CD14. J Biol Chem 269:11425-11434
20. Zhang DE, Hetherington CJ, Gonzalez DA, Chen HM, Tenen DG (1994) Regulation of CD14 expression during monocytic differentiation induced with 1alpha,25-Dihydroxyvitamin D3. J Immunol 153:3276-3284

21. Hagemeier C, Bannister AJ, Cook A, Kouzarides T (1993) The activation domain of transcription factor PU.1 binds the retinoblastoma (RB) protein and the transcription factor TFIID in vitro: RB shows sequence similarity to TFIID and TFIIB. Proc Natl Acad Sci U S A 90:1580-1584
22. Eichbaum QG, Iyer R, Raveh DP, Mathieu C, Ezekowitz AB (1994) Restriction of interferon gamma responsiveness and basal expression of the myeloid human FcgammaR1b gene is mediated by a functional PU.1 site and a transcription initiator consensus. J Exp Med 179:1985-1996
23. Moreau-Gachelin F, Tavitian A, Tambourin P (1988) Spi-1 is a putative oncogene in virally induced murine erythroleukaemias. Nature 331:277-280
24. Moreau-Gachelin F (1994) Spi-1/PU.1: an oncogene of the ets family. Biochim Biophys Acta 1198:149-163
25. Klemsz MJ, McKercher SR, Celada A, Van Beveren C, Maki RA (1990) The macrophage and B cell-specific transcription factor PU.1 is related to the ets oncogene. Cell 61:113-124
26. Scott EW, Simon MC, Anastai J, Singh H (1994) The transcription factor PU.1 is required for the development of multiple hematopoietic lineages. Science 265:1573-1577
27. Pahl HL, Scheibe RJ, Zhang DE, Chen HM, Galson DL, Maki RA, Tenen DG (1993) The proto-oncogene PU.1 regulates expression of the myeloid-specific CD11b promoter. J Biol Chem 268:5014-5020
28. Perez C, Coeffier E, Moreau-Gachelin F, Wietzerbin J, Benech PD (1994) Involvement of the transcription factor PU.1/Spi-1 in myeloid cell-restricted expression of an interferon-inducible gene encoding the human high-affinity Fc gamma receptor. Mol Cell Biol 14:5023-5031
29. Feinman R, Qiu WQ, Pearse RN, Nikolajczyk BS, Sen R, Sheffery M, Ravetch JV (1994) PU.1 and an HLH family member contribute to the myeloid-specific transcription of the Fc-gammaRIIIA promoter. EMBO J 13:3852 3860
30. Ahne B, Stratling WH (1994) Characterization of a myeloid-specific enhancer of the chicken lysozyme gene - Major role for an Ets transcription factor-binding site. J Biol Chem 269:17794-17801
31. Rosmarin AG, Caprio D, Levy R, Simkevich C (1995) CD18 (β_2 leukocyte integrin) promoter requires PU.1 transcription factor for myeloid activity. Proc Natl Acad Sci U S A 92:801-805
32. Moulton KS, Semple K, Wu H, Glass CK (1994) Cell-specific expression of the macrophage scavenger receptor gene is dependent on PU.1 and a composite AP-1/ets motif. Mol Cell Biol 14:4408-4418
33. Kominato Y, Galson DL, Waterman WR, Webb AC, Auron PE (1995) Monocyte-specific expression of the human prointerleukin 1β gene (IL1β) is dependent upon promoter sequences which bind the hematopoietic transcription factor Spi-1/PU.1. Mol Cell Biol 15:58-68
34. Srikanth S, Rado TA (1994) A 30-base pair element is responsible for the myeloid-specific activity of the human neutrophil elastase promoter. J Biol Chem 269:32626-32633
35. Nuchprayoon I, Meyers S, Scott LM, Suzow J, Hiebert S, Friedman AD (1994) PEBP2/CBF, the murine homolog of the human myeloid AML1 and PEBP2beta/CBFbeta proto-oncoproteins, regulates the murine myeloperoxidase and neutrophil elastase genes in immature myeloid cells. Mol Cell Biol 14:5558-5568
36. Carvalho M, Derse D (1993) The PU.1/Spi-1 proto-oncogene is a transcriptional regulator of a lentivirus promoter. J Virol 67:3885-3890
37. Maury W (1994) Monocyte maturation controls expression of equine infectious anemia virus. J Virol 68:6270-6279
38. Landschulz WH, Johnson PF, McKnight SL (1988) The leucine zipper: a hypothetical structure common to a new class of DNA binding proteins. Science 240:1759-1764
39. Johnson PF, Landschulz WH, Graves BJ, McKnight SL (1987) Identification of a rat liver nuclear protein that binds to the enhancer core element of three animal viruses. Genes Dev 1:133-146

40. Katz S, Kowenzleutz E, Muller C, Meese K, Ness SA, Leutz A (1993) The NF-M transcription factor is related to C/EBP-beta and plays a role in signal transduction, differentiation and leukemogenesis of avian myelomonocytic cells. EMBO J 12:1321-1332
41. Natsuka S, Akira S, Nishio Y, Hashimoto S, Sugita T, Isshiki H, Kishimoto T (1992) Macrophage differentiation-specific expression of NF-IL6, a transcription factor for interleukin-6. Blood 79:460-466
42. Scott LM, Civin CI, Rorth P, Friedman AD (1992) A novel temporal expression pattern of three C/EBP family members in differentiating myelomonocytic cells. Blood 80:1725-1735
43. Tsukada J, Saito K, Waterman WR, Webb AC, Auron PE (1994) Transcription factors NF-IL6 and CREB recognize a common essential site in the human prointerleukin 1β gene. Mol Cell Biol 14:7285-7297
44. Tanaka T, Akira S, Yoshida K, Umemoto M, Yoneda Y, Shirafuji N, Fujiwara H, Suematsu S, Yoshida N, Kishimoto T (1995) Targeted disruption of the NF-IL6 gene discloses its essential role in bacteria killing and tumor cytotoxicity by macroophages. Cell 80:353-361
45. Miyoshi H, Shimizu K, Kozu T, Maseki N, Kaneko Y, Ohki M (1991) t(8;21) breakpoints on chromosome 21 in acute myeloid leukemia are clustered within a limited region of a single gene, AML1. Proc Natl Acad Sci U S A 88:10431-10434
46. Ogawa E, Maruyama M, Kagoshima H, Inuzuka M, Lu J, Satake M, Shigesada K, Ito Y (1993) PEBP2/PEA2 Represents a Family of Transcription Factors Homologous to the Products of the Drosophila Runt Gene and the Human AML1 Gene. Proc Natl Acad Sci U S A 90:6859-6863
47. Wang S, Wang Q, Crute BE, Melnikova IN, Keller SR, Speck NA (1993) Cloning and characterization of subunits of the T-cell receptor and murine leukemia virus enhancer core-binding factor. Mol Cell Biol 13:3324-3339
48. Kamachi Y, Ogawa E, Asano M, Ishida S, Murakami Y, Satake M, Ito Y, Shigesada K (1990) Purification of a mouse nuclear factor that binds to both the A and B cores of the polyomavirus enhancer. J Virol 64:4808-4819
49. Levanon D, Negreanu V, Bernstein Y, Baram I, Avivi L, Groner Y (1994) AML1, AML2, and AML3, the human members of the runt domain gene-family: cDNA structure, expression, and chromosomal localization. Genomics 23:425-432
50. Meyers S, Downing JR, Hiebert SW (1993) Identification of AML-1 and the (8;21) Translocation Protein (AML-1/ETO) as Sequence-Specific DNA-Binding Proteins - The Runt Homology Domain Is Required for DNA Binding and Protein-Protein Interactions. Mol Cell Biol 13:6336-6345
51. Satake M, Nomura S, Yamaguchpiiwai Y, Takahama Y, Hashimot Y, Niki M, Kitamura Y, Ito Y (1995) Expression of the runt domain-encoding PEBP2 alpha genes in T cells during thymic development. Mol Cell Biol 15:1662-1670
52. Miyoshi H, Kozu T, Shimizu K, Enomoto K, Maseki N, Kaneko Y, Kamada N, Ohki M (1993) The t(8-21) Translocation in Acute Myeloid Leukemia Results in Production of an AML1-MTG8 Fusion Transcript. EMBO J 12:2715-2721
53. Nucifora G, Rowley JD (1994) The AML1 and ETO genes in acute myeloid leukemia with a t(8;21). Leuk Lymphoma 14:353-362
54. Downing JR, Head DR, Curcio-Brint AM, Hulshof MG, Motroni TA, Raimondi SC, Carroll AJ, Drabkin HA, Willman C, Theil KS, et al (1993) An AML1/ETO fusion transcript is consistently detected by RNA-based polymerase chain reaction in acute myelogenous leukemia containing the (8;21)(q22;q22) translocation. Blood 81:2860-2865
55. Tsai SF, Strauss E, Orkin SH (1991) Functional analysis and in vivo footprinting implicate the erythroid transcription factor GATA-1 as a positive regulator of its own promoter. Genes Dev 5:919-931
56. Zon LI, Youssoufian H, Mather C, Lodish HF, Orkin SH (1991) Activation of the erythropoietin receptor promoter by transcription factor GATA-1. Proc Natl Acad Sci U S A 88:10638-10641

57. Chen HM, Ray-Gallet D, Zhang P, Hetherington CJ, Gonzalez DA, Zhang D-E, Moreau-Gachelin F, Tenen DG (1995) PU.1 (Spi-1) autoregulates its expression in myeloid cells. Oncogene, accepted for publication pending revision
58. Christy RJ, Kaestner KH, Geiman DE, Lane MD (1991) CCAAT/enhancer binding protein gene promoter: binding of nuclear factors during differentiation of 3T3-L1 preadipocytes. Proc Natl Acad Sci U S A 88:2593-2597
59. Timchenko N, Wilson DR, Taylor LR, Abdelsayed S, Wilde M, Sawadogo M, Darlington GJ (1995) Autoregulation of the human C/EBP alpha gene by stimulation of upstream stimulatory factor binding. Mol Cell Biol 15:1192-1202
60. Schuetze S, Paul R, Gliniak BC, Kabat D (1992) Role of the PU.1 transcription factor in controlling differentiation of Friend erythroleukemia cells. Mol Cell Biol 12:2967-2975
61. Visvader JE, Elefanty AG, Strasser A, Adams JM (1992) GATA-1 but not SCL induces megakaryocytic differentiation in an early myeloid line. EMBO J 11:4557-4564

Regulation of Immature Myeloid Cell Differentiation by PEBP2/CBF, MYB, C/EBP and ETS Family Members

A.D. Friedman
Pediatric Oncology Division, Johns Hopkins Oncology Center, Baltimore, Maryland, USA

Transcriptional Regulation of Hematopoiesis

The differentiation potential of the pluripotent hematopoietic stem cell becomes stochastically restricted until unipotent progenitors arise (Till and McCulloch 1980). A simplified diagram of the cellular and hormonal basis of hematopoiesis is depicted in Fig. 1.

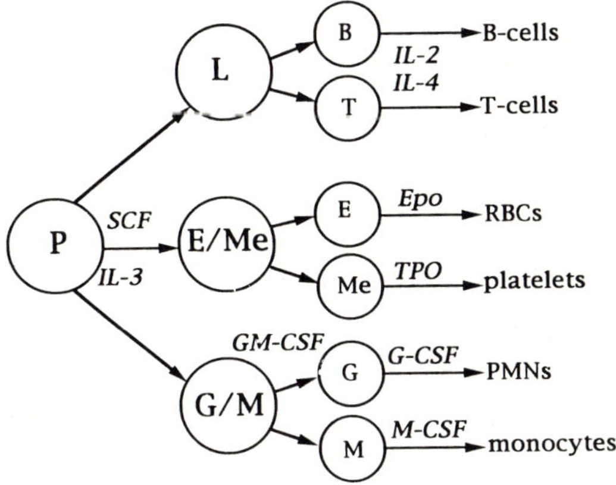

Fig. 1. A model of hematopoiesis. The pluripotent hematopoietic stem cell (P) gives rise to progenitor cells with more restricted potential, through a series of cellular intermediates which are not diagrammed. Three main branches of P cell differentiation are the lymphoid (L) branch, the Erythroid/Megakaryocytic (E/Me) branch, and the Granulocyte/Monocyte (G/M), Myeloid branch. The Basophil and Eosinophil branches are not shown. The proliferation and survival of the progenitors is thought to require hematopoietic hormones such as Stem Cell Factor (SCF) and Interleukin-3 (IL-3), which act predominantly on multipotent progenitors, IL-2 and IL-4 which act on lymphoid cells, Erythropoietin (EPO), which acts on erythroid cells, Thrombopoietin (TPO), which acts on megakaryocitic cells, Granulocyte Monocyte-Colony Stimulating Factor (GM-CSF), which acts predominantly on immature myeloid cells, G-CSF, which acts on neutrophilic cells, and M-CSF, which acts on monocytic cells.

The events which regulate the successive restriction in differentiation potential of hematopoietic progenitors are thought to be cell autonomous, with hematopoietic growth factors serving only to promote the survival and proliferation of particular progenitors. This generality was true for the myeloid and erythroid differentiation of a stem cell line (Fairbairn et al 1993), but has not been established for all growth factors and progenitors. Whether the stochastic events which regulate the successive restriction of progenitor differentiation potential primarily involve transcription factors, kinases, chromatin structure, or other mechanisms is not known. Transcription factors are, however, clearly involved in the expression of these commitment decisions. For example, GATA-1 is required for erythroid differentiation (Simon et al 1992), and Oct-2 is required for lymphoid differentiation (Corcoran et al 1993). The involvement of these transcription factors in hematopoiesis came to light through investigation of the transcriptional regulation of erythroid- and lymphoid-restricted genes. Similarly, my research group began investigating the transcriptional regulation of two genes expressed specifically in immature myeloid cells, myeloperoxidase and neutrophil elastase, with the intent of identifying transcription factors which regulate the generation of commited myeloid progenitors. Perhaps not surprisingly, three of the factors we identified have already been implicated as myeloid leukemia proto-oncoproteins.

DNA Elements Which Regulate The Myeloperoxidase and Neutrophil Elastase Genes

Myeloperoxidase (MPO) and Neutrophil Elastase (NE) are microbicidal proteins present in the primary granules of myeloid cells. These proteins, and their corresponding mRNAs, are not found in other cell lineages (Tobler et al 1988). Transcriptional regulation is in part responsible for the activation of the murine MPO gene during 32D cl3 cells differentiation (Friedman et al 1991) and for shut-off of the human NE gene during HL-60 cell differentiation (Yoshimura et al 1992).

MPO and NE are early markers of both the granulocytic and monocytic myeloid lineages (Pryzwansky et al 1978; Fouret et al 1989). The expression of both murine MPO and NE mRNAs increases markedly during the first days of 32D cl3 cell differentiation (Friedman et al 1991; Nuchprayoon et al 1994). Myeloid cytokine receptor genes may be activated before MPO and NE in myeloid progenitors, but this remains to be established. Interestingly, the elegant work of D. Tenen and colleagues has shown that the GM-CSF, G-CSF, and M-CSF Receptor genes are at least in part regulated by factors similar to those which regulate MPO and NE (see further discussion below and elsewhere in this volume).

The Myeloperoxidase Gene

Both the murine and human MPO genes lack a TATAA box, and their 5' flanking sequences are highly homologous (60%) over 2 kb (as first noted by G. Rovera, pers. communic.). The murine MPO gene initiates transcription from two major sites which are separated by approximately 400 bp (Friedman et al 1991). Two groups identified a DNAse I hypersensitivity sites in the human MPO gene in a region homologous to that containing the distal murine MPO start site (Jorgenson et al 1991; Lubbert et al 1991). Dr. Rovera generously provided a murine MPO genomic clone and the valuable 32D cl3 myeloid cell line (Valtieri et al 1987), enabling us to analyze this region for functional DNA elements.

32D cl3 cells are a diploid, non-leukemic murine cell line which requires IL-3 for survival and cell proliferation. In G-CSF-containing media, they differentiate along the granulocytic lineage (Valtieri et al 1987). Morphologically, their differentiation closely mimics normal granulocytic differentiation, and unlike most myeloid lines they even produce secondary granules. 32D cl3 cells also retain the ability to differentiate along the monocytic pathway (Kreider et al 1990). Thus, 32D cl3 cells retain many characteristics of normal myeloblasts.

We developed a DEAE-dextran procedure which allows reproducible transient transfection of uninduced 32D cl3 cells (Suzow and Friedman 1993), which can then be cultured in G-CSF. Transfection of 32D cl3 cells is at least 10 times less efficient than of fibroblastic lines. Therefore, we relied on the sensitive luciferase assay to detect expression of transfected DNA constructs. When murine MPO 5' flanking region segments were linked to a luciferase reporter and transfected into 32D cl3 cells, luciferase activities barely above background were detected. We reasoned that the TATAA-less murine MPO promoters may not function well in our constructs and so inserted a TATAA box and cap site from the herpes thymidine kinase (TK) gene in place of the murine MPO start sites. We then found that a 1300 bp murine MPO 5' flanking region segment would enhance transcription from the heterologous TK promoter with 100-fold greater potency in induced 32D cl3 cells compared with mouse L cells (Suzow and Friedman 1993). 5' deletional analysis further showed that a 414 bp DNA segment located in the vicinity of the distal MPO mRNA initiation site was similarly a potent myeloid enhancer. Additional analysis of this segment identified several functional elements (Fig. 2), which are apparently conserved in the murine and human MPO genes.

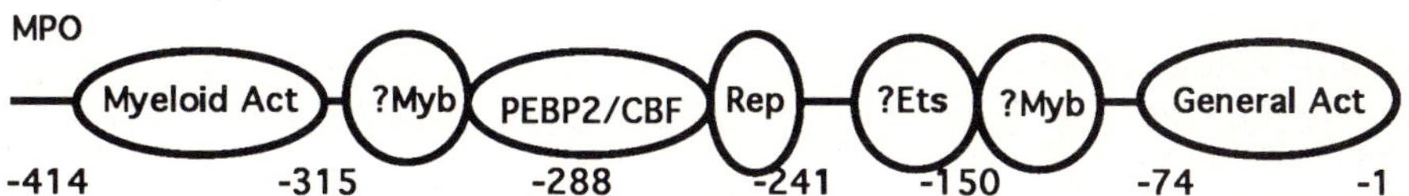

Fig. 2. Functional elements identified in the murine MPO 5' flanking region enhancer (Suzow and Friedman 1993). Deletion of segment -414 to -315 diminished enhancer activity 5-fold only in 32D cl3 cells but not in L cells. Clustered point mutations at -302/-297 diminished activity 4-fold, and this site contains a c-Myb consensus. Mutation of -293/-289 diminished activity 30-fold, and this site contains a PEBP2/CBF consensus. Deletion of bases -269 to -241 or -281 to -241 increased activity 10-fold, suggesting the binding of a transcriptional repressor. Deletion of segment -74 to -1 was markedly deleterious in both 32D cl3 and mouse L cells, suggesting that this segment is important for general promoter activity. In addition, we recently identified a functional Myb consensus sequence and a functional GA-rich sequence, which could bind a member of the Ets family, at approximately -150 (M. Britos-Bray and A.D.F., not shown).

The Neutrophil Elastase Gene

Using the human NE cDNA, provided by R. Crystal, we cloned the murine NE gene. Both the human and murine NE genes contain a TATAA homology and single mRNA start site. We linked the luciferase reporter downstream of the murine NE cap site. Functional, analysis of an 1800 bp murine NE 5' flanking region segment showed that only the proximal 91 bps contained functional elements. This murine NE enhancer was several-hundred fold more active in induced 32D cl3 myeloid cells compared with mouse L cells. Comparison of the human and murine sequences of this 91 bp segment revealed three areas of striking conservation. The underlying sequences predicted the binding of members of the Ets, C/EBP, and Myb families. Introduction of clustered point mutations separately into each of these sites verified that they were functional in 32D cl3 cells. An additional site, located at -70 was shown to be functional in the murine NE enhancer. This site weakly fits the consensus for PEBP2/CBF, but is not conserved in the human NE gene (Nuchprayoon et al 1994).

The human NE gene was recently shown to be adjacent to genes encoding two other myeloid-specific serine proteases, azurocidin and myeloblastin (Zimmer et al 1992). Those authors noted that the immediate 5' region of these three genes contains potential binding sites for members of the Ets, CAAT box-binding, and Myb families of transcription factors. These conserved sites, and the functional sites we identified in the murine NE gene are shown in Fig. 3.

mNE:	**GGAGAGGAAG**	**TGGGGCAAT**	**CAACGG**
hNE:	**AGGGAGGAAG**	**TGGGGCAAT**	**CAACGG**
hAZU:	**GAGAGGGAAA**	**TTGGGCAAT**	**CAACTG**
hMBN:	**AAGGAGGAAG**	**CTGGGCATT**	**CAACGG**
CONSENSUS:	**RRRGAGGAAG**	**T(T/G)NNGNAA(T/G)**	**(T/C)AAC(T/G)G**
	ETS	**C/EBP**	**MYB**

Fig. 3. Conserved binding sites in murine NE (mNE), human NE (hNE), human azurocidin (hAZU) and human myeloblastin (hMBN) 5' flanking regions. The consensus sequences for PU.1 (an Ets factor), C/EBP, and c-Myb are shown. Note that whereas the mNE and hNE genes conserve the spacing between these binding sites, neither their spacing nor their orientation relative to the DNA helix are conserved between the NE, hAZU, and hMBN genes.

Transcription Factors Which Regulate The Myeloperoxidase and Neutrophil Elastase Genes

PEBP2/CBF

Polyoma Enhancer Binding Protein 2/Core Binding Factors (PEBP2/CBF) were purified based on their ability to bind the core sites of polyomavirus or Moloney murine leukemia virus (see references in Nuchprayoon et al 1994). The PEPB2/CBFs are a family of heterodimeric factors, each of which contains a common PEBP2/CBFβ subunit and a specific PEBP2/CBFα subunit. The three identified α subunits are encoded by separate genes. The PEBP2/CBFαB subunit was also termed AML1, because its corresponding gene is present at the breakpoint of t(8;21) present in some cases of acute myeloid leukemia. The PEBP2/CBFβ subunit does not bind DNA, but strengthens binding by the α subunits. The tissue distribution of the PEBP2/CBFs has not been established, but evidence suggesting that they are restricted to lymphoid and myeloid cells has been presented.

The consensus binding sequence for PEBP2/CBFα subunits is 5'-ACCPuCA-3'. The strong functional element located at -290 to -284 of the murine MPO gene, 5-AACCACA-3' fits these consensus sequences, whereas our initial down-mutation, 5'-TAGCACA-3', did not. We detected two gel shift species in 32D cl3 nuclear extracts which bound the wild-type site, but not the mutant site, and termed these MyNF1 (Suzow and Friedman 1993). As C/EBP and several other simian virus 40 enhancer core binding proteins were known to bind the related sequence 5'-TTCCACA-3', we determined whether the MyNF1s could bind that site - they could not. But they did bind and activate transcription via 5'-GACCGCA-3', which fits the PEBP2/CBF consensus. Finally, together with S.

Hiebert and colleagues we showed that the MyNF1 gel shift species were super-shifted by an N-terminal PEBP2/CBFα subunit antisera, establishing that PEBP2/CBF binds and regulates the murine MPO proximal enhancer. Which PEBP2/CBFα subunit activates murine MPO in 32D cl3 cells remains to be determined, as the antisera employed would recognize both αA and αB, and we detected both αA and αB mRNAs in these cells (Nuchprayoon et al 1994).

We similarly showed that PEBP2/CBF weakly binds a functional site located at -70 in the murine NE gene (Nuchprayoon et al 1994). Conceivably, such weak *in vitro* binding might translate into strong *in vivo* cooperativity with other factors. In addition to MPO and murine NE, PEBP2/CBF has been shown to activate a third early myeloid gene, the M-CSF receptor (Zhang et al 1994a).

C/EBP

CCAAT/Enhancer binding proteins (C/EBP) are a family of bZIP DNA-binding proteins which bind DNA as homo- or hetero-dimers. The founding member, C/EBPα, was originally purified and cloned from rat liver. The C/EBP family also includes C/EBPβ, C/EBPγ, and C/EBPδ (see references in Scott et al 1992). C/EBPα is present only in terminally differentiated hepatocytes, where it activates lineage-specific genes (e.g. Friedman et al 1989). Also, C/EBPs help regulate terminal adipocyte differentiation (Yeh et al 1995), and C/EBPα even inhibits proliferation of 3T3-L1 preadipocytes (Umek et al 1991).

Within hematopoietic cells we detected C/EBPα, C/EBPβ, and C/EBPδ in all myeloid cell lines examined, but not in lymphoid or erythroid lines, by Western blotting. C/EBPα was detected as well in myeloid cells prepared from normal marrow and in *de novo* myeloid leukemia cells, but not in normal lymphoid or erythroid cells or in *de novo* lymphoid leukemias (Scott et al 1992). These findings were consistent with the detection of a C/EBP gel-shift activity in avian myeloid but not lymphoid or erythroid cell extracts (Sterneck et al 1992).

In contrast to immature hepatocytes and adipocytes, immature myeloid cells were found to tolerate high-level C/EBPα expression, and expression of this C/EBP isoform diminished rather than increased with myeloid maturation (Scott et al 1992). On the other hand, we also observed that C/EBPβ levels increased during 32D cl3 maturation and were higher in a macrophage than in monoblastic cell lines. These findings are constistent with the macrophage functional defects present in the a C/EBPβ-knockout mouse (Tanaka et al 1995).

We are collaborating with B. Lüscher and colleagues to further determine the role that C/EBPs, as well as c-Myb and Ets family members, play in regulating the murine NE proximal enhancer. Both of our groups have found that the murine NE C/EBP site, located between -61 and -53, can bind C/EBPα, and that the murine NE enhancer is activated by a C/EBPα expression vector in non-myeloid cells only if that site is intact (not shown). Two genes expressed specifically in immature myeloid cells had previously been shown to be activated by NF-M (avian C/EBPβ), the avian specific mim-1 gene and lysozyme (Ness et al 1993). Another immature myeloid gene, the M-CSF receptor contains a functional site which also fits the C/EBP consensus (Zhang et al 1994a).

c-Myb

c-Myb is predominantly expressed in immature hematopoietic cells (Sheiness and Gardinier 1984). Mice lacking c-Myb are defective for fetal hematopoiesis (Mucenski et al 1991). c-Myb cooperated with avian C/EBPβ to activate the mim-1 and lysozyme genes in myeloid cells (Ness et al 1993). We have detected functional elements which fit the c-Myb consensus in both the MPO and NE genes. In collaboration with B. Lüscher and colleagues, we have found that the murine NE proximal promoter is activated by c-Myb in a non-myeloid cell line only if the c-Myb site located between -51 and -45 is intact; and, as with mim-1 and lysozyme, c-Myb and C/EBPs activate murine NE synergistically (not shown).

c-Myb and PEBP2/CBF cooperate to regulate the T cell receptor δ enhancer in T cells (Hernandez-Munain and Krangel 1994), and these two factors may also help regulate MPO and murine NE expression.

Ets Family Members

The Ets family member PU.1 has been shown to regulate the CD11b and M-CSF receptor genes in monocytic cells (Pahl et al 1993; Zhang et al 1994b), and was recently found to also regulate the G-CSF receptor and GM-CSF receptor genes (Smith et al 1994; Hohaus and Tenen 1994). Also, we recently found using gel shift and super-shift assays that PU.1 can bind the functional element located between -88 and -78 in the murine NE enhancer (I. Nuchprayoon and A.D.F., not shown). PU.1 knockout mice indeed lack myeloid, as well as lymphoid, cells (Scott et al 1994). However, other Ets family members are also present in myeloid cells, and several of these may well bind the same sites as PU.1 in these myeloid genes (Klemsz et al 1993 and references therein). Indeed, we recently found that an Ets-2 expression vector activated the murine NE enhancer in NIH3T3 cells, whereas a PU.1 expression vector did not (I. Nuchprayoon and A.D.F., not shown). Ets-2 and c-Myb can combine to regulate the hematopoietic progenitor-specific CD34 promoter (Melotti and Calabretta 1994).

Combinatorial Activation and Lineage Specificities

The transcriptional activator MyoD can activate a program of muscle-specific gene expression in a variety of non-muscle cells (Davis et al 1987). However, the majority of cell lineages may be specified by a combination of transcription factors, none of which may be restricted in expression to that lineage. A. Leutz and colleagues first proposed that the combination of a C/EBP and c-Myb may specify early myeloid differentiation (Ness et al 1993). The results described above led us to suggest that a combination of two or more factors from the PEBP2/CBF, C/EBP, c-Myb, and Ets family of factors may be required to activate genes in immature myeloid cells (Nuchprayoon et al 1994).

Of these four groups of transcription factors, only the C/EBPs have been found to be restricted in their expression to myeloid cells, within the hematopoietic lineages (Scott et al 1992), although subsets of the PEBP2/CBF or Ets families might be myeloid specific as well. As C/EBPα expression predominates in early myeloid cells, it will be of great interest to determine whether this protein is essential for production of early myeloid cells. We employed a naturally occuring dominant-negative C/EBP, CHOP10 (Ron and Habener 1992), to inhibit C/EBP activities in 32D cl3 myeloid cells. In two cell lines expressing high-levels of CHOP10 from the transduced retroviral vector, C/EBP activities were markedly reduced, as assessed using a reporter construct containing two C/EBP DNA-binding sites upstream of a TK promoter and a luciferase reporter. These cell lines proliferated no differently than control cell in IL-3, but apoptosed in G-CSF-containing media, whereas the control cells differentiated (Friedman 1993). Thus, C/EBP activities may be required in the early stages of myeloid differentiation.

Links Between Transcriptional Regulation in Immature Myeloid Cells and Leukemogenesis

PEBP2/CBF

The PEBP2/CBFαB, or AML1, gene is located at the breakpoint of t(8;21)(q22;q22), associated with some cases of FAB M2 acute myeloid leukemia (AML, Miyoshi et al 1991). The AML1 gene is also present at the breakpoint of t(3;21)(q26;q22), in a subset of myeloid leukemias (e.g. Sacchi et al 1994). Finally, the PEBP2/CBFβ gene is disrupted by inv(16)(p13q22), associated with FAB M4eo cases of AML (Liu et al 1993). The inv(16) oncoprotein, CBFβ-MYH11, has been found capable of inhibiting binding of PEBP2/CBF to DNA (Hajra et al 1995), and the t(8;21) oncoprotein, AML1-ETO, has been shown to be a transcriptional repressor (Meyers et al 1995). These results, together with the finding that PEBP2/CBF regulates early myeloid-specific genes such as MPO, suggest a model in which these two oncoproteins act by inhibiting early myeloid differentiation. The FAB M3 myeloid oncoprotein, PML-RARα, may act analogously (Tsai et al 1992, Fig 4). The differentiation block in M2 and M4eo AMLs may be leaky, as some granulocytic differentiation is evident in these leukemias. Also, the eosinophilic blasts present in M4eo AMLs may reflect activation of an eosinophilic differentiation program by CBFβ-MYH11.

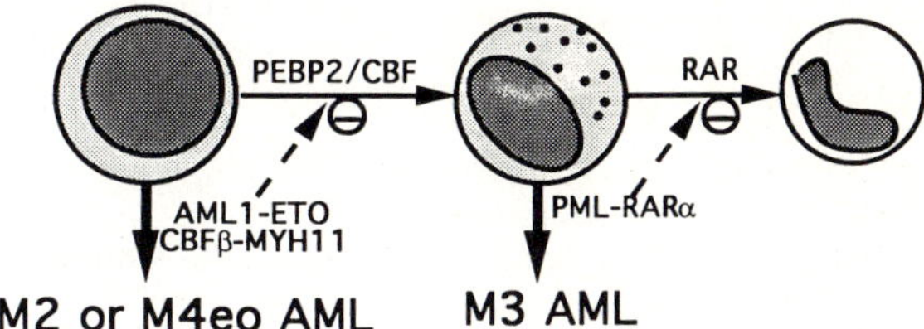

Fig. 4. Proposed inhibition of myeloid differentiation by AML1-ETO, CBFβ-MYH11, and PML-RARα. PEBP2/CBF is required for activation of MPO and other genes leading to the promyelocyte stage. RARα is required for differentiation past this stage.

c-Myb

Although c-Myb has not been implicated in human leukemias, activated forms of c-Myb have been implicated in both myeloid and erythroid leukemogenesis in avian and murine systems (reviewed in Introna et al 1994). c-Myb may regulate both proliferation and differentiation of immature myeloid cells. This dual role may be true for members of the C/EBP, Ets, and PEBP2/CBF families as well.

Ets and C/EBP families

Over-expression of Ets family members PU.1 or fli-1 can contribute to erythroleukemia, and activated ets-1 can contribute to both myeloid and erythroid leukemia (Hromas and Klemsz 1994). Perhaps aberrant expression of transcription factors normally present only in myeloid and lymphoid cells contributes to inhibition of erythroid differentiation and thus to leukemogenesis.

C/EBPs have not been implicated as proto-oncoproteins.

References

Corcoran LM, M.Karvelas, GJV Nossal, Z-S Ye, T Jacks, D Baltimore (1993) Oct-2, although not required for early B-cell development, is critical for later B-cell maturation and for post-natal survival. Genes & Dev 7:570-582

Davis RL, H Weintraub, AB Lassar (1987) Expression of a single transfected cDNA converts fibroblasts to myoblasts. Cell 51:987-1000

Fairbairn LJ, GJ Cowling, BM Reipert TM Dexter (1993) Suppression of apoptosis allows differentiation and development of a multipotent hematopoietic cell line in the absence of added growth factors. Cell 74:823-832

Fouret P, RM duBois, J-F.Bernaudin, H.Takahashi, VJ Ferrans, RG Crystal (1989) Expression of the neutrophil elastase gene during human bone marrow cell differentiation. J Exp Med 169:833-845

Friedman AD, WH Landschulz, SL McKnight (1989) CCAAT/enhancer binding protein activates the promoter of the serum albumin gene in cultured hepatoma cells. Genes & Dev 3: 1314-22

Friedman AD, BL Krieder, D Venturelli, G.Rovera (1991) Transcriptional regulation of two myeloid-specific genes, myeloperoxidase and lactoferrin, during differentiation of the murine cell line 32D Cl3. Blood 78: 2426-2432

Friedman AD (1993) Expression of CHOP10 in 32D cl3 cells inhibits endogenous C/EBP activities and decreases cell survival and proliferation. Blood 82:108a

Hajra A, PP Liu, Q Wang, CA Kelley, T Stacy, RS Adelstein, NA Speck, FS Collins (1995) The leukemic CBFb smooth muscle myosin heavy chain chimeric protein requires both CBFb and myosin heavy chain domains for transformation of NIH3T3 cells. Proc Natl Acad Sci, USA 92:1926-1930

Hernandez-Munain C, MS Krangel (1994) Regulation of the T-cell receptor d enhancer by functional cooperation between c-Myb and core-binding factors. Mol. Cell. Biol. 14:473-483

Hohaus S, DG Tenen (1994) The transcription factor PU.1 regulates the expression of the GM-CSF receptor alpha promoter. Blood 84:510a.

Hromas R, M Klemsz (1994) The ETS oncogene family in development, proliferation and neoplasia. Int J Hematol 59:257-265

Introna M, M Luchetti, M Castellano, M Arsura, J Golay (1994) The myb oncogene family of transcription factors: potent regulators of hematopoieitc cell proliferation and differentiation. Cancer Biology 5:113-124

Jorgenson, KF, GR Antoun, IF Zipf (1991) Chromatin structural analysis of the 5' end and contiguous flanking region of the myeloperoxidase gene. Blood 77:159-164

Klemsz MJ, RA Maki, T Papayannopoulou, J Moore, R Hromas (1993) Characterization of the ets oncogene family member fli-1. J Biol Chem 268:5769-5773

Kreider BL, PD Phillips, MB.Prystowsky, N Shirsat, JH Pierce, R Tushinski, G Rovera (1990) Induction of the GM-CSF receptor by G-CSF increases the differentiative options of a murine hematopoietic progenitor cell. Mol Cell Biol 10: 4846-4853

Liu P, SA Tarle, A.Hajre, DF Claxton, P Marlton, M Freedman, MJ Siciliano, FS Collins (1993) Fusion between transcription factor CBFb/PEBP2b and a myosin heavy chain in acute myeloid leukemia. Science 261:1041-1044

Lubbert, M, CW Miller, HP Koeffler (1991) Changes of DNA methylation and chromatin structure in the human MPO gene during myeloid differentiation. Blood 78:345-356

Melotti P, B Calabretta (1994) Ets-2 and c-Myb act independently in regulating expression of the heamatopoietic stem cell antigen CD34. J Biol Chem 269:25303-25309

Meyers S, N Lenny, SW Hiebert (1995) The t(8;21) fusion protein interferes with AML-1B-dependent transcripional activation. Mol Cell Biol 15:1974-1982

Miyoshi H, K Shimizu, T Kozu, N Maseki, Y Kaneko, M Ohki (1991) t(8;21) breakpoints on chromosome 21 in acute myeloid leukemia are clustured within a limited region of a single gene, AML1. Proc Natl Acad Sci, USA 88:10431-10434

Mucenski ML, K McClain, AB Kier, SH Swerdlow, CM Schreiner, TA Miller, DW Pietryga, WJ Scott, SS Potter (1991) A functional c-myb gene is required for normal murine fetal hepatic hematopoiesis. Cell 65: 677-689

Ness SA, E Kowenz-Leutz, T Casini, T Graf, A Leutz (1993) Myb and NF-M: combinatorial activators of myeloid genes in heterologous cell types. Genes & Dev 7:749-459

Nisson PE, PC.Watkins, N.Sacchi (1992) Transcriptionally active chimeric gene derived from the fusion of the AML1 gene and a novel gene on chromosome 8 in t(8;21) leukemic cells. Cancer Genet Cytogenet 63:81-88

Nuchprayoon I, S Meyers, LM Scott, J Suzow, S Hiebert, AD Friedman (1994) PEBP2/CBF, the murine homolog of the human myeloid AML1 and PEBP2/CBFb oncoproteins, regulates the murine MPO and neutrophil elastase genes in immature myeloid cells. Mol Cell Biol 14:5558-5568

Pahl HL, RJ Scheibe, H-M Chen, D-E Zhang, DL Galson, RA Maki, DG Tenen (1993) The proto-oncogene PU.1 regulates expression of the myeloid-specific CD11b promoter. J Biol Chem 61:1053-1095

Pryzwansky KB, LE Martin, JK Spitznagel (1978) Immunocytochemical localization of myeloperoxidase, lactoferrin, lysozyme, and neutral proteases in human monocytes and neutrophilic granulocytes. J Reticuloendoth Soc 24:295-310

Ron D, JF Habener (1992) CHOP-10, a novel developmentally regulated nuclear protein that dimerizes with transcription factor C/EBP and LAP and functions as a dominant-negative inhibitor of gene transcription. Genes & Dev 6: 439-453

Scott, LM, CI Civin P Rorth, AD Friedman (1992) A novel expression pattern of three C/EBP family members in differentiating myelomonocytic cells. Blood, 80:1725-1735

Scott EW, MC Simon, J Anastasi, H Singh (1994) Requirement of transcription factor PU.1 in the development of multiple hematopoietic lineages. Science 265:1573-1577

Sheiness D, M Gardinier (1984.)Expression of a proto-oncogene (proto-myb) in hematopoietic tissues of mice. Mol Cell Biol 4: 1206-1212

Simon, MC, L.Pevny, MV Wiles, G Keller, F Costantini, SH Orkin (1992) Rescue of the block to erythropoiesis in murine ES cells lacking a functional GATA-1 gene: Analysis of a transcription factor in a developmental program. Nature 336:688-690.

Smith LT, D Gonzalez, DG Tenen (1994) The myeloid specific G-CSF receptor promoter contains a functional site for the myeloid transcription factor PU.1. Blood 84:372a

Sterneck E, C Muller, S Katz, A Leutz (1992) Autocrine growth induced by kinase type oncogenes in myeloid cells requires AP-1 and NF-M, a myeloid specific, C/EBP-like factor. Embo J 11: 115-126

Suzow J, AD Friedman (1993) The murine myeloperoxidase promoter contains several functional elements - one element binds a cell type-restricted transcription factor, myeloid nuclear factor 1 (MyNF1). Mol Cell Biol 13:2141-2151

Tanaka T, S Akira, M Yoshida, Y Umemoto, N Yoneda, H Shirafuji, S Suematsu, N Yoshida, T Kishimoto (1995) Targeted disruption of the NF-IL6 gene discloses its essential role in bacteria killing and tumor cytotoxicitiy by macrophages. Cell 80:353-361

Till JE, EA McCulloch (1980) Hemopoietic stem cell differentiation. Bioch Biophys Acta 605:43-109

Tobler A, CW Miller, KR.Johnson, ME Selsted, G Rovera, P Koeffler (1988) Regulation of gene expression of myeloperoxidase during myeloid differentiation. J Cell Phys 136:215-225

Tsai, S, S Bartelmez, R Heyman, K Damm, R Evans, and SJ Collins (1992) A mutated retinoic acid receptor-a exhibiting dominant-negative activity alters the lineage development of a multipotent hematopoietic cell line. Genes & Dev 6:2258-2269

Umek, RM, AD Friedman, SL McKnight (1991) CCAAT/enhancer binding protein: A component of a differentiation switch. Science 251:288-292

Valtieri M, DJ Tweardy D Caracciolo, K Johnson, F Mavilio, S Altman, D Snatoli, G Rovera (1987) Regulation of proliferative and differentiative responses in a murine progenitor cell line. J. Immunol 138: 3829-3835

Yeh W-C, Z Cao, M Classon, SL McKnight (1995) Cascade regulation of terminal adipocyte differentiation by three members of the C/EBP family of leucine zipper proteins. Genes Dev 9:168-181

Yoshimura K, RG Crystal (1992) Transcriptional and posttranscriptional modulation of human neutrophil elastase gene expression. Blood 79:2733-2740

Zhang D-E, K-I Fujioka, CJ Hetherington, LH Shapiro, H-M Chen, AT Look, DG Tenen (1994a) Identification of a region which directs the monocytic activity of the CSF-1 (M-CSF) receptor promoter and binds PEBP2/CBF (AML1). Mol Cell Biol 14:8085-8095

Zhang D-E, CJ Hetherington H-M Chen, DG Tenen (1994b) The macrophage transcription factor PU.1 directs tissue-specific expression of the M-CSF receptor. Mol Cell Biol 14:373-381

Zimmer M, RL Medcalf, TM Fink, C Mattmann, P Lichter, DE Jenne (1992) Three human elastase-like genes coordinately expressed in the myelomonocyte lineage are organized as a single genetic locus on 19pter. Proc Natl Acad Sci, USA 89:8215-8219

Hematopoietic Transcriptional Regulation by the Myeloid Zinc Finger Gene, MZF-1.

R. Hromas[1], B. Davis[2], F. J. Rauscher III[3], M. Klemsz[1], D. Tenen[4], S. Hoffman[5], Dawei Xu[1], and J. F. Morris

1-Hematology/Oncology and the Walther Oncology Center, Indiana University Medical Center, 975 W. Walnut St., Indianapolis, IN 46202; 2-Cancer Center, UT Medical Branch at Galveston, Galveston, TX; 3-Molecular Oncology, Wistar Institute, Philadelphia, PA; 4- Hematology, Beth Israel Hospital, Harvard Medical School, Boston, MA; 5- Miami University, Oxford, OH; 6-Pediatric Hematology/Oncology, Medical College of Wisconsin, Milwaukee, WI.

Summary- Transcriptional regulators control much of hematopoiesis. One such transcriptional regulator is the myeloid zinc finger gene MZF-1. MZF-1 has been localized to the telomere of chromosome 19q, where a large number of related zinc finger genes reside. It has been found to be essential in granulopoiesis. It is a bi-functional transcriptional regulator, repressing transcription in non-hematopoietic cells, and activating transcription in cells of hematopoietic origins. Its consensus DNA binding site has been isolated, and sites in several promoters of myeloid-specific genes, such as CD34, lactoferrin, and myeloperoxidase, have been defined. In co-transfection experiments MZF-1 has been found to regulate transcription from the CD34 promoter.

Cloning of the Myeloid Zinc Finger Gene, MZF-1

Most embryonic gene expression is regulated at the transcriptional level. Since hematopoiesis may be viewed as a continuous model of development, it can be inferred that much of the genetic regulation of hematopoiesis may be under transcriptional control. This hypothesis has been validated by 1) the vast majority of leukemic translocations isolated involve transcription factors, 2) the genetic regulators of hematopoietic lineage phenotypes that have been cloned have been transcription factors [1-3].

The present study was based on two assumptions: 1) novel transcription factors controlled much of the lineage commitment and stage progression in hematopoiesis, and 2) these transcription factors would fall into known gene families based on conserved DNA-binding domain structures. Thus, using a degenerate oligonucleotide to the conserved H-C link (HTGERKP) in the Krupple family of zinc finger genes, we cloned MZF-1 from a cDNA library made from the peripheral leukocytes of a patient with chronic myelogenous leukemia. These peripheral leukocytes contained a wide variety of myeloid cells, from myeloblasts through granulocytes.

As shown below, MZF-1 has thirteen zinc finger domains, divided in two groups [4]. The amino-terminal group has four fingers, and the carboxy-terminal group has nine fingers. This structure is reminiscent of KID-1 and YY1, two other zinc finger genes important in development [5,6]. MZF-1 zinc finger domains are also related to PLZF-1, which is part of a fusion protein with RAR-alpha in t(11;17) promyelocytic leukemias [7]. There are acidic regions in the amino-terminal non-zinc finger portion of the protein which may be part of the transcriptional regulatory domain. Interestingly, there is a 24 amino acid glycine-proline rich link between the first and second zinc finger regions. Glycine-proline-rich domains have also been implicated in transcriptional regulation in a large number of transcription factors, including Oct-3, Rex-1, and CP-1 [4]. This glycine-proline-rich region, therefore, may serve as a hinge, allowing the two finger domains to bind DNA more efficiently, or as a separate transcriptional regulatory domain.

High stringency northern analysis using the zinc finger region of MZF-1 found multiple cross-hybridizing mRNA species. Likewise, high stringency Southern blotting of human DNA using the same zinc finger region of MZF-1 found a large number of cross-hybridizing bands. These findings lead us to postulate that MZF-1 was the prototype of a large family of related C2-H2 zinc finger genes.

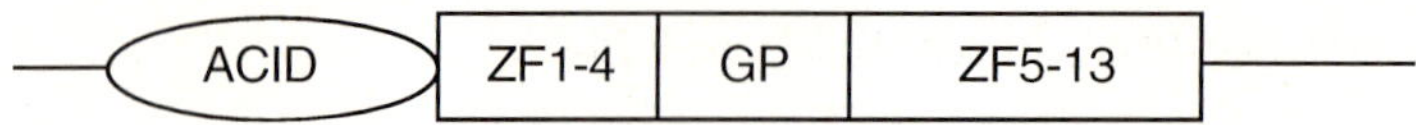

Figure 1. Schematic of the functional domains of the myeloid zinc finger gene MZF-1. The acidic region in the amino-terminal portion of the protein is designated by ACID. The first four zinc fingers (ZF1-4) are separated by a glycine-proline-rich region (GP) from the last nine fingers (ZF5-13)

By northern analysis MZF-1 was found to be expressed HEL, HL60, K562, and KG-1 cells. It was not present in Daudi, C10, Eskol, JY, FHS 738, U937, rat brain, Panc-1, melanoma, SJ4, HepG2, placenta, smooth muscle, or endothelium [4]. By cRNA in situ hybridization of normal marrow MZF-1 was found to be expressed in myeloid cells [8], from myeloblasts to metamyelocytes, but in no other marrow cell types. Thus, the expression pattern of MZF-1 implied a role in myeloid development.

Chromosomal Localization of MZF-1

Using in situ chromosomal hybridization MZF-1 was localized to human chromosome 19q13.4 [4]. Since at that time about a third of the known zinc finger genes had been mapped to chromosome 19, we thought that perhaps other members of the MZF family might also be on chromosome 19. Thus, we probed a chromosome 19-specific cosmid genomic library with the H-C link oligomer described above. We isolated over 490 cosmids that had Krupple-like zinc finger sequences within them. These were arrayed robotically on nitrocellulose filters, and have served as templates for many investigators to assess whether a given zinc finger gene was on chromosome 19. Since the location for most of these cosmids on chromosome 19 was known, once a zinc finger gene was found to hybridize to this filter, in almost every case its exact chromosomal 19 location was also known.

However, the large number of zinc finger -containing cosmids prohibited individual analysis. So, we concentrated on those zinc finger-cosmids that were located close to MZF-1 on chromosome 19q13.4. Using a combination of yacs, bacs, and high resolution FISH, the telomeric region of 19q was mapped with respect to these zinc finger containing cosmids. Cosmids containing zinc finger sequences were identified within a megabase of the telomere. The DNA fragments that contained these zinc finger sequences were subcloned, and tested for tissue patterns of expression using northern analysis. The zinc finger sequences we analyzed all had ubiquitous patterns of expression.

In our analysis of the telomeric region of chromosome 19q we found that MZF-1 was the most telomeric gene yet mapped. The entire coding region of MZF-1 is present in a 6.6 kB Hind3-EcoR1 fragment, with the EcoR1 site being most telomeric. In addition, MZF-1 is oriented with the 5' end being closest to the telomere. The 5' end of the MZF-1 gene is within 13 kB of a 3 kB Hind3 fragment that contains the 19q telomeric repeats. These repeats cross-hybridized by FISH to telomeres of the other chromosomes, indicating that they represent the true end of 19q. This data is being prepared for submission.

The finding that MZF-1 is located at the telomere of chromosome 19q is significant because there is evidence that the aging hematopoietic stem cell loses telomeric size with time [9]. With age the incidence of stem cell disorders such as myelodysplasia and leukemia increases markedly. It is possible that the regulatory region of MZF-1 or the transcribed sequence is disrupted with age, which could play a role in the increased incidence of hematopoietic stem cell disorders with age. Thus, one important future study will be to assess whether MZF-1 genomic structure or expression is disrupted in both normal and aberrant hematopoietic cells as they age.

Myeloid Transcriptional Regulation by MZF-1

MZF-1 encodes a zinc finger gene that is a putative regulator of transcription [10]. To evaluate the transcriptional regulatory function of MZF-1 co-transfection studies were performed in various cell lines.

The coding region of MZF-1 was cloned in frame into the GAL4 expression plasmid pSG424 to produce a GAL4-MZF fusion protein, where MZF-1 was fused to the carboxy-terminus of the GAL4 zinc finger region. To assay for the effect of MZF-1 on transcription mediated through the GAL4 binding sites, pSG424 was co-transfected with a CAT reporter plasmid, pGAL45/tk-CAT, where 5 GAL4 binding sites are just 5' of a minimal thymidine kinase promoter driving chloramphenicol acetyl transferase.

To assess whether MZF-1 was a transcriptional activator or repressor GAL4-MZF was co-transfected with pGAL45/tk-CAT into two non-hematopoietic cell lines, 3T3, 293, and two hematopoietic cell lines, Jurkat and K562. In this fusion protein the transcriptional regulatory region is provided by MZF-1 and the DNA binding domain by the GAL4 zinc fingers.

In the two non-hematopoietic cell lines GAL4-MZF-1 repressed CAT activity. In 3T3 cells GAL4-MZF repressed CAT activity an average of 9-fold from GAL4 alone. In 293 cells with the same transduction procedure GAL4-MZF repressed CAT activity by an average of 10-fold.

Surprisingly, GAL4-MZF behaved differently in hematopoietic cells. In contrast to the non-hematopoietic cells, it was found to activate CAT expression in both cell lines. In K562 cells GAL4-MZF increased CAT activity by an average of 12.3-fold. In Jurkat cells GAL4-MZF increased CAT activity an average of 31.0-fold.

Therefore, based on the GAL4 assays, MZF-1 had bi-functional transcriptional regulatory activity. In one intracellular environment it functions as a transcriptional repressor, and in another as an activator.

To assess transcriptional regulation mediated by the DNA binding of the MZF-1 zinc finger domains the coding region of MZF-1 was cloned into the expression vector CB6. This plasmid was termed CB6-MZF. Three copies of the ZN1-4 consensus DNA binding site were cloned in forward orientation into pBL2CAT, which has chloramphenicol acetyl transferase driven by a minimal thymidine kinase promoter. This plasmid was termed M3/tk-CAT. To assay for the effect of MZF-1 on transcription initiated from a heterologous promoter with 3 copies of the MZF-1 consensus binding site, CB6-MZF was co-transfected with M3/tk-CAT into various cell types, and CAT activity assessed.

These co-transfection experiments allowed examination of the transcriptional regulatory activity of MZF-1 mediated by its own DNA binding domains uninfluenced by a promoter with which MZF-1 could interact. This reporter was co-transfected with CB6-MZF into non-hematopoietic and hematopoietic cell lines. In the hematopoietic cell line K562 CB6-MZF activated CAT by an average of 9.8-fold. In the hematopoietic cell line Jurkat CB6-MZF activated CAT by an average 8.3-fold. These data also indicated that in hematopoietic cell lines MZF-1 was functioning as a transcriptional activator.

The consensus DNA binding sites of MZF-1 were isolated by mobility shift selection of degenerate oligonucleotides and PCR amplification [11]. The first finger domain ZF1-4 bound to a consensus sequence of 5'-AGTGGGGA-3', while ZF5-13 bound to a consensus of 5'-CGGGnGAGGGGGAA-3'. These guanine-rich sites resemble NF-kB binding sites. The full length recombinant MZF-1 could bind to either sequence. Indeed, the two MZF-1 finger domains, ZF1-4 and ZF5-13 could bind to each others consensus binding sequence. In addition, it was discovered that ZF5-13 probably needs all nine fingers to bind to its consensus sequence. Deletion of as few as 3 fingers from either end of ZF5-13 could abrogate DNA binding [11].

Computer searches of data bases for promoter sequences that contained these consensus MZF-1 binding sites found several possible matches. The promoters of the myeloid genes CD34, myeloperoxidase, and lactoferrin all contained consensus MZF-1 binding sites. Three MZF-1 binding sites were found in the CD34 promoter, at -575, -519, and -142 from the start site, with the numbering according to He et al [12]. Recombinant MZF-1 will bind to each of these sites by mobility shift analysis. This binding can be appropriately competed off by cold duplex oligonucleotides containing the MZF-1 consensus DNA binding sites. There was one MZF-1 binding site in the human myeloperoxidase promoter, at -278 according to Morishita et al [13]. Recombinant MZF-1 will also bind to this site in a mobility shift assay. There was one binding site within the human lactoferrin promoter that perfectly matched the consensus for ZF1-4, at -82 according to Johnston et al [14]. Full length recombinant MZF-1 will also bind to this site.

We have not yet explored the potential regulation of the lactoferrin or myeloperoxidase promoters by MZF-1. However, we have extensively analyzed the regulation of the CD34 promoter by MZF-1 in co-transfection experiments.

The CD34 promoter was cloned in front of a luciferase reporter gene to assess whether MZF-1 could regulate the CD34 promoter. When CB6-MZF was co-transfected into Hela cells with CD34-Luc it repressed luciferase activity by an average of 5.2-fold over vector alone at the highest CB6-MZF level. In 3T3 cells at the highest CB6-MZF amounts CB6-MZF repressed CD34-mediated luciferase activity by an average of 10.1-fold. Consistent with the GAL4 and heterologous promoter studies, MZF-1 also functioned as a transcriptional repressor when transcription was initiated by the CD34 promoter in non-hematopoietic cells.

In the hematopoietic cell lines, however, MZF-1 activated transcription from the CD34 promoter. In Jurkat cells at the highest transfected amounts of CB6-MZF CD34 luciferase activity was increased by an average of 7.5-fold. In K562 cells at the highest transfected amounts of CB6-MZF luciferase activity was increased by an average of 18.1-fold. In KG-1 cells the average increase of luciferase activity produced by CB6-MZF over vector alone was 6.1-fold. Mutating the three MZF-1 binding sites in the CD34 promoter prevented activation by MZF-1 in hematopoietic cells. However, it did not prevent repression in Hela cells, implying that the repression in non-hematopoietic cells is not mediated by the MZF-1 binding sites, but may be indirect, perhaps by interacting with other transcriptional repressors. The data in the CD-Luc study was also consistent with the previous GAL4 and heterologous promoter experiments. This study has been submitted for publication.

The presence of MZF-1 binding sites in several myeloid promoters and the explicit regulation of the CD34 promoter by MZF-1 point out that MZF-1 may have a general role in the regulation of hematopoietic gene expression. Lending evidence to the hypothesis that MZF-1 may be an important regulator of hematopoietic development are studies showing that blocking MZF-1 expression inhibits granulopoiesis [8]. Anti-sense but not sense oligonucleotides to MZF-1 decreased G-CSF-driven granulocyte colony formation by 5 to 20-fold, depending on the anti-sense oligonucleotide [8]. The colonies that did form were dysplastic in cytospin morphology. However, erythropoietin-driven BFU-E were not affected by the MZF-1 anti-sense oligonucleotides. This study was recently replicated by Licht and colleagues, in comparing MZF-1 to PLZF-1 (J. Licht, personal communication).

It was not clear from these experiments how blocking MZF-1 expression inhibited G-CFU. It is possible that MZF-1 may be important in stimulating the developing myeloid phenotype, as described in this section. However, as shown in the next section, there is an equal amount of data that implies that MZF-1 may be important in myeloid cell proliferation. Thus, blocking its expression would prevent the proliferation needed to form a colony.

MZF-1 in Cell Proliferation

To study the effects of disrupted regulation of MZF-1 on myeloid growth characteristics, MZF-1 was continuously over-expressed in the IL-3-dependent myeloid cell line FDCP.1. The complete ORF of MZF-1 was subcloned into the retroviral expression vector pLXSN (LXSN-MZF), packaged in PA317 cells, and retroviral supernatants used to infect FDCP.1 cells. FDCP.1 cells were also infected with LNL6, a related retrovirus that carries only neomycin phosphotransferase, as a control. Both cell types were selected in G418 for stable integrants. The FDCP.1 cells infected with LXSN-MZF (FDCP-MZF) had generally the same morphology as those infected with LNL-6 (FDCP-LNL). Immunohistochemistry analyses were performed to assess whether MZF-1 expression altered the phenotype of the FDCP.1 cells. Both cell lines expressed lysozyme to the same extent. Neither expressed myeloperoxidase, factor VIII, or hemoglobin. Of interest is that the FDCP-MZF cells expressed esterase and proliferating cell nuclear antigen at two-fold higher levels than the control cells. In cytochemical stains the FDCP-MZF cells were also markedly more PAS positive than controls. These changes suggested that forced MZF-1 expression may alter the phenotype of FDCP.1 cells.

To investigate whether MZF-1 over-expression affected the apoptosis that occurs with IL-3 withdrawal in FDCP.1 cells, both cell lines were washed three times, and resuspended in growth media without IL-3. No outgrowth of IL-3 independent FDCP-MZF cells occurred in either cell line. However, by 24 hours after cytokine removal, 84% of the FDCP-LNL cells had apoptotic nuclear changes, while none of the FDCP-MZF cells did. At 24 hours the FDCP-MZF cells had little nucleosomal DNA cleavage, a characteristic of cells undergoing apoptosis, while the control cells had all cellular DNA cleaved into nucleosomal fragments. After 6 days without IL-3 there were no viable control cells, but 18% of FDCP-MZF cells were still viable. Injection of 10,000 FDCP-MZF or FDCP-LNL cells into congenic DBA2 mice produced injection site tumors in 70% of mice given FDCP-MZF cells but none in the mice given the FDCP-LNL cells. These tumors had the histology of granulocytic sarcomas. PCR analysis of these tumors found that neo DNA sequences were amplified, indicating that LXSN-MZF was still present. These data suggest that over-expression of MZF-1 can disrupt the normal proliferative behavior of myeloid cells. This study is being prepared for publication.

MZF-1 can also disrupt the normal proliferative behavior of embryonic fibroblasts. When MZF-1 was retrovirally transduced and over-expressed in NIH 3T3 cells, where it is not normally expressed, it rapidly transformed those cells [15]. However, two other myeloid transcriptional regulators, PU.1 and HOX12, did not transform 3T3 cells when they were similarly over-expressed.

The loss of contact inhibition was assayed in 3T3/LNL versus 3T3/MZF cells by focus formation. When 3T3/LNL cells were grown to confluency on a 100 mm culture dish and stained with crystal violet they had an average 3.2 foci per plate. 3T3 cells infected with LXSN-HOX or LXSN-PU and selected with G418 had an average of 3.5 and 12 foci per plate, respectively. These values were not significantly different from the 3T3/LNL cells. 3T3 cells transduced with HOX12 or PU.1 did not change in morphology from normal NIH 3T3 cells. In addition, there was no difference in focus formation between normal 3T3 and 3T3/LNL cells.

The uncloned, heterogenous population of 3T3/MZF cells averaged 234 foci per plate, the clonal 3T3/MZF#1 cells 231 foci per plate, and the clonal 3T3/MZF#2 174 foci per plate. Large foci of 3T3/MZF cells formed quickly in culture, even before the adherent layer of cells neared confluency. Thus, 3T3 cells transduced with MZF-1 but not LNL6, HOX12 or PU.1 produced a loss of contact inhibition.

Loss of substrate dependence by 3T3/MZF cells was assayed by colony growth on a layer of soft agar. When 3T3/MZF cells were plated in soft agar, they formed macroscopic colonies in two weeks. Normal 3T3 or control 3T3/LNL cells did not produce any colonies in soft agar. However, 3T3/MZF parent cells had an average soft agar cloning efficiency of 43%, the 3T3/MZF#1 cells 35%, and the 3T3/MZF#2 cells also 35%. 3T3 cells over-expressing MZF-1 are able to grow without adhering to a substrate. 3T3/HOX and 3T3/PU cells did not form any colonies in soft agar. The 3T3/LNL, 3T3/HOX, and 3T3/PU cell lines served as an internal control for the possibility that retroviral integration and G418 selection might have produced the 3T3-MZF oncogenic transformation.

The uncloned and clonally-derived 3T3 cell lines transduced with MZF-1 formed tumors in athymic mice. These tumors had the characteristics of aggressive fibrosarcomas. They had a high nuclear to cytoplasmic ratio, pleomorphic nuclei with prominent nucleoli, and a high grade of mitotic indices. These tumors appeared within four weeks of injection, and expanded rapidly. On sacrifice at eight weeks the mice were found to have tumor invading through the peritoneum and into the abdominal cavity. Malignant cells studded the inner peritoneum, and involved bowel and liver. Normal 3T3 cells did not form tumors in these mice while 3T3/MZF formed tumors in 5 of 5 mice injected, 3T3/MZF#1 in 4 of 5 mice, and 3T3/MZF#2 in 5 of 5 mice.

The MZF-1-transformed 3T3 cells traversed the cell cycle at twice as fast a rate as the untransformed cells. There were approximately twice as many cells in S-phase by flow cytometry in the 3T3/MZF populations. In addition, the 3T3/MZF cells also markedly overexpressed cyclin A and E. This raises the possibility that MZF-1 may stimulate cell cycle progression, perhaps by increasing the expression of cyclins.

These data imply that the aberrant expression of nodal regulators of development can be neoplastic. It lends evidence for the hypothesis that malignancy is a normal development gone awry.

Conclusion

MZF-1 is a Krupple-class zinc finger gene that is expressed preferentially in myeloid cells and is essential in the development of granulocyte colonies in vitro. It is located at the telomere of human chromosome 19q, at the end of a cluster of other zinc finger genes. MZF-1 is a DNA-binding protein, and recognizes a consensus sequence present in many myeloid promoters. In non-hematopoietic cells MZF-1 functions as a transcriptional repressor and as a transcriptional activator in hematopoietic cells. Over-expression of MZF-1 will aggressively transform 3T3 cells, and inhibit apoptosis and promote leukemogenesis in the IL-3-dependent myeloid leukemia line FDCP.1.

References

1. Hromas R, Zon L, and Friedman A (1992) Hematopoietic transcriptional regulators and the origins of leukemia. Crit rev Oncol Hematol 12:167-190.
2. Rabbits TH (1994) Chromosomal translocations in cancer. Nature 372:143-149.
3. Cline M (1994) The molecular basis of leukemia. New Engl J Med 330:328-336.
4. Hromas R, Collins SJ, Hickstein D, et al (1991) A retinoic-acid responsive human zinc finger gene, MZF-1, preferentially expressed in myeloid cells. J Biol Chem 266:14183-14187.
5. Witzgall R, O'Leary E, Gessner R, et al (1993) Kid-1, a putative renal transcription factor: regulation during ontogeny and in response to ischemia and toxic injury. Mol Cell Biol 13:1933-1942
6. Shi Y, Seto E, Chang LS, Shenk T (1991): Transcriptional repression by YY1, a human GLI-Krupple-related protein, and relief of repression by adenovirus E1A protein. Cell 67:377-388.
7. Zhen C, Brand NJ, Chen A, et al (1993) Fusion between a novel Krupple-like zinc finger gene and the retinoic acid receptor alpha locus due to a variant t(11;17) translocation associated with acute promyelocytic leukemia. EMBO J 12:1161-1171.
8. Bavisotto L, Kaushansky K, Lin N, and Hromas R (1991) Anti-sense oligonucleotides from the stage-specific myeloid zinc finger gene MZF-1 inhibit granulopoiesis in vitro. J Exp Med 174:1097-1101.
9. Vaziri H, Dragowska W, Allsopp RC, et al (1994) Evidence for a mitotic clock in human hematopoietic stem cells: loss of telomeric DNA with age. Proc Natl Acad Sci USA 91:9857-9860.
10. Evans R, and Hollenberg SM (1988) Zinc fingers: Gilt by association. Cell 52:1-3.
11. Morris J, Hromas R, and Rauscher F III (1994) Characterization of the DNA-binding properties of the myeloid zinc finger protein MZF-1: Two independent DNA-binding domains recognize two DNA consensus sequences with a common G-rich core. Mol Cell Biol 14:1786-1795.
12. He XY, Antao VP, Basila D, Marx JC, and Davis BR (1992) Isolation and characterization of the human CD34 gene. Blood 79:2296-2302.
13. Morishita K, Tsuchiya M, Asano S, Kaziro Y, and Nagata S (1987) Chromosomal gene structure of human myeloperoxidase and regulation of its expression by granulocyte colony stimulating factor. J Biol Chem 262:15208-15213.
14. Johnston J, Rintels P, Chung J, et al (1992) Lactoferrin gene promoter: structural integrity and non-expression in HL60 cells. Blood 79:2998-3006.
15. Hromas R, Morris J, Cornetta K, et al (1995) Aberrant expression of the myeloid zinc finger gene MZF-1 in NIH 3T3 cells is oncogenic. Cancer Res, in press.

Coordinate Regulation of Neutrophil Secondary Granule Protein Gene Expression

A. Khanna-Gupta, T. Zibello, and N. Berliner
Yale University School of Medicine

Introduction

Evidence from study of both normal and leukemic cells suggests that a crucial step in neutrophil maturation occurs in the transition from the promyelocyte to the myelocyte stage. The transition from the promyelocyte to the myelocyte in normal marrow cells is accompanied both by the loss of proliferative capacity associated with terminal maturation [1], and by the loss of the capacity for alternative maturation [2]. Normal promyelocytes respond to stimuli of both granulocyte and monocyte differentiation, but myelocytes are restricted to terminal granulocyte maturation [2]. In this regard, it is striking that acute myeloid leukemias invariably involve proliferation of cells arrested in development at or before the promyelocyte stage. Consequently, an understanding of the transition from the promyelocyte to the myelocyte stage should provide crucial insights into both the control mechanisms governing normal hematopoietic cell differentiation and the ways in which disruption of the control mechanisms can contribute to leukemic transformation.

The acquisition of different types of granules marks progressive stages of granulocyte differentiation. The promyelocyte stage is characterized by primary ("nonspecific") granules which persist in decreasing numbers in the later phases of both neutrophil and monocyte maturation [3]. As a promyelocyte undergoes the transition to the myelocyte stage it acquires secondary ("specific") granules [4]. Specific granules contain four major identified proteins, namely, transcobalamin I (TCI), lactoferrin (LF), human neutrophil collagenase (HNC), and human neutrophil gelatinase (HNG). The appearance of secondary granules, in association with the expression of their content proteins, provides a unique marker of commitment to terminal neutrophil differentiation. Thus, the expression of genes responsible for the formation of secondary granule membranes and their contents should provide a convenient probe for the processes controlling normal neutrophil differentiation.

Although many structural and functional features of primary and secondary granules are defined, little else is known about their biogenesis. It is known that primary granules are the first clear marker of granulocyte differentiation. It is also known that secondary granules are seen only in granulocytes at and beyond the myelocyte stage; hence, they are invariably absent from leukemic cells. A clear function has not been found for any of the major secondary granule proteins, although two of them, lactoferrin and TCI, are expressed in a wide range of tissues. Specific granule deficiency, a rare syndrome characterized by absence of secondary granules and failure of gene expression of the secondary granule content proteins, is associated with increased infections, but the functional defect in the granulocytes associated with the syndrome is as yet undefined [5]. The clinical manifestations of isolated R-binder deficiency are minimal [6]. Lactoferrin deficiency without other defects in secondary granules has not been encountered [7].

Although granule formation is clearly closely tied to normal neutrophil differentiation, the processes governing the stage-specific development of these crucial structural and functional components of developing granulocytes remain to be elucidated. Two lines of evidence summarized here suggest indirectly that the regulation of the genes encoding for the secondary granule proteins is dependent on *trans*-acting factors: evidence suggested by studies of the leukemic cell lines, HL60 and NB4, and analysis of patients with specific granule deficiency. Based on the results, we would propose that a factor governing the coordinate regulation of neutrophil secondary granule protein gene expression may be involved in early events in the neutrophil differentiation pathway. Further analysis of shared regulatory features of these late-stage differentiation markers may therefore provide important insights into the more proximal events in neutrophil differentiation.

Secondary Granule Protein mRNA Expression is Coordinate

In order to investigate the regulation of neutrophil secondary granule protein (SGP) gene expression as a probe for the transcriptional regulation of neutrophil maturation, we isolated cDNA and genomic clones for the four major SGP genes, and examined their expression during neutrophil maturation [8-12]. As described below, leukemic cell lines are an inappropriate system in which to study this question, as they fail to express SGP genes, even upon chemical induction of neutrophil maturation. Consequently, we have examined the stage-specific expression of SGP genes in two other systems. We have investigated the pattern of primary and secondary granule protein gene expression in a factor-dependent murine stem cell line, 32Dcl3, which is considered to be one of the best available cell culture models of "normal" neutrophil maturation. We have also looked at granule protein gene expression in isolated human bone marrow progenitors induced to undergo neutrophil maturation *in vitro*.

32DCl3 Cells
The 32Dcl3 cell line (32D) is an IL3-dependent murine stem cell line which can be induced to undergo differentiation to mature neutrophils upon induction with G-CSF. We initially determined that this cell line shows a sequential pattern of primary and secondary granule protein gene expression by analyzing expression of myeloperoxidase (MPO) and lactoferrin (LF). We then undertook to clone one of the two mouse neutrophil metalloproteinase genes (neutrophil gelatinase or collagenase). Using highly conserved sequences for functional domains of those genes which are shared by nearly all metalloproteinases across wide species barriers, we synthesized degenerate primers for PCR amplification of cDNA from the 32D cell line, and successfully isolated a PCR fragment encoding a portion of the mouse homolog of the 92 kDa gelatinase expressed in both neutrophils and monocytes. This was then used to obtain a full length mouse neutrophil gelatinase (MNG) clone from a mouse monocyte library. We then demonstrated that upon induction of neutrophil differentiation with G-CSF, the expression of the MNG gene, as determined by Northern blot analysis, did in

fact parallel that of the LF gene. Nuclear run-on assays confirmed that control of expression was exercised at the level of mRNA transcription [13].

G-CSF induction of normal marrow progenitors *in vitro*
We have examined the stage-specific expression of the primary and secondary granule protein genes in purified bone marrow progenitors induced with growth factors to undergo *in vitro* differentiation to mature neutrophils. A fraction of marrow cells enriched for hematopoietic progenitor cells (CD34+ HLA-DR+) was isolated from normal human bone marrow by monoclonal antibody staining and fluorescence activated cell sorting. Cells were cultured in a suspension system for three days in the presence of stem cell factor and IL3, following which G-CSF was added. Cells were harvested daily and analyzed for phenotypic maturation by morphologic criteria, and total RNA was obtained for analysis of myeloid gene expression. Maturation was observed to progress to the late metamyelocyte and band stage over a period of 10-12 days. Neutrophil-specific gene expression was assayed by RT-PCR. Induction with G-CSF resulted in sequential expression of primary and secondary granule proteins. Interestingly, primary granule protein genes showed surprisingly asynchronous expression. MPO expression was detectable from the time of isolation of CD34+ cells, despite the fact that the cells did not yet contain visible granules. However, the expression of lysozyme and defensin, two other primary granule proteins, was first detectable at about 2 days and 4-5 days respectively. Secondary granule protein genes, however, showed synchronous mRNA expression, as determined by RT-PCR of LF and TCI. Expression of secondary granule protein genes was detected from day 7-8 [14]. This recapitulation of a program of sequential expression of primary and secondary granule protein genes suggests that *in vitro* marrow culture suspensions with appropriate growth factors can mimic normal granulocytic maturation, and may be a useful model of transcriptional regulation of neutrophils.

Leukemia is Associated with a Failure of SGP Gene Expression

HL60 is a human leukemia cell line which has been observed to undergo phenotypic maturation upon exposure to a variety of chemical agents. Upon induction with dimethyl sulfoxide (DMSO) or retinoic acid, the cells undergo transition to a more mature neutrophil phenotype; when stimulated with phorbol esters, they acquire the phenotypic characteristics of more mature monocytes. Uninduced HL60 cells have the morphologic appearance of promyelocytes; they contain primary granules and express high levels of myeloperoxidase. The granulocytic maturation induced by DMSO and other agents is associated with the acquisition of many properties associated with normal mature neutrophils. Morphologically, the cells resemble predominantly metamyelocytes and bands; like mature neutrophils, they reduce nitroblue tetrazolium, and display phagocytosis [15]. However, they do not acquire secondary granules [16].

Similarly, studies of patients with leukemia have shown abnormalities in SGP gene expression in apparently normal neutrophils in patients with newly-diagnosed acute leukemia [17]; other studies suggest that these circulating neutrophils are in fact part of the malignant clone [18]. These studies suggest that one abnormality of the leukemic cell is a disruption of the differentiation program which affects more distal events in neutrophil maturation, as reflected in the coordinate absence of expression of secondary granule proteins.

Having obtained probes for all of the SGP genes, we examined the defect in SGP gene expression in leukemic cell lines induced toward mature neutrophils to determine the nature of their defect in SGP gene expression. For these experiments, we studied HL60 cells induced with DMSO, as well as NB4 cells induced with all-*trans* retinoic acid.

1. HL60 cells
HL60 cells are differentially inducible with TPA or DMSO to monocytes or granulocytes respectively. HL60 cells induced with DMSO undergo defective neutrophil maturation, manifested by a coordinate failure of secondary granule protein gene expression. We have looked at the defect in SGP gene expression in uninduced and DMSO-induced HL60 cells. Northern blot analysis reveals a total failure to accumulate mRNA for any of the SGP genes upon induction with DMSO. Nuclear run-on assays further confirmed that the SGP genes are uniformly untranscribed in these cells. In addition, we fully sequenced over 600bp of the HL60 lactoferrin promoter region and showed that it is structurally intact. This suggests that the failure of gene expression in these cells reflects a regulatory defect [11].

2. NB4 cells
NB4 is an acute promyelocytic leukemia cell line that has been shown to be inducible to terminal neutrophil maturation with all-*trans* retinoic acid (ATRA). Because initial reports of the maturation of this cell line induced by ATRA suggested that it achieved a more "normal" differentiated phenotype, we undertook to examine secondary granule protein gene expression in that cell line. We found that, in a manner parallel to HL60, NB4 cells displayed limited phenotypic differentiation, although they acquired the ability to reduce NBT. Upon induction with ATRA, they did not express mRNA for any of the secondary granule proteins [19].

Coordinate Regulation of SGP Genes is Neutrophil-Specific

Human neutrophil gelatinase (HNG) is unique among the four major content proteins of the neutrophil secondary granule in that it is also expressed in the monocyte/macrophage lineage. We have examined HNG expression during differentiation of both HL60 and NB4 along monocytoid lines induced with TPA. Both lines showed striking phenotypic maturation along the monocytoid-macrophage lineage with TPA induction, with a change to an adherent phenotype. In both cell lines, Northern blot analysis showed induction of expression of CD18, *c-fos*, and human neutrophil gelatinase (HNG). We have observed that HNG expression is absent in myeloblast (HL60) cells and NB4 promyelocytic cells induced toward the neutrophil lineage with DMSO or ATRA respectively (which do not express any secondary granule proteins), but present in both cell lines when induced along the monocyte/macrophage lineage with TPA [12,19]. This suggests that the regulatory pathways governing HNG expression in the two granulocytic lineages are distinct, and that while the neutrophil-specific pathway is defective in leukemic cells, the monocyte-specific regulatory pathway remains intact. Further evaluation of the control of HNG expression could therefore provide a useful probe of the factors determining lineage differentiation between neutrophils and monocytes.

Specific Granule Deficiency is a Transcriptional Defect

Specific granule deficiency (SGD) is a rare congenital disorder characterized by recurrent infections. SGD is characterized by heterogeneous developmental defects in the neutrophil, complete absence of neutrophil secondary granules, and marked deficiency or absence of their content proteins. Evaluation of the cells at the protein level confirms that they contain none of the proteins found in secondary granules. Work summarized here, from our laboratory as well as our collaborators, has shown that leukocytes from patients with specific granule deficiency exhibit almost total absence of secondary granule mRNA [5,20].

We obtained bone marrow from a patient with SGD, and isolated total mRNA. We then performed Northern blot analysis on normal and SGD marrow, to determine the mRNA accumulation for all four SGP genes, as well as the primary granule protein genes MPO and defensins. We demonstrated absent or negligible mRNA for all of the SGP genes, as well as for defensins. These levels correlated well with previously determined levels of secondary granule proteins in this patient's neutrophils. This suggests that SGD may well represent a defect in a common *trans*-acting factor governing SGP gene expression.

In the absence of these results implicating a defect in mRNA accumulation, one might postulate that failure to make secondary granule proteins would most likely result from a defect in protein processing or transport into granules. However, the observed absence of all of the secondary granule proteins (the genes for which are not linked) in association with a deficiency of the mRNAs encoding those genes, suggests that the abnormality instead resides in a failure of gene transcription of all of the SGP genes. We postulate that this failure could result from the absence or abnormal function of a shared *trans*-acting factor regulating their transcription.

Conclusion

In summary, we have identified the neutrophil secondary granule protein genes as a potential probe for the events determining lineage commitment and terminal maturation to neutrophils. Toward this end, we have cloned the cDNAs and genomic genes for the four secondary granule content proteins, and studied their expression in a variety of hematopoietic cell lines, in a patient with specific granule deficiency, and in normal marrow progenitors. Our results support the hypothesis that secondary granule protein gene expression is coordinately regulated at the level of mRNA transcription.

The abnormalities in secondary granule protein gene expression in leukemic cell lines suggests the possibility that leukemic transformation results in a defect in this regulatory pathway. Given that these cell lines represent cells arrested at or before the promyelocyte stage, it is not surprising that the uninduced cell lines do not express secondary granule proteins. However, the failure to transcribe these genes after the chemical induction of a more mature phenotype implies that the shared factor regulating their coordinate expression may be involved in early events in the neutrophil differentiation pathway. This would suggest that disruption of secondary granule protein gene expression may be a downstream manifestation of the process of malignant transformation.

The finding that the factor dependent cell line 32Dcl3 retains the capacity to express secondary granule proteins offers a potential probe for the further analysis of the defect in leukemia. One striking difference between 32Dcl3 cells and leukemic cell lines is that the latter are growth factor independent. The loss of the capacity for normal late stage gene expression may accompany the changes which render the cell hyperproliferative and growth-factor independent.

In conclusion, the control of expression of the secondary granule protein genes appears to be exercised by factors that may have far-reaching influences on early stages of neutrophil differentiation. Current studies are directed at the analysis of the transacting factors determining that expression, with the long term goal of analyzing the expression of those factors in pathologic states. Further analysis of shared regulatory features of these late-stage differentiation markers may therefore provide important insights into the more proximal events in neutrophil differentiation and into the maturation defect induced by leukemic transformation.

References

1. Janoshowitz H, Moore MAS, Metcalf D (1971) Density gradient segregation of bone marrow cells with the capacity to form granulocytic and macrophage colonies in vitro Exp Cell Res 68: 220-224
2. Warner HR, Athens JW (1964) Analysis of granulocyte kinetics in blood and bone marrow. Ann NY Acad Sci 113: 523-535
3. Bainton DF (1975) Neutrophil granules Br J Hematol 29: 17-22
4. Bainton DF, Ullyot JL, Farquhar MG (1971) The development of the neutrophilic polymorphonuclear leukocytes in human bone marrow. J Exp Med 134: 907-34
5. Lomax KJ, Gallin JI, Benz EJ, Rado TA, Boxer LA, Malech HL (1989) Selective defect in myeloid cell lactoferrin gene expression in neutrophil specific granule deficiency J Clin Invest 83: 514-519
6. Carmel R (1983) R-Binder deficiency: A clinically benign cause of cobalamin pseudodeficiency. J Amer Med Assoc 250: 1886-1890
7. Parmley RT, Tzeng DY, Baehner RL, Boxer LA (1983) Abnormal distribution of complex carbohydrates in neutrophils of a patient with lactoferrin deficiency. Blood 62: 538-548
8. Johnston J, Bollekens J, Allen RH, Berliner N (1989) Structure of the cDNA encoding transcobalamin I, a neutrophil granule protein J Biol Chem 264:15754-15757
9. Devarajan P, Mookhtiar K, Van Wart H, Berliner N (1991) Structure and expression of the cDNA encoding human neutrophil collagenase Blood 77:2731-2738

10. Johnston J, Yang-Feng T, Berliner N (1991) Structure and mapping of the chromosomal gene encoding human transcobalamin I: Comparison to human intrinsic factor Genomics 12: 459-464
11. Johnston J, Rintels P, Chung J, Sather J, Benz EJ, Berliner N (1992) The lactoferrin gene promoter: Structural integrity and non-expression in HL60 cells Blood 79: 2998-3006
12. Devarajan P, Johnston J, Ginsberg S, Van Wart H, Berliner N (1992) Structure and expression of neutrophil gelatinase cDNA: Identity with Type IV collagenase from HT1080 cells J Biol Chem. 267: 25228-25232
13. Graubert T, Johnston J, Berliner N (1993) Cloning and expression of the cDNA encoding mouse neutrophil gelatinase: Demonstration of coordinate secondary granule protein gene expression during terminal neutrophil maturation. Blood 82: 3192-3197
14. Berliner N, Hsing A, Sigurdsson F, Zain M, Graubert T, Bruno E, Hoffman R (1995) G-CSF induction of normal human marrow progenitors results in neutrophil-specific gene expression. Blood 85:799-803
15. Collins SJ, Ruscett FW, Gallagher RE (1977) Normal functional characteristics of cultured human promyelocytic leukemia cells (HL-60) after induction of differentiation by dimethylsulfoxide. Nature 270:347-349
16. Fibach E, Peled T, Treves A, Kornberg A, Rachmilewitz E (1982) Modulation of the maturation of human leukemic promyelocytes (HL-60) to granulocytes or macrophages Leukemia Res 6: 781-790
17. Miyauchi J, Watanabe Y, Enomoto Y, Takeuchi K (1983) Lactoferrin deficient polymorphonuclear leucocytes in leukemias: A semiquantitative and ultrastructural cytochemical study J Clin Pathol 36:1397-1405
18. Fearon ER, Burke PJ, Schiffer CA, Zehnbauer BA, Vogelstein B (1986) Differentiation of leukemia cells to polymorphonuclear leukocytes in patients with acute nonlymphocytic leukemia. N Engl J Med 315: 15-24
19. Khanna-Gupta A, Kolibaba K, Berliner N (1994) NB4 cells exhibit bilineage potential and an aberrant pattern of neutrophil secondary granule protein gene expression Blood 84: 294-302
20. Johnston J, Boxer LA, Berliner N (1992) Correlation of RNA levels with protein deficiencies in specific granule deficiency. Blood 80: 2088-2091

Interleukin-5 Receptor α Subunit Gene Regulation in Human Eosinophil Development: Identification of a Unique Cis-Element that acts lie an Enhacer in Regulating Activity of the IL-5Rα Promoter

Z. Sun[1], D. A. Yergeau[1], I.C. Wong[1], T. Tuypens[2], J. Tavernier[2], C. C. Paul[3], M. A. Baumann[3], P. E. Auron[4], D. G. Tenen[5], and S. J. Ackerman[1]

From the Divisions of Infectious Diseases[1] and Hematology-Oncology[5], Department of Medicine, Beth Israel Hospital and Harvard Medical School, Boston, MA 02215, Roche Research Gent, Hoffman-La Roche, Gent, Belgium[2], Research Service, Dayton Veterans Affairs Medical Center and Hematology-Oncology Division[3], Wright State University Medical School, Dayton OH 45435, and Center for Blood Research and Harvard Medical School[4]

1 Introduction

1.1 IL-5 and eosinophil development. The molecular basis for the commitment of multipotential myeloid progenitors to the eosinophil lineage, and the transcriptional mechanisms by which eosinophil-specific genes are subsequently expressed and regulated during eosinophil development are currently unknown. IL-5, produced primarily by activated T cells [1] and mast cells [2], is integral to both the differentiation and functional maturation of the human eosinophil lineage [3-6]. The development and maturation of eosinophils in the bone marrow, and their post-mitotic functional activation in tissues occurs in response to a number of cytokines in addition to IL-5, including GM-CSF and IL-3 [7-9]. In humans, both IL-3 and GM-CSF have activities on other hematopoietic lineages, whereas IL-5 is eosinophil-specific and plays a crucial role in regulating the differentiation and development of the eosinophil lineage [3]. Although IL-3 and GM-CSF participate in the proliferation and committment of progenitors to the eosinophil lineage, IL-5 is both necessary and sufficient for eosinophil development to proceed[3, 10]. In humans, the high affinity receptor for IL-5 is apparently restricted to eosinophils and hematopoietically related basophils [11]; in contrast to murine B cells, the activity of IL-5 on human B cells is controversial [3, 12] and is still being delineated [13]. Thus, the expression of the high-affinity receptor for IL-5 is an important prerequisite and very early lineage-specific event in the hematopoietic program for these granuloctyes. Overexpression of IL-5 is observed in many eosinophil-associated diseases [14-16] and IL-5 transgenic mice develop profound eosinophilia [4, 5], indicating that IL-5 plays critical roles in promoting the production and function of eosinophils *in vivo*. Of note, IL-5 is also active *in vitro* both in the production of eosinophils from bone marrow, umbilical cord and peripheral blood progenitors, as well as in the priming, activation and enhanced survival of mature eosinophils.

1.2 The IL-5 receptor. A heterodimeric receptor for IL-5 has been identified that is comprised of a unique α subunit (IL-5Rα) and a β subunit ($ß_c$) which is shared with the α subunits of the receptors for IL-3 and GM-CSF [17-19]. The unique IL-5Rα subunit is the primary ligand-binding polypeptide of the receptor, in contrast to the $β_c$ subunit which does not independently bind IL-5; the combination of α and β subunits forms the high affinity receptor. Although signaling through the IL-5R, as well as the receptors for IL-3 and GM-CSF, was thought to occur primarily via the $ß_c$ subunit, intact α and β subunits of the IL-5R were both shown to be required for optimal signal transduction [20, 21]. Further, a critical cytoplasmic domain of the α subunit has recently been shown to play an essential role in protein tyrosine phosphorylation of the $ß_c$ chain, SH2/SHS-containing proteins and consequent cell proliferation [22]. The isolated α subunit if the GM-CSFR has likewise been shown to participate in signaling, albeit via a phosphorylation independent pathway [23]. The gene encoding the IL-5Rα subunit is located on chromosome 3 in the region 3p26 [24]. The organization of this gene reflects the functional domains of the protein [24] and shares many characteristics with other members of the cytokine/hemopoietin receptor gene family [9, 19, 24]. Several alternatively spliced transcripts have been identified in the mRNA and reflect the membrane versus soluble isoforms [25]. Aside from its ability to bind IL-5 *in vitro* [26], the *in vivo* function of the soluble form of the IL-5R has not been elucidated.

1.3 IL-5Rα expression and eosinophil development. In contrast to its expression on both eosinophils and B cells in the mouse [27], expression of the IL-5Rα gene appears to be restricted to the eosinophil and hematopoietically related basophil lineages in humans. The IL-5Rα subunit is expressed at the mRNA level in purified human bone marrow-derived CD34+ progenitors, and is strongly upregulated during IL-3 and/or IL-5 induced eosinophil differentiation of these cells (ZJS and SJA, unpublished results). In addition to mature eosinophils and basophils, the high affinity IL-5R is constitutively expressed on a number of eosinophil-inducible and eosinophil-differentiated myeloid leukemic cell lines [19, 28-30] (see below). Early expression of the IL-5Rα gene may be a critical step in the process of commitment and differentiation of multipotential myeloid progenitors to the eosinophil lineage. Recent studies with transgenic mice that ubiquitously express the murine IL-5Rα gene constitutively through the phosphoglycerate kinase promoter suggest that the lineage specificity of IL-5 is mainly a function of the restricted expression of the IL-5Rα subunit [31]. Like other cytokine receptor genes analyzed thus far, regulation of IL-5Rα gene expression in this process is likely controlled in part at the transcriptional level in a cell-type specific and temporal manner [32-35]. However, the transcriptional mechanisms and lineage specific cis-elements and trans-acting factors critical to the expression and regulation of the IL-5Rα subunit have not been determined.

1.4 Cell lines for analysis of IL-5Rα gene regulation. Studies of human eosinophil differentiation and gene regulation have been aided significantly by the development of eosinophil-committed subclones of the promyelocytic leukemia cell line HL-60 [38, 39] and the acute myeloid leukemia line AML14 [28, 29], which proliferate in suspension cultures, continue to resemble very immature myeloid precursors, and are uniquely capable of being driven towards eosinophil differentiation (Fig. 1). The AML14 line has the unusual ability to differentiate into eosinophils with a nearly mature phenotype (AML14.eos) in response to stimulation by IL-3, GM-CSF and IL-5 [28, 29], and to upregulate the expression of the genes encoding the eosinophil granule cationic proteins (EPO, MBP, EDN and ECP) and eosinophil lysophospholipase (CLC protein) [29]. AML14 is committed to the eosinophil lineage [28], and AML14.eos, the cytokine-induced fully differentiated eosinophilic myelocyte subline of AML14, continues to proliferate and maintain a differentiated phenotype with cytokine (IL-3, GM-CSF, IL-5) supplementation [29]. AML14.3D10 is a recently cloned factor-independent subline of AML14.eos that continues to proliferate as a granulated eosinophilic myelocyte [40]. The eosinophil-committed HL-60-C15 line was cloned following induction of HL-60 by alkaline culture and butyric acid [38]. These cell lines either constitutively express the IL-5Rα subunit gene and a functional high affinity IL-5R

(AML14, AML14.eos, AML14.3D10 [28, 29, 40], or express the IL-5Rα subunit and IL-5R after induction towards eosinophil differentiation with butyrate (HL-60-C15) [17, 30]. These eosinophilic cell lines have proven their utility as models for analyzing the mechanisms regulating expression of the IL-5Rα subunit gene [36, 37] and other eosinophil genes [41-43], including characterization of the cis-acting elements and nuclear factors that may be critically involved in their myeloid and/or eosinophil lineage-specific expression during eosinophil development.

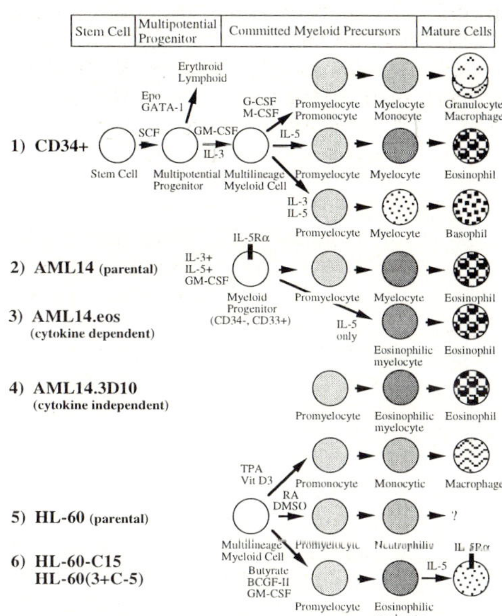

Fig. 1. Models of normal myeloid/eosinophil differentiation of (CD34+) stem cells/progenitors and eosinophil-inducible and eosinophil-differentiated cell lines used for IL-5Rα promoter characterization.

2 Characterization of the IL-5Rα Gene Promoter

In our initial studies of the regulation of IL-5Rα gene expression, we have mapped the transcriptional start site, cloned and characterized the promoter, and identified a 34 bp upstream region critical for functional promoter activity *in vitro* [36]. We have subsequently characterized this functional region of the gene, identifying a unique enhancer-like cis-element (EOS1) 421bp upstream of the transcriptional start site [37]. This element is important both for full promoter function and for lineage-specific activity in myeloid and eosinophilic cell lines. Further, the EOS1 element exhibits characteristics of a lineage-specific enhancer for which we have not detected any significant similarities to consensus sequences for known transcription factor binding sites. The EOS1 element binds a novel nuclear factor(s) that is expressed by eosinophilic as well as other myeloid leukemic cell lines that can be driven towards eosinophilic differentiation, but is absent in non-myeloid and non-hematopoietic lines. Interaction between this unique element and its cognate transcription factor(s) is likely critical for regulating expression of the human IL-5Rα gene and may be essential to the development of the eosinophil lineage.

2.1 A unique nuclear protein-binding cis-element is present in the functional IL-5Rα promoter region. We previously showed that the region between bp -432 and -398 of the IL-5Rα gene was both necessary and sufficient for maximal promoter activity *in vitro* [36]. To detect interactions between potential transcriptional activators and the cis-element(s) in this region, we performed electrophoretic mobility shift assays using a ^{32}P-end labeled 93bp DNA probe which encompassed the region from bp -469 to -377. Two major DNA-protein complexes, identified as C1 and C2, were obtained using nuclear extracts isolated from certain myeloid cell lines, including eosinophil-committed HL-60-C15, promyelocytic HL-60, undifferentiated eosinophil-committed AML-14, fully differentiated AML14.eos, and myeloid U937. In contrast, these complexes were not detected using nuclear extracts from lymphoid BJA-B or non-hematopoietic HeLa cell lines (Fig. 9 in reference [36]). To precisely identify the nuclear protein binding element(s) in this region of promoter sequence, we analyzed both the C1 and C2 protein-DNA complexes by methylation interference using a partially methylated 93 bp DNA fragment between bp -469 and -377 (Fig. 2). On the coding strand, methylation of the G residues at bp -430 and -427 strongly blocked nuclear factor(s) binding, and at bp -422 and -421 blocked complex formation as well, though slightly less effectively. On the non-coding strand, methylation of the two G residues at bp -426 and -425 likewise strongly blocked complex formation. These results indicated that a GTTGCCTAGG sequence between bp -430 and -421, now referred to as EOS1, comprised a cis-element which comes in direct contact with the DNA-binding protein(s) forming both the C1 and C2 complexes. Since both the C1 and C2 complexes showed precisely the same methylation interference pattern on the coding and non-coding strands, both complexes are likely formed by the same nuclear factor, with the difference in mobility potentially due to limited proteolysis, binding of monomer versus dimer, or post-translational modifications such as phosphorylation.

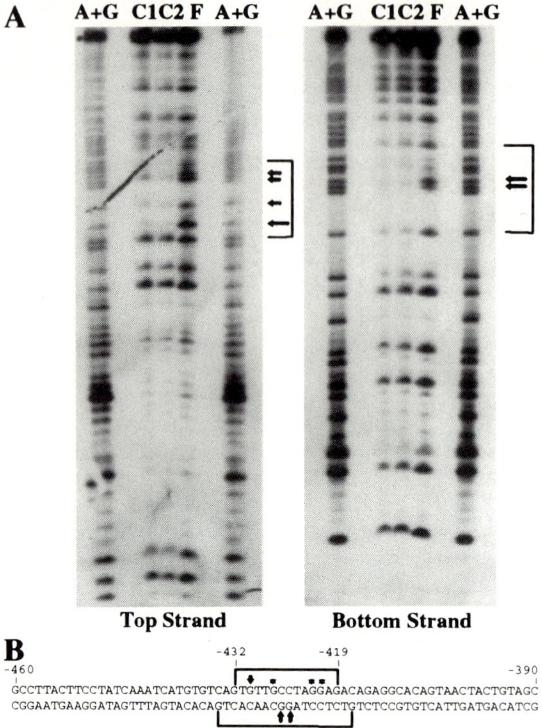

Fig. 2. Methylation interference analysis localizes the nuclear factor binding site in the functional region of the IL-5Rα promoter to bp -430 to -421. (A) A 93 bp probe from bp -469 to -377 of the IL-5Rα promoter was single end labeled with γ-32P-ATP. The labeled probe was partially methylated by treatment with dimethylsulfate and incubated with nuclear extract from the AML14 eosinophil-committed cell line. Protein-DNA complexes were separated on a low ionic strength preparative 4% polyacrylamide gel. Bands comprising free probe not exposed to nuclear proteins (F) and C1 and C2 complexes as shown in Figure 3, were visualized by autoradiography, excised from the gel, isolated by electroelution, extracted with phenol-chloroform, precipitated with ethanol, and treated with 1M piperidine. Equal amounts of radioactivity were subjected to electrophoresis on 8% polyacrylamide-urea gels. Maxam and Gilbert sequencing reactions for A+G were used as markers. The G residues for which DNA methylation interfered with nuclear factor binding on the coding or non-coding strands are indicated (arrows) and sequence bracketed below in panel (B) for the bp -432 to -420 region containing the 10bp GTTGCCTAGG element. Reproduced from reference [37].

2.2 Mutation of the EOS1 site abolishes nuclear factor binding.

Three 18 bp oligonucleotides from bp -434 to -417 were synthesized with either the wild type binding motif, GTTGCCTAGG, or the binding motif with mutations in either 3 (<u>A</u>TTG<u>GG</u>TAGG, Mut A) or 2 bases (GTTGCCTA<u>CC</u> Mut B) in the EOS1 element.

Double stranded wild type or mutant oligonucleotides were prepared and used as competitors or probes for electrophoretic mobility shift assays (Fig. 3). A 50 or 100 fold molar excess of wild type oligonucleotide completely blocked formation of both the C1 and C2 complexes, whereas the mutant oligos failed to block complex formation with either the wild type DNA or oligonucleotide probes. Gel shift analysis with the oligonucleotide probe was much cleaner than with the 93 bp DNA fragment, providing essentially no non-specific bands (Fig. 3, a versus b). A third minor but specific complex (C3) was detected using the bp -434 to -417 oligonucleotide probe (Fig. 3b), which was not visualized using the 93 bp DNA probe. Finally, the mutant B oligonucleotide did not bind the AML14-derived EOS1 factor(s) when used directly as the probe for gel shift analysis (Fig. 3b). Thus, mutations at either three bases (<u>G</u>TTG<u>CC</u>TAGG) or two bases (GTTGCCTA<u>GG</u>) in the EOS1 element were sufficient to completely destroy the binding site.

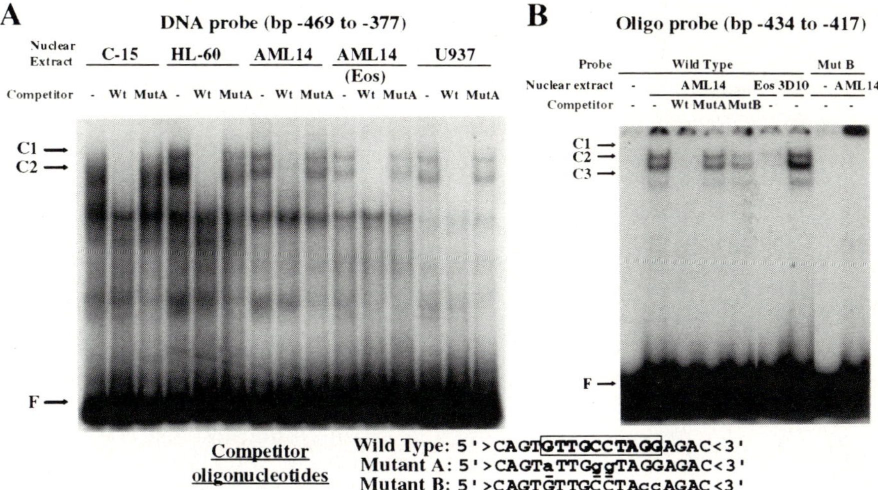

Fig. 3. Nuclear proteins of eosinophil and other myeloid cell lines bind specifically to a unique 10 bp cis-element required for activity of the IL-5Rα promoter. In (A), a 93 bp probe from bp -469 to -377 of the IL-5Rα promoter was labelled and used for gel shift analysis. Two low mobility DNA-protein complexes (C1 and C2, arrows) were identified using nuclear extracts from both the eosinophilic (HL-60-C15, AML14, AML14.eos) and other myeloid (HL-60, U937) cell lines. The specificity of formation of the C1/C2 complexes with the bp -430 to -421 element of the promoter is indicated by the inhibition of complex formation by a 50-fold molar excess of the wild type but not mutant A oligonucleotide covering this region. In (B), the doublestranded wild type or mutant B oligonucleotides were labeled and used as probes for the gel shift analysis. In addition to the specific C1 and C2 complexes identified using the longer 93 bp DNA fragment in panel A, a minor but specific third complex (C3) was detected. All three complexes formed using the AML14 and AML14.eos nuclear extracts were likewise obtained using an AML14.3D10 nuclear extract. A 100-fold molar excess of mutant oligonucleotides A and B failed to block complex formation, and the mutant B oligonucleotide probe did not bind the AML14 EOS1 nuclear factor(s). Reproduced from reference [37].

2.3 Mutation of the EOS1 site markedly reduces IL-5Rα promoter activity. To determine whether the EOS1 cis-element was functionally important for promoter activity and possessed enhancer-like activity, we generated IL-5Rα promoter constructs which contained the wild type or mutated element in the pXP2 and pT81 luciferase vectors (Fig. 4a), and analyzed reporter gene activity in transient tranfections of various eosinophil, myeloid and non-myeloid/non-hematopoietic cell lines (Fig. 4b). Results from these experiments showed that promoter activity of the mutant, -436Mut-luc construct was 75-99% lower compared to its wild type counterpart in the various cell lines tested. Comparison of the luciferase activities of the mutant -436Mut-luc and -179-luc deletion constructs [36] showed essentially identical reductions in promoter activity. These results indicate that the EOS1 element, as the only nuclear factor binding site and functional motif we have been able to identify in this region, accounts for the majority of IL-5Rα gene promoter activity *in vitro* in the eosinophilic and other myeloid cell lines.

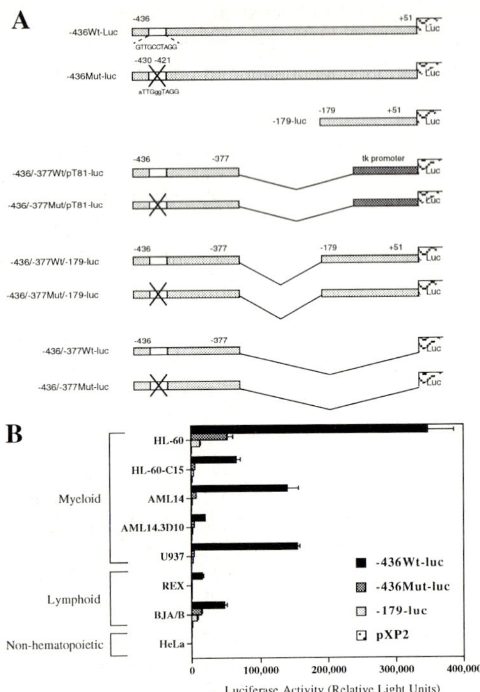

Fig. 4. Mutation of the EOS1 element (GTTGCCTAGG to aTTGggTAGG) destroys IL-5Rα promoter activity. (A) Wild type or mutant IL-5Rα promoter constructs containing the indicated regions of sequence were generated in the promoterless pXP2 luciferase vector or pT81 vector containing the Herpes simplex minimal thymidine kinase promoter. (B) The wild type or mutant constructs were transiently transfected into the various eosinophil, myeloid, lymphoid and non-hematopoietic cell lines as indicated. Mutation of the 10 bp element shown to bind nuclear factors in HL-60-C15 and other eosinophil and myeloid cell lines (Fig. 3) functionally inhibited promoter activity of the -436/luc construct by >90%, as well as significantly decreased enhancer-like activity in the bp -436 to -377/pT81/luc construct (shown in Fig. 5b). Reproduced from reference [37].

2.4 The EOS1 sequence is an enhancer-like element which requires a basal promoter for its full activity and lineage specificity. The EOS1 element, located in the region between bp -430 to -421, is ~400bp upstream of the putative TATA box for this gene [36]. To determine whether this element is capable of functioning independently without a proximal basal promoter region, we inserted a 60bp sequence (bp -436 to -377), containing the GTTGCCTAGG element, directly upstream of the luciferase reporter gene in the promoterless pXP2 vector (Fig. 4a). Results from transient transfections showed that the -436/-377Wt-luc construct, like the -179-luc deletion construct, expressed <8% of the promoter activity of the maximally active -436Wt-luc construct containing this element (Fig. 5a). These results suggest that EOS1 acts like most other enhancers, requiring a proximal basal promoter for initiation of transcription. However, the combination of IL-5Rα sequences between bp -179 and +51 with the EOS1 upstream element produced only 15% to 35% of the activity of the most active full length promoter construct in the

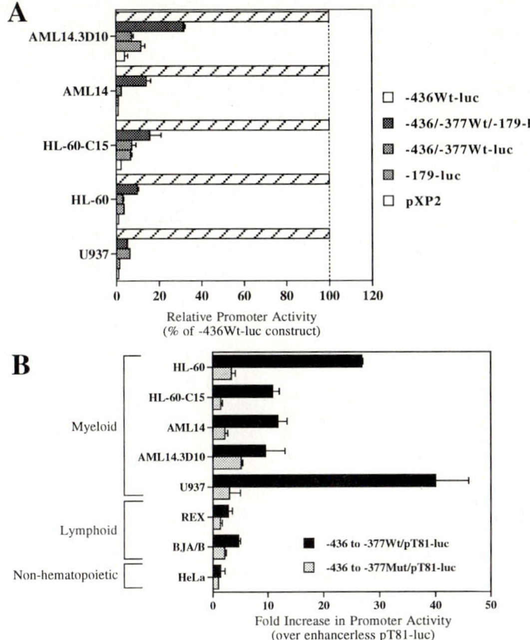

Fig. 5. The relative activities of different IL-5Rα promoter constructs in the pT81 plasmid demonstrates lineage specific and enhancer-like activity of the EOS1 element. (A) Relative promoter activities of -436/-377wt/-179-luc, -436/-377wt-luc and -179-luc constructs were compared to the maximally active wild type -436wt-luc construct (100%) in several eosinophilic and myeloid cell lines. (B) The 60 bp region between bp -436 to -377, containing a wild type (GTTGCCTAGG) or mutant (aTTGggTAGG) binding site, was cloned into pT81 luciferase vector containing a basal thymidine kinase promoter. The fold increase in promoter activity was measured relative to the enhancerless pT81 parent vector in the various cell lines. Reproduced from reference [37].

various myeloid cell lines (Fig. 5a), suggesting there may be additional functionally relevant sequences in the bp -377 to -179 region. To further characterize the enhancer-like activity of the EOS1 element, we inserted a 60bp sequence (bp -436 to -377), containing either the wild type or mutant sequence, into the pT81-luc vector upstream of its basal thymidine kinase promoter. These two constructs, identified as -436/-377Wt/pT81-luc and -436/-377Mut/pT81-luc (Fig. 4a), and the parent pT81-luc plasmid were transiently transfected into the various eosinophil, myeloid, lymphoid and non-myeloid cell lines (Fig. 5b). In comparison to the enhancerless pT81-luc plasmid, promoter activity of the -436/-377Wt/pT81 construct was increased 10-fold in both the eosinophilic HL-60 -C15 and AML14 cell lines, and 40-fold in the myeloid U937 line, versus only an ~5 fold or no increase in the lymphocytic BJA-B and non-hematopoietic HeLa cell lines, respectively. The -436/-377Mut/pT81-luc construct with a three base mutation in the functional EOS1 element showed significantly less promoter activity than its wild type counterpart in all cell lines analyzed (Fig. 5b). These results imply that the EOS1 element functions as a myeloid lineage-specific enhancer with a critical role in the cell-specific expression of the IL-5Rα gene. Further, since this enhancer element was highly active with a basal tk promoter, its enhancer activity is not likely promoter specific.

3 Discussion

Little is currently known regarding the mechanisms by which human cytokine and growth factor receptor genes are expressed and regulated during the commitment and differentiation of hematopoietic progenitors to the myeloid lineages in general or eosinophil lineage in particular. IL-5 is a late-acting, lineage-specific cytokine that demonstrates maximum activity on an eosinophil progenitor pool expanded by the earlier-acting, multipotential cytokines such as IL-3 or GM-CSF [3]. The unique α subunit of the IL-5R is required for both ligand binding [20] and for signal transduction [21, 22] and is therefore critical to the specific action of IL-5 on the eosinophil lineage. Expression of the IL-5Rα gene, like other hematopoietic growth factor receptor genes including those encoding the M-CSF (CSF-1) [32-34], G-CSFR [35, 44] and GM-CSFRα [45, 46] receptors, may be critical to the entry of multipotential myeloid progenitors into a particular hematopoietic developmental program. Thus, elucidation of the processes controlling IL-5Rα gene expression will be extremely pertinent to understanding the mechanisms involved in regulating eosinophil lineage commitment and differentiation. Since activation and prolonged survival of the mature eosinophil is also regulated in part by IL-5, the mechanisms regulating expression of the IL-5Rα subunit gene are likewise pertinent to the activities of IL-5 and other eosinophil-active cytokines on eosinophil function [9, 47], especially with regard to the mechanisms for IL-5-mediated signal transduction pathways in eosinophil activation [48, 49]. These regulatory mechanisms are also likely to be important both to the development of eosinophilia and to the functional activation of eosinophils in tissues in eosinophil-associated allergic, parasitic, inflammatory and other diseases.

We previously isolated the 5' upstream region of the human IL-5Rα gene and identified a 34 bp promoter region that was highly functional in the eosinophil lineage [36]. We have subsequently identified a unique and previously uncharacterized cis-element (EOS1) in this 34bp region of the promoter, identified a nuclear factor or factors that binds specifically to the EOS1 element, characterized the functional importance of EOS1 for promoter activity in eosinophil versus other myeloid, lymphoid or non-myeloid lineages, and defined the element as an eosinophil/myeloid lineage-specific enhancer-like sequence [37]. Our results suggest that full tissue specificity of the IL-5Rα promoter may depend on both the distal EOS1 element and the more proximal basal promoter region of this gene. Constructs containing the IL-5Rα enhancer element driving the thymidine kinase promoter in the pT81 vector expressed markedly increased levels of activity compared to the enhancerless pT81 parent construct in the various eosinophil and myeloid lines tested, significantly greater increases than in the non-myeloid lines. These results suggested that the EOS1 element is myeloid and possibly eosinophil lineage specific in its function. However, mutation of the EOS1 site in the context of the -436 bp promoter further decreased the low levels of promoter activity detected in some of the non-myeloid lines (REX and BJA-B), suggesting these transformed lymphoid lines may contain low but functional levels of factors capable of transactivating the IL-5Rα promoter via the EOS1 element. Of interest, the -436/-377Wt/-179-luc construct was somewhat more active than the-436/-377Wt/pT81-luc construct in the AML14 cell line. AML14 constitutively expresses the IL-5Rα subunit and a functional high affinity IL-5R [28] and culture with eosinophil-active cytokines (IL-5, IL-3 and GM-CSF) induces it into a fully differentiated eosinophilic myelocyte subline (AML14.eos) [29]. The differentiation potential of this line probably reflects the fact that expression of a number of regulatory genes that control eosinophil differentiation can be induced by growth factors such as IL-5, IL-3 and/or GM-CSF. The IL-5Rα gene is likely to be an important component of this process and studies of IL-5Rα gene regulation will likely result in greater understanding of the processes involved in the induction and/or maintenance of eosinophilia and tissue damage in syndromes such as HES, in infections with parasitic helminths, and in the pathogenesis of eosinophil-associated allergic diseases such as asthma.

We have identified a nuclear factor or factors in the eosinophilic cell lines studied that binds specifically to the EOS1 enhancer-like element and that likely regulates expression of the IL-5Rα gene. Whether the two DNA-nuclear factor complexes formed by the interaction of the EOS1 cis-element and its factor represent monomer versus homodimer or heterodimer binding, proteolytic cleavage or postranslation modification has not been determined. Consensus sequences for known transcription factor binding sites were not detected in searches of a 1994 update of the Ghosh transcription factor data base [50]. Recently, Zhang and colleagues [33] reported that a functional element, TGTGGTTGCCT, in the M-CSF receptor promoter interacts with the PEBP2/CBF (AML1) transcription factor. Comparison of this AML1 element to the EOS1 (GTTGCCTAGG) element in the IL-5Rα promoter showed identity in 7 of 11 3' bp of the AML1 sequence. To determine whether the PEBP2/CBF (AML1) factor binds to the IL-5Rα EOS1 enhancer, we used two double stranded oligonucleotides containing either a perfect AML1 binding motif (TGTGGT) or the M-CSF receptor element itself (TGTGGTTGCCT) that binds AML1 [33], to compete with the IL-5Rα promoter sequence for formation of the C1 and C2 complexes. Results from electrophoretic mobility shift assays showed that C1 and C2 complex formation was not inhibited by either of the AML1-binding sequences (data not shown). In addition, a specific anti-AML1 antibody (kindly provided by Dr. Scott Hiebert) was used to try to supershift the IL-5Rα C1/C2 complexes, but this antibody likewise failed to recognize the nuclear factor(s) forming these complexes (data not shown). From these results, we conclude that PEBP2/CBF (AML1) is not involved in the formation of complexes with the IL-5Rα EOS1 element nor in its function.

The structure of the murine IL-5Rα chain gene was published recently by Imamura and colleagues [51]. In marked contrast to the selective expression of the human gene in the eosinophil and basophil lineages [3, 52], the murine gene is also expressed in B cells, which have a functional IL-5R [51]. Identification of the promoter regions and transcription factors that control the differential expression of the human gene in eosinophils and basophils only versus the murine gene in eosinophil and B cell lineages [3, 53], should provide insights into their respective mechanisms of lineage-specific regulation. The upstream, 5' flanking region of the murine IL-5Rα gene contains consensus sequences for Ap1, AP-1, GATA-1 and PU.1, but these sequences have not been characterized for their functional relevance [51]. In functional analyses of the murine promoter published thus far [51], 256bp of 5' flanking region (-96 to +160bp) had minimal promoter activity in a pCAT vector in fibroblast (NIH3T3), FDC-P1 and IL-5 dependent pre-B cell (Y16) lines. Of interest, functional characterization of an additional ~1.3kb of upstream sequence to identify a region that directs B cell-selective expression, detected no promoter activity in murine cell lines, suggesting suppressive 5' elements in the gene. Similarity searches of the murine IL-5Rα upstream region for the EOS1 element we have characterized in the human promoter did not identify a similar element within the 740bp of murine upstream sequence published thus far [51]. However, these searches have identified a single region of marked sequence similarity for the human versus murine IL-5Rα genes (bp -76 to -38 versus bp -212 to -174, respectively; 85% identity), a region containing an NF-IL6 (C/EBPß) consensus site [51] for which the functional importance has not been determined.

In comparisons of the human IL-5Rα upstream sequence to the 5' flanking sequences of the human GM-CSFRα gene [45], we did not detect any regions of significant similarity to the functionally active 34 bp region of the IL-5Rα promoter identified previously [36]. Likewise, similarity searches for the EOS1 element itself have failed to identify a similar element in the functionally active promoter region of the GM-CSFRα gene [46]. Recent studies of human GM-CSFRα subunit gene expression have demonstrated that PU.1(Spi-1) and C/EBPα transcription factors bind to and regulate the functional activity of the GM-CSFRα promoter in myelomonocytic cell lines such as U937 in a cell-type specific manner [46]. We have not identified similar binding sites for the PU.1 or C/EBP factors in the human IL-5Rα promoter, nor does *in vitro* translated PU.1 bind to the promoter in electrophoretic mobility shift assays (data not shown). Furthermore, antibodies to C/EBPα, C/EBPβ, and C/EBPδ (kindly provided by Dr. Steven McKnight) as well as oligonucleotides containing functional C/EBP binding sites did not inhibit EOS1 complex formation (data not shown), indicating that EOS1 is not likely one of the C/EBP family of transcription factors. Overall, these findings indicate important differences in the regulatory regions and trans-acting factors for the human IL-5Rα subunit versus murine IL-5Rα and human GM-CSFRα subunit genes, differences likely relevant to their selective expression in eosinophil versus B cell lineages in the two species, and granulocyte/macrophage lineages in humans, respectively. The promoters for a number of other genes preferentially expressed in the myeloid series, including CD11b [54], M-CSF [32], and CD14 [55] have likewise been shown to require either PU.1 and/or Sp1 [55, 56] binding for myeloid-specific expression. The lack of functionally active or consensus binding sites for PU.1 or Sp1 in the IL-5Rα promoter further emphasizes important differences in the lineage-specific regulation of genes in the eosinophil versus other myeloid lineages.

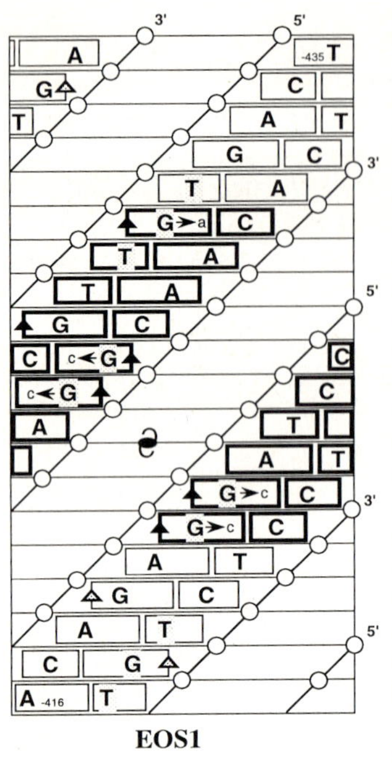

Fig. 6. Modeling of EOS1 nuclear factor binding to the DNA helix based on methylation interference. The DNA from bp -416 to -435 is shown as a cylindrical projection with phosphate backbone indicated by the small open circles. Nucleotide bases representing the core recognition residues in the EOS1, GTTGCCTAGG element are in bolded boxes in the major groove and the methylated residues that strongly or weakly interferred with nuclear factor binding are indicated by the black or shaded triangles, respectively. Mutations of G residues in the binding site that blocked both nuclear factor binding (Fig. 3) and functional activity of the enhancer element (Fig. 4 & 5) are indicated by the arrowheads. The center of potential diad symmetry of the EOS1 element between bp -423 and -424 is indicated by the symbol (§) and corresponding symmetrically arranged bases (shaded). Reproduced from reference [37].

We have recently characterized functional promoter regions for a number of other genes expressed selectively in the eosinophil lineage including those for eosinophil peroxidase [41] and Charcot-Leyden crystal protein (eosinophil lysophospholipase) [42]. Searches of the upstream functional promoter regions of these genes for the IL-5Rα EOS1 element, as well as the published 5' sequences for genes encoding eosinophil granule major basic protein [57], eosinophil cationic protein [58] (SJA, unpublished sequence data) and eosinophil-derived neurotoxin [58], did not detect an identical sequence element. However, since the genes encoding the eosinophil granule-associated proteins are likely expressed later in eosinophil development, subsequent to signaling through an IL-5R expressed by early multipotential myeloid progenitors, these granule protein genes may be coordinately regulated in an entirely different fashion than the IL-5Rα subunit during eosinophil differentiation and granulogenesis.

4 Summary and Conclusions

Further functional and biochemical characterization of the nuclear factor(s) which interacts with the EOS1 enhancer-like element in the IL-5Rα promoter is currently in progress. Since different transcription factors recognize and interact with DNA in distinct fashions and with distinct structural motifs [59], we have modeled potential binding of the EOS1 factor to its cis-element based upon its methylation interference pattern (Fig. 2), using a cylindrical DNA helical projection [59, 60] (Fig. 6). Over a length of two helical turns, all nuclear protein contacts indicated by methylation interference map to one side of the DNA helix, suggesting that EOS1 binds in the major groove, across the minor groove, and on only one side of the helix. Further review of the model also reveals a potential diad symmetry for the binding site, suggestive of binding by a homodimer and consistent with the formation of the two DNA-protein complexes in our electrophoretic mobility shift experiments that could represent interactions with monomer versus dimer. Comparison of the EOS1 binding motif to similar models for the binding of other transcription factor families for which structural crystallographic and/or binding data is available suggests a similarity of the EOS1 complex to that of the bacterial helix-turn-helix phage λ and 434 repressor-operator complexes [61], and the Cys_4 zinc finger glucocorticoid response element (GRE) [62] DNA-binding motifs, all of which show similar diad symmetry and binding in the major groove on one side of the DNA [59]. The possibility that EOS1 functions as a GRE is being investigated, especially since there is a consensus AP-1 site at bp -440 to -432 of the IL-5Rα promoter, immediately adjacent to the EOS1 binding site (see Fig. 5 in reference [36]) and AP-1/GRE interactions have been identified for composite response elements in the regulation of a number of different genes [63, 64]. The identification or cloning of EOS1, a potentially novel and eosinophil lineage-active transcription factor, should enhance our understanding of the processes involved in eosinophil development in particular and myeloid lineage committment and differentiation in general.

Acknowledgements. The authors thank Dr. Scott Hiebert for providing the anti-AML1 antiserum, Dr. Steven McKnight for providing the antisera to C/EBPα,β, and δ, and Mary Singleton for secretarial and administrative assistance. This work was supported by National Institute of Allergy and Infectious Disease grant AI33043 (to SJA), National Cancer Institute grant CA41456 (to DGT), American Lung Association research grant (to ZJS), and Department of Veteran Affairs Merit Review grants (to CCP and MAB). ZJS is an Edward Livingston Trudeau Scholar of the American Lung Association.

References

1. Takatsu, K., Tominaga, A., Harada, N., Mita, S., Matsumoto, M., Takahashi, T., Kikuchi, Y., Yamaguchi, N. (1988) T cell-replacing factor (TRF)/interleukin 5 (IL-5): molecular and functional properties. Immunol Rev 102, 107-35.
2. Galli, S.J., Gordon, J.R., Wershil, B.K., Elovic, A., Wong, D.T., Weller, P.F. (1994) Mast cell and eosinophil cytokines in allergy and inflammation. In Eosinophils in Allergy and Inflammation, Volume 2 (A. B. Kay and G. J. Gleich, eds), Marcel Dekker, New York 255-280.

3. Sanderson, C.J. (1990) Eosinophil differentiation factor (IL-5). Immunol Ser 49, 231-56.
4. Dent, L.A., Strath, M., Mellor, A.L., Sanderson, C.J. (1990) Eosinophilia in transgenic mice expressing interleukin 5. J Exp Med 172, 1425-31.
5. Tominaga, A., Takaki, S., Koyama, N., Katoh, S., Matsumoto, R., Migita, M., Hitoshi, Y., Hosoya, Y., Yamauchi, S., Kanai, Y., et, a. (1991) Transgenic mice expressing a B cell growth and differentiation factor gene (interleukin 5) develop eosinophilia and autoantibody production. J Exp Med 173, 429-37.
6. Takatsu, K., Takaki, S., Hitoshi, Y. (1994) Interleukin-5 and its receptor system: implications in the immune system and inflammation. Adv Immunol 57, 145-90.
7. Saito, H., Hatake, K., Dvorak, A.M., Leiferman, K.M., Donnenberg, A.D., Arai, N., Ishizaka, K., Ishizaka, T. (1988) Selective differentiation and proliferation of hematopoietic cells induced by recombinant human interleukins. Proc Natl Acad Sci U S A 85, 2288-92.
8. Clutterbuck, E.J., Hirst, E.M., Sanderson, C.J. (1989) Human interleukin-5 (IL-5) regulates the production of eosinophils in human bone marrow cultures: comparison and interaction with IL-1, IL- 3, IL-6, and GMCSF. Blood 73, 1504-12.
9. Goodall, G.J., Bagley, C.J., Vadas, M.A., Lopez, A.F. (1993) A model for the interaction of the GM-CSF, IL-3 and IL-5 receptors with their ligands. Growth Factors 8, 87-97.
10. Sanderson, C.J. (1991) Control of eosinophilia. Int Arch Allergy Appl Immunol 94, 122-6.
11. Lopez, A.F., Shannon, M.F., Chia, M.M., Park, L., Vadas, M.A. (1992) Regulation of human eosinophil production and function by interleukin- 5. Immunol Ser 57, 549-71.
12. Clutterbuck, E., Shields, J.G., Gordon, J., Smith, S.H., Boyd, A., Callard, R.E., Campbell, H.D., Young, I.G., Sanderson, C.J. (1987) Recombinant human interleukin 5 is an eosinophil differentiation factor but has no activity in standard human B cell growth factor assays. Eur J Immunol 17, 1743-50.
13. Baumann, M.A., Tolbert, M., Mahrer, S., Paul, C.C. (1993) Evidence for an alternative functional low affinity IL-5 receptor on human B-lymphocytes. Blood 82, 241a.
14. Owen, W.F., Rothenberg, M.E., Petersen, J., Weller, P.F., Silberstein, D., Sheffer, A.L., Stevens, R.L., Soberman, R.J., Austen, K.F. (1989) Interleukin 5 and phenotypically altered eosinophils in the blood of patients with the idiopathic hypereosinophilic syndrome. J Exp Med 170, 343-8.
15. Owen, W., Jr., Petersen, J., Sheff, D.M., Folkerth, R.D., Anderson, R.J., Corson, J.M., Sheffer, A.L., Austen, K.F. (1990) Hypodense eosinophils and interleukin 5 activity in the blood of patients with the eosinophilia-myalgia syndrome. Proc Natl Acad Sci U S A 87, 8647-51.
16. Sanderson, C.J. (1992) Interleukin-5, eosinophils, and disease. Blood 79, 3101-9.
17. Tavernier, J., Devos, R., Cornelis, S., Tuypens, T., Van der Heyden, J., Fiers, W., Plaetinck, G. (1991) A human high affinity interleukin-5 receptor (IL5R) is composed of an IL5-specific alpha chain and a beta chain shared with the receptor for GM-CSF. Cell 66, 1175-84.
18. Murata, Y., Takaki, S., Migita, M., Kikuchi, Y., Tominaga, A., Takatsu, K. (1992) Molecular cloning and expression of the human interleukin 5 receptor. J Exp Med 175, 341-51.
19. Miyajima, A., Mui, A.L., Ogorochi, T., Sakamaki, K. (1993) Receptors for granulocyte-macrophage colony-stimulating factor, interleukin-3, and interleukin-5. Blood 82, 1960-74.
20. Takaki, S., Murata, Y., Kitamura, T., Miyajima, A., Tominaga, A., Takatsu, K. (1993) Reconstitution of the functional receptors for murine and human interleukin 5. J Exp Med 177, 1523-9.
21. Takaki, S., Kanazawa, H., Shiiba, M., Takatsu, K. (1994) A critical cytoplasmic domain of the interleukin-5 (IL-5) receptor alpha chain and its function in IL-5-mediated growth signal transduction. Mol Cell Biol 14, 7404-13 1994.
22. Sato, S., Katagiri, T., Takaki, S., Kikuchi, Y., Hitoshi, Y., Yonehara, S., Tsukada, S., Kitamura, D., Watanabe, T., Witte, O., Takatsu, K. (1994) IL-5 receptor-mediated tyrosine phosphorylation of SH2/SHS-containing proteins and activation of Bruton's tyrosine and Janus 2 kinases. J. Exp. Med. 180, 2101-2111.
23. Ding, D.X., Rivas, C.I., Heaney, M.L., Raines, M.A., Vera, J.C., Golde, D.W. (1994) The

alpha subunit of the human granulocyte-macrophage colony- stimulating factor receptor signals for glucose transport via a phosphorylation-independent pathway. Proc Natl Acad Sci U S A 91, 2537-41.

24. Tuypens, T., Plaetinck, G., Baker, E., Sutherland, G., Brusselle, G., Fiers, W., Devos, R., Tavernier, J. (1992) Organization and chromosomal localization of the human interleukin 5 receptor alpha-chain gene. Eur Cytokine Netw 3, 451-9.
25. Tavernier, J., Tuypens, T., Plaetinck, G., Verhee, A., Fiers, W., Devos, R. (1992) Molecular basis of the membrane-anchored and two soluble isoforms of the human interleukin 5 receptor alpha subunit. Proc Natl Acad Sci U S A 89, 7041-5.
26. Devos, R., Guisez, Y., Cornelis, S., Verhee, A., Van der Heyden, J., Manneberg, M., Lahm, H.W., Fiers, W., Tavernier, J., Plaetinck, G. (1993) Recombinant soluble human interleukin-5 (hIL-5) receptor molecules. Cross-linking and stoichiometry of binding to IL-5. J Biol Chem 268, 6581-7.
27. Barry, S.C., McKenzie, A.N., Strath, M., Sanderson, C.J. (1991) Analysis of interleukin 5 receptors on murine eosinophils: a comparison with receptors on B13 cells. Cytokine 3, 339-44.
28. Paul, C.C., Tolbert, M., Mahrer, S., Singh, A., Grace, M.J., Baumann, M.A. (1993) Cooperative effects of interleukin-3 (IL-3), IL-5, and granulocyte- macrophage colony-stimulating factor: a new myeloid cell line inducible to eosinophils. Blood 81, 1193-9.
29. Paul, C.C., Ackerman, S.J., Mahrer, S., Tolbert, M., Dvorak, A.M., Baumann, M.A. (1994) Cytokine induction of granule protein synthesis in an eosinophil-inducible human myeloid cell line, AML14. J Leuk Biol 56, 74-79.
30. Plaetinck, G., Van der Heyden, J., Tavernier, J., Fache, I., Tuypens, T., Fischkoff, S., Fiers, W., Devos, R. (1990) Characterization of interleukin 5 receptors on eosinophilic sublines from human promyelocytic leukemia (HL-60) cells. J Exp Med 172, 683-91.
31. Takagi, M., Hara, T., Ichihara, M., Takatsu, K., Miyajima, A. (1995) Multi-colony stimulating activity of interleukin 5 (IL-5) on hematopoietic progenitors from transgenic mice that express IL-5 receptor alpha subunit constitutively. J Exp Med 181, 889-99.
32. Zhang, D.E., Hetherington, C.J., Chen, H.M., Tenen, D.G. (1994) The macrophage transcription factor PU.1 directs tissue-specific expression of the macrophage colony-stimulating factor receptor. Mol Cell Biol 14, 373-81.
33. Zhang, D.-E., Fujioka, K.-I., Hetherington, C.J., Shapiro, L.H., Chen, H.-M., Look, A.T., Tenen, D.G. (1994) Identification of a region which directs the monocytic activity of the colony stimulating factor 1 (macrophage colony-stimulating factor) receptor promoter and binds PEBP2/CBF (AML1). Mol. Cell Biol. 14, 8085-8095.
34. Roberts, W.M., Shapiro, L.H., Ashmun, R.A., Look, A.T. (1992) Transcription of the human colony-stimulating factor-1 receptor gene is regulated by separate tissue-specific promoters. Blood 79, 586-93.
35. Smith, L.A., Dziennis, S., Hohaus, S., Tenen, D.G. (1993) The granulocyte colony stimulating factor receptor promoter directs tissue specific expression in myeloid cells. Blood 82, 184a (abstract #720).
36. Sun, Z., Yergeau, D.A., Tuypens, T., Tavernier, J., Paul, C.C., Baumann, M.A., Tenen, D.G., Ackerman, S.J. (1995) Identification and characterization of a functional promoter region in the human eosinophil IL-5 receptor alpha subunit gene. J Biol Chem 270, 1462-1471.
37. Sun, Z., Yergeau, D.A., Wong, I.C., Paul, C.C., Baumann, M.A., Tavernier, J., Tenen, D.G., Ackerman, S.J. (1995) The unique cis-element, EOS1 (GTTGCCTAGG), acts like an enhancer in regulating expression of the human eosinophil interleukin-5 receptor α subunit gene. J. Immunol., submitted for publication .
38. Fischkoff, S.A. (1988) Graded increase in probability of eosinophilic differentiation of HL- 60 promyelocytic leukemia cells induced by culture under alkaline conditions. Leuk Res 12, 679-86.
39. Tomonaga, M., Gasson, J.C., Quan, S.G., Golde, D.W. (1986) Establishment of eosinophilic sublines from human promyelocytic leukemia (HL-60) cells: demonstration of

multipotentiality and single- lineage commitment of HL-60 stem cells. Blood 67, 1433-41.

40. Paul, C.C., Mahrer, S., Tolbert, M., Elbert, B.L., Ackerman, S.J., Baumann, M.A. (1995) Changing the differentiation program of hematopoietic cells: Retinoic Acid-induced shift of eosinophil-committed cells to neutrophils. Blood , Submitted for publication.
41. Yamaguchi, Y., Zhang, D.-E., Sun, Z.-J., Albee, E.A., Nagata, S., Tenen, D.G., Ackerman, S.J. (1994) Functional characterization of the promoter for the gene encoding human eosinophil peroxidase. J Biol Chem 269, 19410-19419.
42. Gomolin, H.I., Yamaguchi, Y., Paulpillai, A.V., Dvorak, L.A., Ackerman, S.J., Tenen, D.G. (1993) Human eosinophil Charcot-Leyden crystal protein: cloning and characterization of a lysophospholipase gene promoter. Blood 82, 1868-74.
43. Zon, L.I., Yamaguchi, Y., Yee, K., Albee, E.A., Kimura, A., Bennett, J.C., Orkin, S.H., Ackerman, S.J. (1993) Expression of mRNA for the GATA-binding proteins in human eosinophils and basophils: potential role in gene transcription. Blood 81, 3234-41.
44. Seto, Y., Fukunaga, R., Nagata, S. (1992) Chromosomal gene organization of the human granulocyte colony- stimulating factor receptor. J Immunol 148, 259-66.
45. Nakagawa, Y., Kosugi, H., Miyajima, A., Arai, K.-I., Yokota, T. (1994) Structure of the gene encoding the α subunit of the human granulocyte-macrophage colony stimulating factor receptor. J Biol Chem 269, 10905-10912.
46. Hohaus, S., Voso, M.T., Sun, Z., Tenen, D.G. (1995) PU.1 (Spi-1) and C/EBPα regulate the expression of the granulocyte-macrophage colony-stimulating factor receptor α. Mol. Cell. Biol. Accepted for publication with revision.
47. Lopez, A.F., Vadas, M.A., Woodcock, J.M., Milton, S.E., Lewis, A., Elliott, M.J., Gillis, D., Ireland, R., Olwell, E., Park, L.S. (1991) Interleukin-5, interleukin-3, and granulocyte-macrophage colony- stimulating factor cross-compete for binding to cell surface receptors on human eosinophils. J Biol Chem 266, 24741-7.
48. van der Bruggen, T., Caldenhoven, E., Kanters, D., Coffer, P., Raaijmakers, J.A., Lammers, J.W., Koenderman, L. (1995) Interleukin-5 signaling in human eosinophils involves JAK2 tyrosine kinase and Stat1 alpha. Blood 85, 1442-8.
49. Pazdrak, K., Schreiber, D., Forsythe, P., Justement, L., Alam, R. (1995) The intracellular signal transduction mechanism of Interleukin 5 in eosinophils: The involvement of lyn tyrosine kinase and the Ras-Raf-1- MEK microtubule-associated protein kinase pathway. J. Exp. Med. 181, 1827-1834.
50. Ghosh, D. (1992) TFD: the transcription factors database. Nucleic Acids Res 11, 2091-3.
51. Imamura, F., Takaki, S., Akagi, K., Ando, M., Yamamura, K.-I., Takatsu, K., Tominaga, A. (1994) The murine interleukin-5 receptor α-subunit gene: Characterization of the gene structure and chromosome mapping. DNA and Cell Biol 13, 283-292.
52. Denburg, J.A., Silver, J.E., Abrams, J.S. (1991) Interleukin-5 is a human basophilopoietin: induction of histamine content and basophilic differentiation of HL-60 cells and of peripheral blood basophil-eosinophil progenitors. Blood 77, 1462-8.
53. Sanderson, C.J., Campbell, H.D., Young, I.G. (1988) Molecular and cellular biology of eosinophil differentiation factor (interleukin-5) and its effects on human and mouse B cells. Immunol Rev 102, 29-50.
54. Pahl, H.L., Scheibe, R.J., Zhang, D.E., Chen, H.M., Galson, D.L., Maki, R.A., Tenen, D.G. (1993) The proto-oncogene PU.1 regulates expression of the myeloid-specific CD11b promoter. J Biol Chem 268, 5014-20.
55. Zhang, D.E., Hetherington, C.J., Tan, S., Dziennis, S., Gonzalez, D.A., Chen, H.-M., Tenen, D.G. (1994) Sp1 is a critical factor for the monocyte specific expression of CD14. J Biol Chem 269, 11425-11434.
56. Chen, H.M., Pahl, H.L., Scheibe, R.J., Zhang, D.E., Tenen, D.G. (1993) The Sp1 transcription factor binds the CD11b promoter specifically in myeloid cells in vivo and is essential for myeloid-specific promoter activity. J Biol Chem 268, 8230-9.
57. Barker, R.L., Loegering, D.A., Arakawa, K.C., Pease, L.R., Gleich, G.J. (1990) Cloning and sequence analysis of the human gene encoding eosinophil major basic protein. Gene 86,285-9.
58. Hamann, K.J., Ten, R.M., Loegering, D.A., Jenkins, R.B., Heise, M.T., Schad, C.R.,

Pease, L.R., Gleich, G.J., Barker, R.L. (1990) Structure and chromosome localization of the human eosinophil-derived neurotoxin and eosinophil cationic protein genes: evidence for intronless coding sequences in the ribonuclease gene superfamily. Genomics 7, 535-46.

59. Pabo, C.O., Sauer, R.T. (1992) Transcription factors: structural families and principles of DNA recognition. Annu Rev Biochem 61, 1053-95.
60. Kissinger, C.R., Liu, B.S., Martin-Blanco, E., Kornberg, T.B., Pabo, C.O. (1990) Crystal structure of an engrailed homeodomain-DNA complex at 2.8 A resolution: a framework for understanding homeodomain-DNA interactions. Cell 63, 579-90.
61. Pabo, C.O., Aggarwal, A.K., Jordan, S.R., Beamer, L.J., Obeysekare, U.R., Harrison, S.C. (1990) Conserved residues make similar contacts in two repressor-operator complexes. Science 247, 1210-3.
62. Luisi, B.F., Xu, W.X., Otwinowski, Z., Freedman, L.P., Yamamoto, K.R., Sigler, P.B. (1991) Crystallographic analysis of the interaction of the glucocorticoid receptor with DNA. Nature 352, 497-505.
63. Miner, J.N., Yamamoto, K.R. (1992) The basic region of AP-1 specifies glucocorticoid receptor activity at a composite response element. Genes Dev 6, 2491-501.
64. Pearce, D., Yamamoto, K.R. (1993) Mineralocorticoid and glucocorticoid receptor activities distinguished by nonreceptor factors at a composite response element [see comments]. Science 259, 1161-5.

Molecular Aspects of Myeloid Leukemia. I. Murine Studies

Retroviral Insertional Mutagenesis in Murine Promonocytic Leukemias: *c-myb* and *Mml*1.

L. Wolff, R. Koller, J. Bies, V. Nazarov, B. Hoffman, A. Amanullah, M. Krall, and B. Mock
Laboratory of Genetics, National Cancer Institute, Bethesda, MD 20892-4255 and Fels Institute for Cancer Research and Molecular Biology, Temple University School of Medicine, Philadelphia, PA 19140

Abstract

Studies have focused on two genetic loci, c-*myb* and *Mml1*, whose activation by retroviral insertional mutagenesis contribute to promonocytic leukemia in our acute monocytic leukemia (AMoL) model. Multiple mechanisms of activation of c-*myb* by retroviral insertional mutagenesis implicate both transcriptional deregulation and protein truncation in conversion of this proto-oncogene to an oncogene. Because transformation by c-Myb can be viewed as a block to differentiation our studies moved into two in vitro systems to evaluate effects of truncated forms of c-Myb on cytokine induced maturation of myeloid progenitors to the granulocyte and macrophage lineages. Deregulated expression of truncated and full length c-Myb did not result in maintenance of the myelomonocytic progenitor state but rather a block in differentiation at intermediate to late steps in the maturation processes of myelomonocytic cells. Our results argue that inhibition of differentiation is due to c-Myb's ability to maintain the proliferative state of cells. Interestingly, the phenotype of continuously proliferating monocytic cells resembles that of the tumor cell phenotype. Recently we identified a new target of integration, *Mml1,* which is rearranged in ten promonocytic leukemias that do not have c-*myb* rearrangements. This locus which was mapped to chromosome 10 is presently being characterized.

The Disease Model

Promonocytic leukemias develop as a result of intravenous inoculation of replication competent retroviruses into adult BALB/c or DBA/2 mice that are undergoing a chronic peritoneal inflammation (1-5). They are referred to collectively as MMLs for murine leukemia virus (MuLV)-induced myeloid leukemia. These leukemias develop as a consequence of retroviral insertional mutagenesis involving either c-*myb* or recently demonstrated genetic target *Mml-1.*

During the early phase of disease, myeloid progenitors undergo transformation in hematopoietic organs by the mutagenic action of the virus. For example, c-*myb* was demonstrated by a sensitive RT-PCR approach to be activated by insertional mutagenesis in cells of the bone marrow and spleen within the first month after virus inoculation (6). In leukemias involving c-*myb,* activation of the proto-oncogene by the virus is the first step in what appears to be a multistep process and neoplastic development is dependent upon the pristane-induced chronic inflammation. Support for this comes from observation that the mutagenic event occurs in 100% of mice inoculated with virus in the absence of the promoting effects of the pristane, but these mice never develop leukemia. Only about 50% of the mice succumb to full blown disease with pristane treatment suggesting that progression beyond this preleukemic phase is rate limiting for disease. One explanation for this is that the immune response becomes active in eliminating cells during this crucial phase. Interestingly, susceptibility and resistance does not correlate with the capacity of mouse strains to support virus replication and activation of c-*myb* but rather the capacity to undergo progression (7). Similarly, pathogenic determinants of the retrovirus, namely the ψ-*gag*-PR and *env* regions do not determine whether c-*myb* will be activated by the virus but whether the mice will develop disease once the proto-oncogene is activated (8).

It is believed that progression involves differentiation, further oncogenic or promoting events, and clonal expansion. The fact that splenectomy decreases disease incidence indicates that this organ, in addition to the peritoneal cavity may function in this progression (9). We have demonstrated increased myelopoiesis in this organ following pristane treatment which may explain its ability to facilitate myeloid tumor development (9).

Our studies have demonstrated that insertional mutagenesis may occur in multiple cell types but selection of transforming oncogenes appears to be cell-type specific. For example, c-*myb* is activated in the thymus shortly after inoculation of Moloney MuLV into newborn mice, however c-*myb* is never activated in the lymphomas that are derived by this widely used method of tumor induction (10).

Mechanisms of c-*myb* Activation

C-*myb* activation in promonocytic leukemia can occur by proviral insertion at more than one site. With integrations at the 5' end of the gene constitutive promotor function is provided by the virus and the protein is truncated at the amino terminus by 20, 47, or 71 aa.(1,3,4) Integration can also occur within the ninth exon which gives rise to carboxyl truncation of 248 aa (4). Similar truncations have also been observed to occur in other murine and avian leukemias and in vitro transformed cell lines. Very recently we also identified novel insertion sites at the distal 3' end the gene in some DBA/2 tumors induced by Friend MuLV(5). It was found that the proviruses in these instances were located within a narrow region in the beginning of the large (greater than 10 kb) intron 14. The integration of the provirus resulted in inhibition of splicing to exon 15 in the affected allele thereby removing 38 amino acids encoded by exon 15 and adding 14 amino acids encoded by the beginning of intron 14.

The c-Myb protein is a nuclear regulator whose transcriptional activity has been reviewed (11,12). Figure 1 shows its overall structure including its DNA binding, transactivation, and negative regulatory domains. In addition, it depicts the various truncated versions the c-Myb either detected or predicted in promonocytic leukemias. Although some type of N-and C-terminal truncation of the protein occurs in all leukemias so far analyzed, it is unclear whether protein truncation is essential for leukemia. It has been hypothesized that truncations at the amino-terminus of c-Myb could affect its DNA binding. Increased binding might result from removal of the casein kinase II phosphorylation site through which negative regulation of this function has been shown to occur (13,14). Alternatively, removal of the R1 region of the binding domain might lower the binding capacity for specific target sequences (15). Deletion of the C-terminal negative regulatory domain has been demonstrated to increase transcriptional activity by c-Myb (16,17,18).

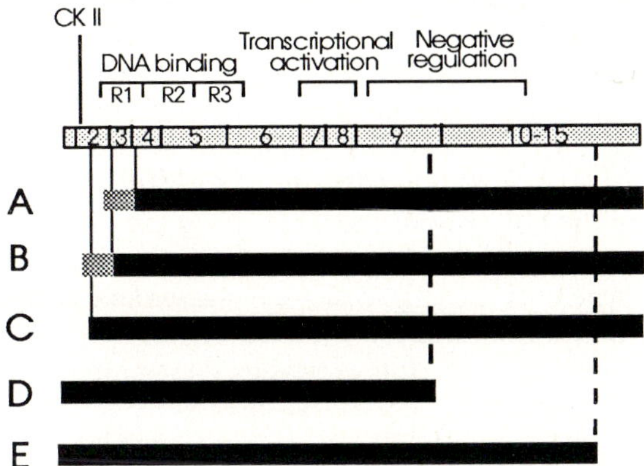

Figure 1. Truncated forms of c-Myb protein in promonocytic leukemias called MML (see ref. 1,3-5). The top bar depicts the normal c-Myb protein with exon numbers inside. Functional regions are shown above. R1,R2, R3 are direct repeat regions within the DNA binding domain. Shaded region at amino termini denote amino acids encode by viral *gag*. CkII, casein kinase II.

Although C-*myb* activation by insertional mutagenesis has been associated with protein truncation, it has also been observed that disregulation of *myb* transcription is an important feature of the activation as well. Since several of our recently examined tumors had only small protein truncations at either the C-or N-terminus, we began to wonder if protein truncation was absolutely required for transformation. Because transformation by c-Myb can be viewed as a block to differentiation our studies moved into two in vitro systems to evaluate effects of normal and truncated forms of c-Myb on cytokine induced maturation of myeloid progenitors. We were particularly interested in determining to what extent maturation inhibition occurs; a widely held view in the field is that c-Myb maintains cells in the progenitor or immature state, however, inconsistent with this is the fact that promonocytic leukemias that overexpress c-Myb have a very mature phenotype (2).

Effects of Full Length c-Myb and Leukemia-specific Truncated Forms on Differentiation of Granulocytes and Macrophages

Granulocyte differentiation. To study the effects of Myb on granulocyte differentiation we utilized the 32Dcl3 cells which differentiate in response to the physiological regulator, granulocyte colony stimulating factor (G-CSF). The results of this study were recently published but will be summarized here (19).

Initial evaluation of G-CSF-induced differentiation of this cell line demonstrated that c-Myb is expressed for the first 8-9 days of the 12-14 day maturation process. We predicted, therefore, that Myb may be essential for the early to intermediate steps of maturation and that overexpression would not inhibit the early stages of differentiation. Our data confirmed this hypothesis; cells infected with c-*myb* expressing retroviruses progressed to the promyelocyte stage and there was some evidence for differentiation to the myelocyte stage as well. These cells, however, were inhibited from entering late maturation to the nonmitotic metamyelocyte and neutrophilic granulocyte stages. Interestingly, truncated versions of c-Myb had exactly the same effect as full length c-Myb on differentiation; no difference in the ratios of the various morphological phenotypes was observed. An early differentiation marker, myeloperoxidase was expressed in all cells ectopically producing c-Myb whereas lactoferrin, a late marker was completely suppressed. The cells continued to proliferate in the presence of the G-CSF and no difference was observed in doubling times of cells overexpressing full length or truncated c-Myb.

Macrophage differentiation. M1 cells treated with IL-6 will differentiate to a mature macrophage phenotype over the course of 4 days. In order to evaluate the effects of normal murine c-Myb and truncated forms on this arm of the differentiation pathway, we infected M1 cells with LXSN vectors expressing these proteins. Helper-free stocks of the viruses were produced in GP + E-86 cells as described in ref. 19. Protein expression in clonal cell lines derived by infection with the viruses is shown in Figure 2. After 24 hrs in the presence of IL-6 endogenously expressed Myb disappeared while exogenously expressed c-Myb levels remained high. All of the cells infected with Myb viruses continued to proliferate in the presence of IL-6 while the cell line infected with empty vector proliferated with a long doubling time for 2-3 days and than ceased to grow (Figure 3).

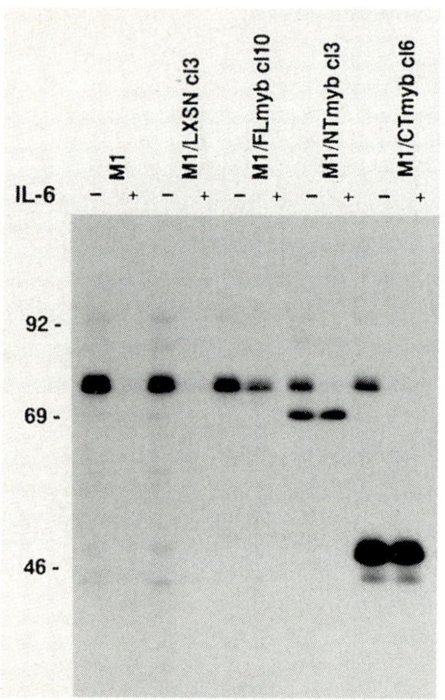

Figure 2. Immune precipitation of c-Myb proteins exogenously expressed in M1 cells. M1 clones infected with vector alone (LXSN) or LXSN expressing full length (FL), C-terminally truncated (CT), or N-terminally truncated c-Myb were either untreated (-) or treated (+) with human(h) recombinant IL-6 (100ng/ml) (PeproTech, Inc.) for 24 hr. Medium for growth and differentiation was RPMI 1640 with 10% horse serum. Immune precipitation was carried using polyclonal rabbit antisera direct toward GST-MybI, amino acids 1-325.

Evaluation of RNA expression by Northern analysis demonstrated that MyD88, a gene shown to be expressed early during differentiation, was upregulated in cells ectopically expressing all forms of the c-Myb protein. Results for the full length protein are shown in Figure 4. With deregulated expression of all forms of c-Myb the cells were able to only partially develop into macrophages and had an intermediate morphology consistent with the promonocytic phenotype observed for MML leukemias (Figure 5). Cell surface markers characteristic of macrophages showed varied expression. For example, expression of Fc and

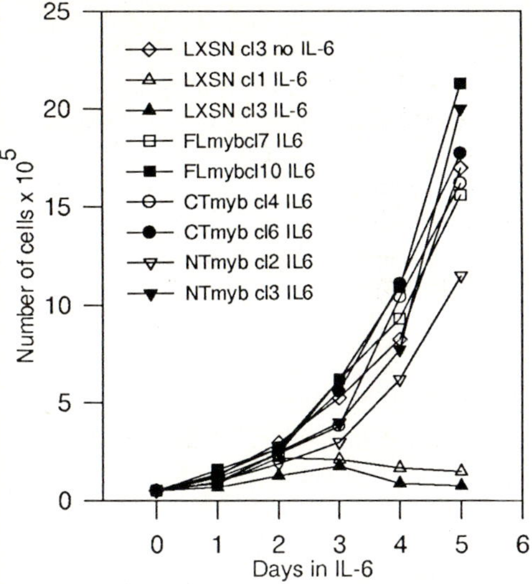

Figure 3. Cell growth curves of M1 cells ectopically expressing FLmyb, CTmyb, or NTmyb. hIL-6 (100 ng/ml) was added at day 0. Cell counts were prepared using trypan blue exclusion and performed in triplicate.

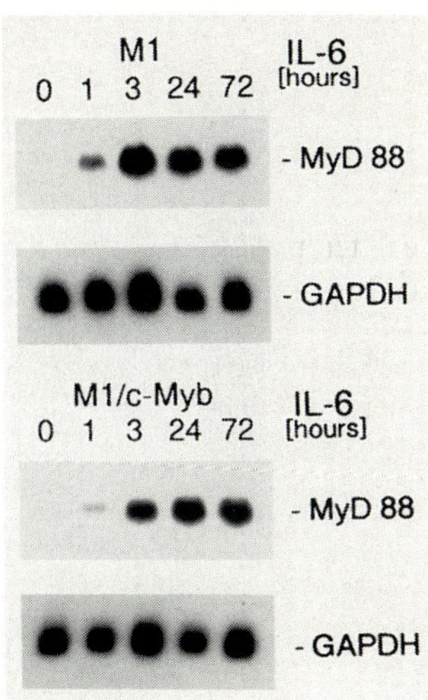

Figure 4. MyD88 expression in M1 cells and M1 ectopically expressing c-Myb. Cells were treated with hIL-6 for times indicated. Total RNA was prepared, separated by electrophoresis and hybridized with a MyD88 cDNA probe (27).

C3 receptors was intermediate compared to untreated M1 cells or M1 cells treated with IL-6 (Figure 6). Mac-1 was suppressed in cells infected with the Myb viruses while Mac-2 expression was unaffected (Figure 7). In all phenotypic marker studies, the expression was determined to be similar for all forms of the c-Myb, leukemia-specific truncated forms or the full length protein.

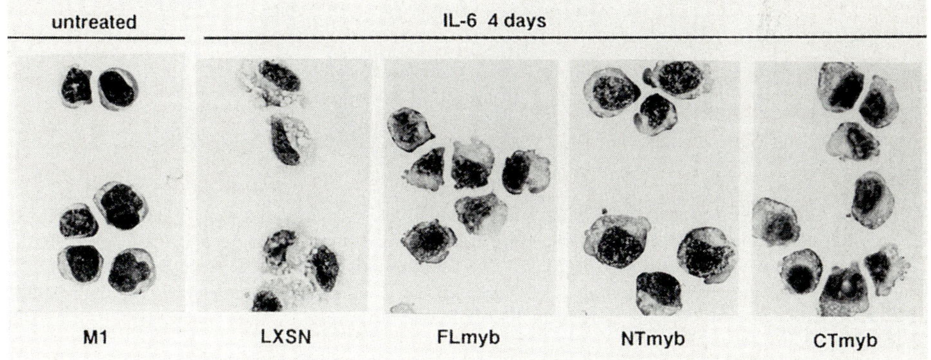

Figure 5. Morphology of M1 cells infected with FLmyb, NTmyb, and CTmyb containing retroviruses. Cells at a concentration of 5 x 10^4 were treated with hIL-6 for 4 days. Cells were prepared by cytospinning and stained with Diff-Quik (Scientific Products).

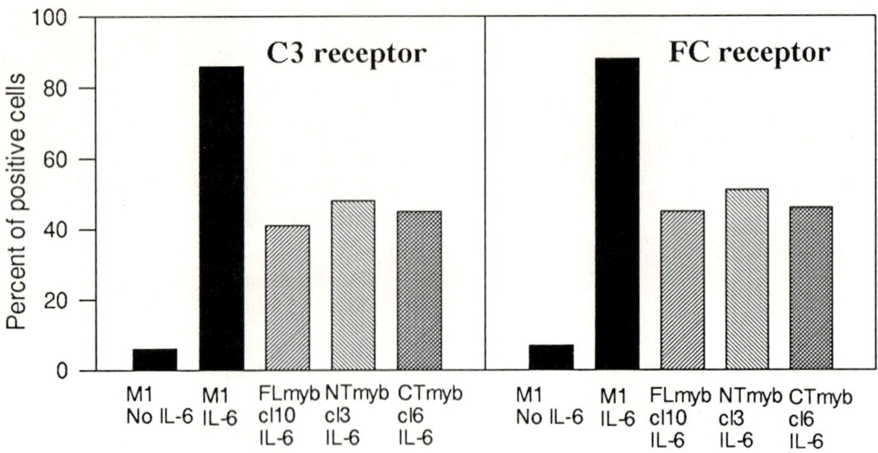

Figure 6. Induction of Fc and C3 receptors following IL-6 treatment of M1 cells constituitively expressing full length and truncated c-Myb proteins. Cells were seeded at 1.5 x 10^5 in the presence of IL-6 and the percentage of cells positive for the receptors was determined as previously reported (20) after 3 days. All values represent the means of three independent determinations.

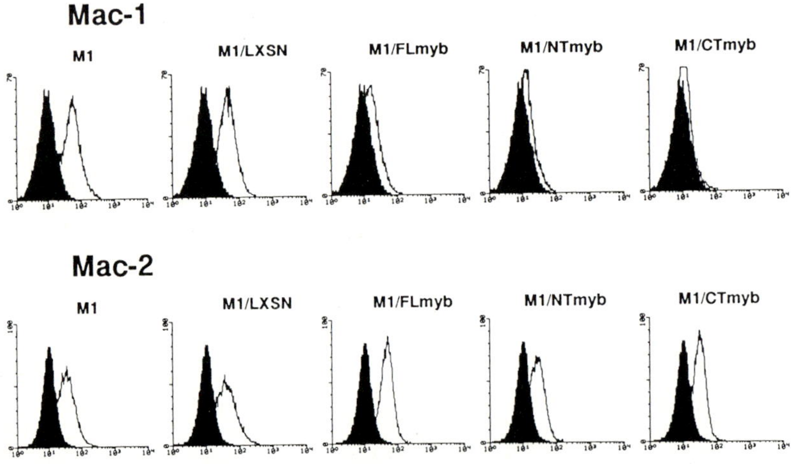

Figure 7. Cell surface expression of Mac-1 and Mac-2 following differentiation of M1 cells infected with c-*myb* containing retroviruses. Cells were prepared and analyzed using a FACSCAN (Becton Dickinson) as previously described (3). Mac-1 was detected with monoclonal antibody M1/70; Mac-2 was detected with monoclonal antibody 2.4G2 (Boehringer Mannheim). Data is shown as the relative number of cells versus fluorescence intensity. Black histograms represent the control M1 cells growing in the absence of IL-6.

Overall conclusions from differentiation studies. These results demonstrate that normal murine c-Myb does not maintain the progenitor state but rather blocks differentiation at intermediate to late steps in the maturation processes of myelo-monocytic cells. This is an observation which is consistent with studies of by Yanagisawa et al. carried out in WEHI 3B cells (21). Our results argue that inhibition of differentiation is due to c-Myb's ability to maintain the proliferative state of cells, but does not preclude c-Myb's possible other roles in driving maturation in early development or in granulocyte differentiation through regulation of lineage or stage-specific genes (22,23,24,25). Interestingly, M1/c-Myb cells treated with IL-6 are phenotypically similar to that of M1 cells overexpressing c-Myc (26,27) in that their differentiation state is intermediate. This correlates with the evidence that c-Myb upregulates c-Myc transcription (28-31). Our previous demonstration that deregulated expression of human c-Myb results in an earlier block in M1 cells (32) could be explained by species differences. Protein levels in our M1 overexpressing murine c-Myb were quite high suggesting effects observed for these genes from different species was not due to differences in protein levels. Although these results at first sight may seem inconsistent with the demonstration that avian v-*myb* can induce retro-differentiation in c-*myc* transformed avian macrophages (33), one must take into account the fact that avian and murine c-Mybs may function somewhat differently and there are mutations in the avian v-*myb* compared to its normal counterpart.

Mml1, a new common locus of retroviral integration in promonocytic leukemias

Although c-*myb* DNA is a rearranged in greater than 90% of promonocytic leukemias induced by Moloney MuLV or its derivatives in BALB/c mice, there are many such leukemias induced by other MuLVs that do not have c-*myb* rearrangements. This is particularly observed in promonocytic leukemias of DBA/2 mice induced by amphotrophic 4070A, FB29, and Friend MuLV C57(3,4,5). We have now identified a common site of integration, *Mml1*. A NheI fragment containing LTR and cellular flanking sequences from this locus was cloned using DNA from a tumor that did not have a *c-myb* rearrangement, but had a single proviral insertion. Flanking cellular sequences segregated from the proviral sequences in the cloned fragment were then used as a probe in Southern blot hybridization to determine if DNA from other leukemias had rearrangements at the same locus. We have now identified rearrangements of *Mml1* in at least ten MuLV-induced leukemias. The same probe was subsequently used to clone a 12kb segment of *Mml1* from normal liver cellular DNA and map the location of the integration sites as shown in Figure 8.

Chromosomal mapping studies, performed using progeny from interspecies backcross mice generated by mating (BALB/cAn x *M. spretus*)F_1 females to BALB/cAN males, has determined that *Mml1* is located on the proximal end of mouse chromsome 10. Interestingly, there were no recombinants between *c-myb* and *Mml1* in 102 backcross progeny. Despite this, our data suggest that Myb is not involved in leukemia development. Fragments from a 40kb region which include the new *Mml1* integration sites do not cross hybridize with probes from the 5' and 3' ends of the *c-myb* locus. Furthermore, c-Myb mRNA and Myb protein expression appears to be unaffected in leukemias with integrations in *Mml1*. Future studies will be aimed at looking for regions which are actively transcribed from this locus.

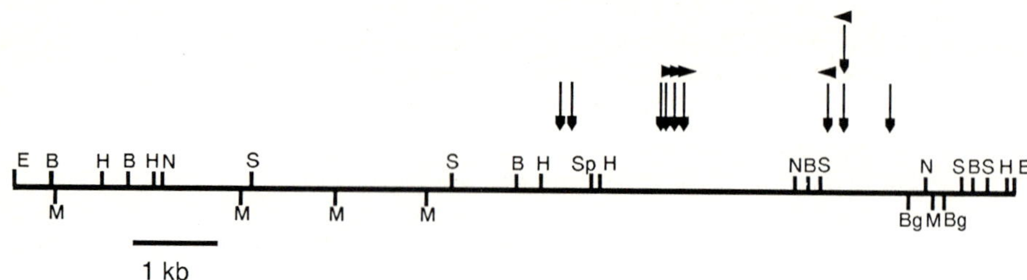

Figure 8. Restriction endonuclease map of cloned region of *Mml1* from normal DBA liver. Positions of integrated retroviruses in 10 leukemias are depicted above. Arrows show orientation of viruses. B,BamHI; E, Bg, BglII;EcoRI; H, HindIII; M, MscI; N,NheI; S,SstI; Sp, SpeI.

References

1. Shen-Ong GLC, Wolff, L (1987) Moloney murine leukemia virus-induced myeloid tumors in adult BALB/c mice: requirement of c-*myb* activation but lack of v-*abl* involvement. J Virol 61:3721-3725
2. Wolff L, Mushinski JF, Shen-Ong GLC, Morse III HC (1988) A chronic inflammatory response: its role in supporting the development of c-*myb* and c-*myc* related promonocytic and monocytc tumors. J Immunol 141:681-689
3. Wolff L, Koller R, Davidson W (1991) Acute myeloid leukmia induction by amphotropic murine retrovirus (4070A): Clonal integrations involve c-*myb* in some but not all leukemias. J Virol 65:3607-3616
4. Mukhopadhyaya R, Wolff L (1992) New sites of proviral integration associated with muine promonocytic leukemias and evidence for alternate mode of c-*myb* activation. J Virol 66:6035-6044
5. Nazarov V, Wolff L (1995) Novel integration sites at the distal 3' end of the c-*myb* locus in retrovirus-induced promonocytic leukemias. J Virol 69:3885-3888
6. Nason-Burchenal K, Wolff L (1993) Activation of c-myb is an early bone marrow event in a murine model for acute promonocytic leukemia. Proc Natl Acad Sci USA 90:1619-1623
7. Nazarov V, Hilbert D, Wolff L (1994) Susceptibility and resistence to Moloney murine leukemia virus-induced promonocytic leukemia. Virology 205:479-485
8. Mukhopadhyaya R, Richardson J, Nazarov V, Corbin A, Koller R, Sitbon M, Wolff L (1994) Different abilities of Friend murine leukmia virus (MuLV) and Moloney MuLV to induce promonocytic leukemia are due to determinants in both Ψ-*gag*-PR and *env* regions. J Virol 68:5100-5107
9. Nason-Burchenal K, and Wolff L (1993) Involvement of the spleen in preleukemic development of a murine retrovirus-induced promonocytic leukemia. Cancer Res 52:5317-5322
10. Belli B, Wolff L, Nazarov V, Fan H (1995) Proviral activation of the c-myb proto-oncogene in preleukemic mice infected neonatally with Moloney murine leukemia virus but not in resulting end-stage T-lymphomas J Virol, in press
11. Graf T (1992) Myb: a transcriptional activator linking proliferation and differentiation in hematopoietic cells. Current Opinion in Genet Devel 2:249-255
12. Introna M, Luchetti M, Castellano M, Arsura M, Golay J (1994) The *myb* oncogene family of transcription factors: potent regulators of hematopoietic cell proliferation and differentiation. Semin in Cancer Biol 5:113-124

13. Luscher B, Christenson E, Litchfield DW, Krebs EG, Eisenman RN (1990) Myb DNA binding is inbitited by phosphorylation at a site deleted during oncogenic activation. Nature 344:517-522
14. Bousset K, Oelgeschlager MHH, Henriksson M, Schreek S, Burkhardt H, Litchfield DW, Luscher-Firzlaff JM, Luscher B (1994) Regulation of transcription factors c-Myc, Max, and C-Myb by casein kinase II. Cell Mol Biol Res 40:501-511
15. Dini P, Lipsick JS (1993) Oncogenic truncation of the first repeat of c-Myb decreases DNA binding in vitro and in vivo. Mol Cell Biol 13:7334-7348
16. Sakura H, Kanei-Ishii C, Nakagoshi H, Gonda TJ, Ishii S (1989) Delineation of three functional domains of the transcriptional activator encoded by the c-*myb* protooncogene. Proc Natl Acad Sci USA 86:5758-5762
17. Kalkbrenner F, Guehmann S, Moelling K (1990) Transcriptional activation by human c-*myb*and v-*myb* genes. Oncogene 5:657-661
18. Dubendorff JW, Whittaker LJ, Eltman JT, Lipsick JS (1992) Carboxy-terminal elements of c-Myb negatively regulate transcriptional activation in *cis* and *trans*. Genes Dev 6:2524-2535
19. Bies J, Mukhopadhyaya R, Pierce J, Wolff L (1995) Only late, nonmitotic stages of granulocyte differentiation in 32cl3 cells are blocked by ectopic expression of murine c-*myb* and its truncated forms. Cell Growth Differ 6:59-68
20. Lord KA, Holffman-Liebermann B, Liebermann D (1990) Dissection of the immediate early response of myeloid leukemic cells to terminal differentiation and growth inhibitory stimuli. Cell Growth Differ 1:637-645
21. Yanagisawa H, Nagasawa T, Kuramochi S, Abe T, Ikawa J, Todokoro K (1991) Constitutive expression of exogenous c-*myb* gene causes maturation block in monocyte-macrophage differentiation. Biochim Biophysica Acta 1088:380-384
22. Burk O, Mink S, Ringwald M, Klempnauer K-H (1993) Synergisic activation of the chicken *mim-1* gene by v-*myb* and C/EBP transcription factors. EMBO 12:2027-2038
23. Ness SA, Kowenz-Leutz E, Casini T, Graf T, Leutz A (1993) Myb and NF-M: combinatorial activators of myeloid gene in heterologous cell types. Genes Devel 7:749-759
24. Nuchprayoon I, Meyers S, Scott LM, Suzow J, Hiebert S, and Friedman AD (1994) PEBP2/CBF, the murine homolog of the human myeloid AML1 and PEB2β/CBFβ proto-oncoproteins, regulates the murine myeloperoxidase and neutrophil elastase genes in immature myeloid cells. Mol Cell Biol 14:5558-5568
25. Shapiro L (1995) Myb and ets proteins cooperate to transactivate an early myeloid gene. J Biol Chem 270:8763-8771
26. Hoffman-Liebermann B, Liebermann DA (1991) Interleukin-6 and leukemia inhibitory factor-induced terminal differentiation of myeloid leukemia cells is blocked at an intermediate stage by constituitive c-*myc*. Mol Cell Biol 11:2375-2381
27. Selvakumaran D, Liebermann D, Hoffman-Liebermann B (1993) Myeloblastic leukemia cells conditionally blocked by Myc-estrogen receptor chimera transgenes for terminal differenteation coupled to growth arrest and apoptosis. Blood 81:2257-2262
28. Evans JL, Moore TL, Kuehl WM, Bender T, Ting FP-Y (1990) Functional analysis of c-Myb protein in T-lymphocytic cell lines shows that it *trans*-activates the c-*myc* promoter. Mol Cell Biol 10:5747-5752
29. Zobel A, Kalkbrenner F, Guehmann S, Nawrath M Vordbrueggen G, Moelling K (1991) Interaction of the v-and cMyb proteins with regulatory sequences of the human c-*myc* gene. Oncogene 6:1397-1407
30. Nakagoshi H, Kanei-Ishii C, Sawazaki T, Mizuguchi G, Ishii S (1992) Transcriptional activation of the c-*myc* gene by the c-*myb* and B-*myb* gene products. Oncogene 7:1233-1240
31. Cogswell JP, Cogswell PC, Kuehl WM, Cuddihy AM, Bender,TM, Engelke U, Marcu KB, Ting JP-Y (1993) Mechanism of c-*myc* regulation by c-Myb in different cell lineages. Mol Cell Biol 13:2858-2869
32. Selvakumaran D, Liebermann DA, Hoffman-Liebermann B (1992) Deregulated c-*myb* disrupts interleukin-6 or leukemia inhibitory factor-induced myeloid differentiation prior to c-*myc*: role in leukemogenesis. Mol Cell Biol 12:2493-2500
33. Burk O, Klempnauer K-H (1991) Estrogen-dependent alterations in differentiation state of myeloid cells cause by a v-*myb*/estrogen receptor. EMBO 10:3713-3719

Molecular Analysis and Characterization of Two Myeloid Leukemia Inducing Murine Retroviruses

E. RASSART, J. HOUDE, C. DENICOURT, M. RU, C. BARAT, E. EDOUARD, L. POLIQUIN and D. BERGERON

Laboratoire de biologie moléculaire, Département des sciences biologiques, Université du Québec à Montréal, Montréal, Canada

Introduction

Murine retroviruses (MuLVs) induce a large variety of hematopoietic tumors through complex mechanisms involving insertional activation of cellular proto-oncogenes in the tumor cell (for review see [1,2]). The type of tumor induced generally depends on sequences located in the viral enhancer [3-5] but also on coding sequences from other regions of the viral genome [6-8]. The Cas-Br-E MuLV is an ecotropic retrovirus that induces hindlimb paralysis in susceptible strains of mice [9]. It also induces mainly non-T, non-B-cell leukemia in infected NIH/Swiss mice [10-12]. The tumor cells show a very undifferentiated blast cell morphology which suggest that they are immature precursors.

The Graffi MuLV was first characterized by Graffi who initially reported in 1957 that cell-free supernatants of Ehrlich sarcoma and derivative tumors induced «chloroleukemias» with a high frequency in various inbred and outbred mouse strains [13]. However, he also observed a large spectrum of leukemic cell types after repeated passages of the putative viral agent; this phenomenon was described as «hematological diversification» [14]. The leukemic cell type observed in these experiments was influenced by the tumorigenic tissue used for the viral extract preparation. These studies which have been limited to morphological and histochemical analysis of the leukemic cell phenotype suggested that the Graffi virus could be a mixture of viral components [15].

1. The Cas-Br-E MuLV

Histopathology of the tumors.

Newborn NIH/Swiss mice were inoculated intraperitoneally with a molecular clone of Cas-Br-E [11,16]. Over the next months, the mice developed several pathologies as described in Table 1. Tumors could be divided into two distinct groups based on gross pathology and latency periods. Tumors classified as non-T, non-B leukemias were characterized by an enlargement of the spleen after a mean latency period of 120 d., accompanied by a severe anemia. Thirty-three tumors were examined and showed no gene rearrangement of the T-cell receptor ß chain neither of the immunoglobulin heavy-chain regions [11]. Histopathological analysis of the tumor cells revealed that they had an immature blast cell morphology with high nuclear-to-cytoplasmic ratios. Analysis of blood smears from 3 tumors showed that the cells were negative for Sudan black, myeloperoxydase and Schiff periodic acid. In one experiment, in which we used a lower titer of Cas-Br-E MuLV, we obtained a larger proportion of myeloid leukemia (50%) with a longer latency. These were characterized by an enlargement of the lymph nodes and/or the thymus with a mean latency period of 210 days. All these findings are consistent with the original description of these tumors as "non-T, non-B cells" [9-11]. They are also very similar to those obtained with the 10A1 MuLV [17,18] and are typical of early, uncommitted precursor cells very close to the stem cell since they still express the c-Kit mRNA [18].

Table 1. Types of leukemia induced by NE-8-Cas-Br-E

	No of mice	Gross pathology	Latency (month)
non-T, non-B leukemia	63	splenomegaly	3-5 / 4-8[a]
T-lymphoma	1	splenomegaly lymphadenopathy gross thymic enlargement	4
myeloid leukemia	5	splenomegaly lymphadenopathy	6 to 8.5
	2	splenomegaly lymphadenopathy gross thymic enlargement	5
	1	lymphadenopathy gross thymic enlargement	7
	1	gross thymic enlargement	9

[a] Two groups of mice were infected with virus at different titers. 53 tumors were obtained with a latency of 3-5 months (virus titer: 5×10^5 PFU/ml) and 10 tumors with a latency of 4-8 months (virus titer: 1×10^5 PFU/ml).

DNA rearrangements in Cas-Br-E-induced tumors

To identify proto-oncogenes associated with transformation, the DNA from 63 non-T, non-B lymphomas were analyzed by Southern blot hybridization for the presence of rearrangement in the genomic regions coding for p53, Evi-1, Evi-2, Pim-1, c-myc, c-myb, EpoR, Spi-1/Pu-1, Fim-1, Fim-2, Fim-3 and Fli-1. We found that Fli-1, p53 and Evi-1 were rearranged in respectively 72%, 23% and 18% of the tumors. As shown in Fig. 1, Evi-1 rearrangements were not associated with p53 or Fli-1 alterations. However, two tumors appeared rearranged for both Fli-1 and Evi-1 but they were oligoclonal, each cell population showing a unique rearrangement. In contrast, p53 and Fli-1 genes are often rearranged within the same tumor. The percentage of double rearrangements could be even higher since we looked for large DNA rearrangements only and not single point mutations.

The Fli-1 locus was originally identified as a viral integration site by Ben-David et al [19] in Friend MuLV-induced erythroleukemias and by us in Cas-Br-E MuLV-induced non-T, non-B lymphomas [11]. The Friend-MuLV-induced tumors were composed of committed erythroid cells as they expressed EpoR and ß-Globin mRNAs. As depicted in Fig. 2A, Friend-MuLV proviruses were integrated within a 2 Kbp genomic region upstream of the Fli-1 coding region and in the opposite transcriptional orientation. In contrast, Cas-Br-E viral integrations were all clustered in a very narrow DNA region (less than 100 bp) in the same transcriptional orientation as the Fli-1 gene. The retrovirus integrates in the Fli-1 exon 1, few nucleotides upstream of the ATG start codon for the Fli-1 gene product [20]. Interestingly, the 10A1 MuLV was recently associated with Fli-1 activation in murine leukemia and it induces tumors remarkably similar to those induced by Cas-Br-E both in their morphology and in the absence of T- or B-cell DNA rearrangements [11,18]. Cell surface analysis has revealed that the morphology of 10A1 MuLV-induced tumors is consistent with proerythroblastic, promyeloblastic, and undifferentiated blastic cell types, negative for 15 lineage-specific markers [18]. The virus was found integrated in the same cluster and orientation as the Cas-Br-E MuLV in Fli-1 exon 1. Most likely, both MuLVs activate Fli-1 transcription from their 3'LTR according to a promoter insertion mechanism even though their LTR sequence is 96% homologous only.

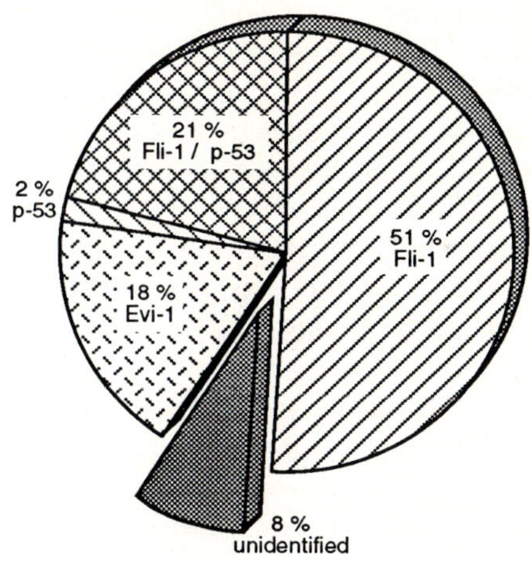

Fig. 1. Distribution of the DNA rearrangements in the Cas-Br-E-induced non-T-, non-B-cell leukemias.

As illustrated in Fig. 2B, Cas-Br-E integrations were mapped in the Evi-1 gene for 7

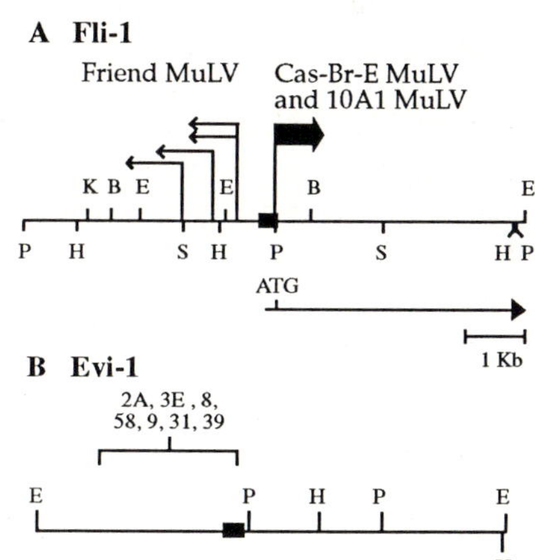

Fig. 2. Position of viral integrations in the Fli-1 and Evi-1 regions. A) Vertical lines indicate the integration sites for Cas-Br-E or Friend MuLVs in the Fli-1 locus. Horizontal arrows indicate one proviral orientation and the thick arrow represents 45 proviral integrations of Cas-Br-E ecotropic and MCF recombinant MuLVs. B) Bracket indicates the integration sites of ecotropic proviruses in the Evi-1 locus. Black boxes indicate the position of the postulated Fli-1 exon 1 and of the Evi-1 exon 2. B, BamHI; E, EcoRI; H, HindIII; K, KpnI; P, PstI; S, SacI.
This Fig. is modified from Fig. 2 in [12].

tumors [20]. They are all located upstream the exon 2, in the main viral integration cluster previously described in myeloid leukemias induced by Cas-Br-E and Moloney MuLVs [21]. In these tumors, Northern blot analysis revealed that Evi-1 transcripts were present, but detected as a smear extending from 1 to 7 Kb [12]. As reported in cell lines derived from Cas-Br-E-induced tumors, the diffuse nature of Evi-1 transcripts may result from differential splicing of retrovirus-Evi-1 RNA sequences [21]. Interestingly, the proviruses found in the Evi-1 locus were all ecotropic in contrast with the Fli-1 locus, where 50% of proviruses were MCF recombinants and 50% were ecotropic MuLV [20].

Proto-oncogene expression in Non-T, non-B lymphomas

We performed Northern blot analysis on the total RNA from Cas-Br-E induced, non-T, non-B tumors to analyse the expression of several proto-oncogenes. Tumors showing DNA rearrangements for Fli-1, p53, Fli-1/p53 and Evi-1 were chosen. As shown in Fig. 3, high expression of Evi-1 was detected only in the Evi-1 rearranged tumors. In these tumors, only low levels of Fli-1, p53, c-myc and Pim-1 were detectable. On the other hand, all Fli-1 and/or p53 rearranged tumors did not synthesize Evi-1 transcripts but expressed much higher levels of Fli-1, p53, c-myc and Pim-1 mRNAs. Interestingly, these four genes have been reported to be rearranged in transplanted erythroid tumors induced by the Friend MuLV [22].

Only 2 out of 63 non-T, non-B tumors showed a DNA rearrangement for p53 in the absence of Fli-1 alteration. Surprisingly, in these two tumors, Northern blot analysis revealed a 2,4 Kb transcript in addition to the normal 2,8 Kb Pim-1 transcript, in the absence of any Pim-1 DNA rearrangement (Fig. 4B). The serine/threonine protein kinase Pim-1 was reported to be frequently activated in Moloney MuLV-induced T cell lymphomas. In these tumors, the major cluster of viral integration was mapped in the non-translated region of the last exon of Pim-1 [23], a region associated

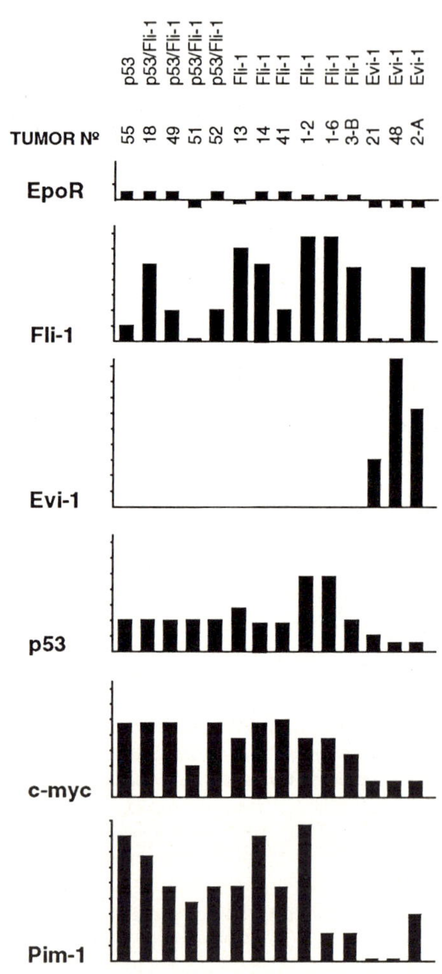

Fig. 3. Densitometric quantification of mRNA expression in tumors rearranged for p53, Fli-1 and Evi-1. The quantification was normalized by hybridization with the ß-actin probe. The data are plotted relative to the level in spleen (base line). This Fig. is from [12].

with sequences confering instability to Pim-1 mRNA [24]. In Cas-Br-E-induced tumors, the activation of Pim-1 expression is not the result of viral integration but rather of an indirect mechanism possibly associated with the activation of Fli-1 expression. In the two tumors rearranged for p53 but not for Fli-1, sufficient activation of Pim-1 may occur through the expression of the shorter mRNA. Indeed, in rat testes germ cells, the shorter 2,4 Kb Pim-1 transcript, which arises through the use of an alternate polyadenylation signal, was shown to be extremely stable [25]. These findings suggest that the expression of Fli-1 and Pim-1 may be related. Consistent

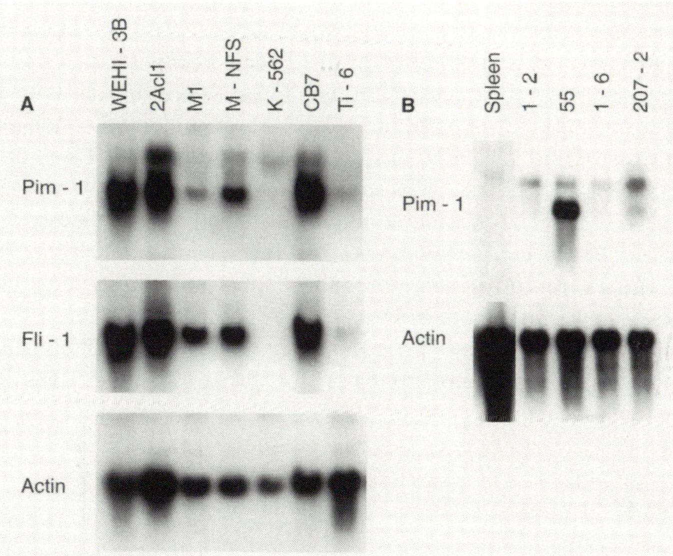

Fig. 4. Northern blot analysis of hematopoietic cell lines (A) and Cas-Br-E-induced tumors (B). Total RNAs from erythroid (K-562, CB7), myeloid (WEHI-3B, M1, M-NFS), lymphoid (Ti6) cell lines, Cas-Br-E-induced tumors (2ACl1, 1-2, 55, 1-6, 207-2), and normal spleen were successively hybrized with the probes indicated in the left of each panel.

with this hypothesis, analysis of various hematopoietic cell lines revealed a perfect correlation in the level of expression of these two genes (Fig. 4). It is tempting to speculate that, in the absence of Fli-1, the shorter Pim-1 transcript is necessary for tumor progression and development.

In conclusion, it seems that our non-T, non-B tumors can be divided in two categories, one with Evi-1 activation and the other with Fli-1 activation accompanied by enhanced expression of Pim-1 and frequently with DNA rearrangements of p53. Evi-1 has been shown to repress GATA-1-dependent transactivation [26] which itself is essential for erythroid differentiation [27]. It is therefore possible that the Cas-Br-E infects immature progenitor cells and when integrated in the Evi-1 locus, enhanced Evi-1 expression may block erythroid factors thus promoting myeloid characteristics of the tumor cells. However, when integration occurs in the Fli-1 locus, both Fli-1 and Pim-1 become activated and the tumor cells maintain their undifferentiated phenotype or progress toward proerythroblast cells.

2. The Graffi MuLV

Molecular cloning and restriction map

We took advantage of the ability of a probe derived from the U3 portion of Moloney-MuLV [28] to detect and clone proviruses from a genomic library of Graffi MuLV newly infected NIH 3T3 cells. Two types of molecular clones, GV-1.2 and GV-1.4, corresponded to complete, apparently non-defective retroviruses [15]. Two other types were defective molecules or MCF recombinants. Clones GV-1.2 and GV-1.4 were digested with several restriction enzymes and analyzed by Southern transfer and hybridization with probes derived from different portions of the Moloney-MuLV genome. A summary of their restriction maps is shown in Fig. 5. Clone GV-1.2 is very similar to clone GV-1.4 except for its LTR which is about 60 nucleotides longer and for a few restriction sites. The nucleotide sequence of the LTR of clone GV-1.2 showed a 60 bp perfect direct repeat in the U3 region which is present in only one copy in clone GV-1.4. The restriction maps of these clones are clearly distinct from those of Moloney, Friend, Rauscher, Cas-Br-E MuLVs, Kaplan RadLV and endogenous viruses (ecotropic, xenotropic and amphotropic) suggesting that the Graffi virus is a novel leukemogenic retrovirus [15]. A comparison of the LTR sequences with those of other retroviruses showed a best homology between clone GV-1.4 and the amphotropic 4070 MuLV and Cas-Br-E MuLV with 92% and 91,5% of homology respectively.

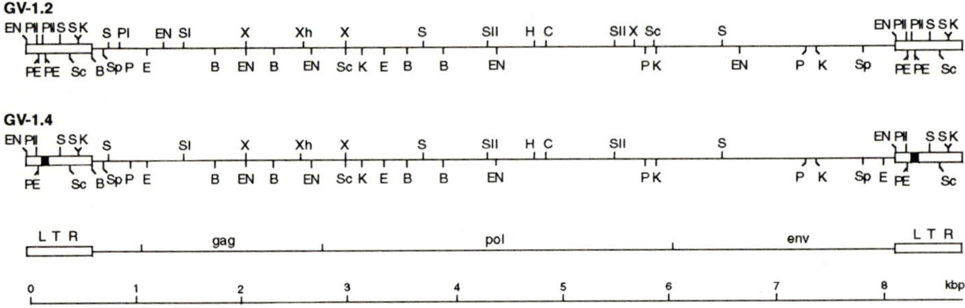

Fig. 5. Restriction endonuclease map of Graffi MuLV clone GV-1.2 and GV-1.4. B: BamHI, C: ClaI, E: EcoRV, EN: EcoNI, H: Hind III, K: KpnI, P: PstI, PI: PvuI, PII: PvuII, S: SmaI, SI: SalI, SII: SacII, Sc: SacI, Sp: SpeI, X: XbaI, Xh: XhoI.

Biological and clinical characteristics of the cloned Graffi-MuLVs

Infectious retroviruses, recovered following transfection of clones GV-1.2 and GV-1.4 DNAs, were ecotropic, XC positive and NB tropic. They also grew well on Rat-1 but not on Mink cells. They were injected intraperitoneally into newborn Balb/c and NFS mice. All mice inoculated with GV-1.2, GV-1.4 and the uncloned parental virus mixture (GV) developed leukemia. Leukemic cells began to appear in peripheral blood by 4 weeks post infection. Gross post mortem examination of the animals revealed splenomegaly and generalized lymphadenopathy often accompanied with thymic and mesenteric lymph node enlargements. Peripheral blood smears contained predominantly myeloid blasts with rare differentiation to the promyelocyte stage or a hiatus to mature neutrophils. Immature forms predominated in tissue imprints of lymph nodes, spleen and thymus. Approximately 15-20% of blasts showed positive staining for myeloperoxidase and 60-70% staining for naphtol AS-D chloroacetate esterase. Interestingly, we obtained chloroleukemias (greenish coloration of lymph nodes reflecting endogenous myeloperoxidase activity) with clone GV-1.4 exclusively in about 10% of the diseased animals.

The cumulative incidence of disease is shown in Fig. 6. Identical results were also obtained in NIH/Swiss mice (data not shown). The latency was significantly longer for clone GV-1.4 in the three strains of mice compared to clone GV-1.2 or to parental GV virus.

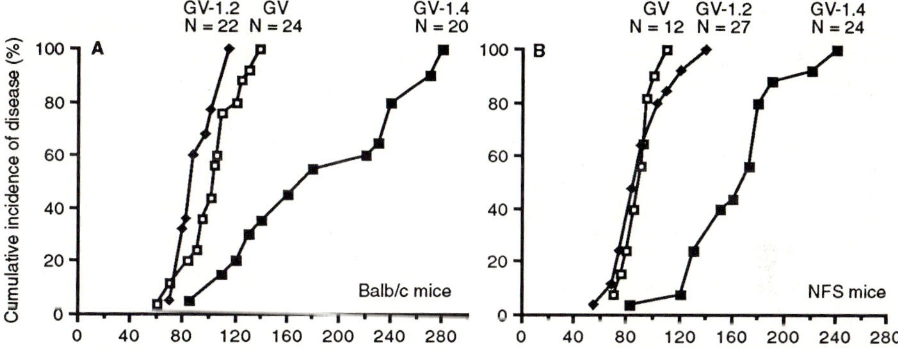

Fig. 6. Cumulative incidence of disease (in days) induced by parental Graffi MuLV (GV), and molecular clones GV-1.2 and GV-1.4. Newborn Balb/c mice (A) and NFS mice (B) were inoculated intraperitoneally with $5X10^4$ PFU. Animals were sacrificed when they showed signs of advanced disease.

Since most mice developed thymic and lymph node enlargement in addition to splenomegaly, the leukemic cell phenotype was further characterized by the analysis of cell surface antigens and molecular markers. The tumor cells were negative for Thy-1.2, surface Ig and Ly-5 markers. Southern blot analysis of tumor DNAs with IgH and TCR-ß gene probes revealed DNA rearrangements in certain tumors with either one or both probes (data not shown). In some cases, the rearranged allele was not equimolar when compared to the germ line allele. However, we confirmed the clonality of the tumors with a Graffi-derived U3LTR probe as we could detect by Southern analysis a discrete pattern of bands for each tumor DNA. All leukemias which demonstrated mixed lineage markers showed a clonal pattern of viral integration despite non-equimolar rearrangement of IgH or TcR genes, similar to findings in human chronic myeloid leukemia [29-31]. Indeed, concomitant expression of myeloid and lymphoid markers has been observed in several human leukemias such as acute and chronic myeloid leukemia and acute lymphocytic leukemia [31].

DNA rearrangements in Graffi-induced tumors

To identify proto-oncogenes associated with Graffi-MuLV-induced leukemias, we analysed 30 tumors for DNA rearrangements of genes that are related to the myeloid lineage. Thus, we have tested Evi-1, Evi-2, Evi-3, Evi-4, c-myb, Fim-1, Fim-2, Fim-3, Fis-1, Meis-1 and Spi-1/Pu-1. One tumor only was found rearranged for Spi-1/Pu-1 and none of the other genes tested was found rearranged. This suggests that potentially new genes are involved in Graffi-induced leukemias.

Analysis of the U3 LTR region of Cas-Br-E and Graffi MuLVs

Both retroviruses are capable of causing myeloid leukemia in mice. However, Cas-Br-E MuLV induces usually a low percentage of myeloid disease in mice. In fact, in two series of experiments, we obtained a higher incidence of myeloid leukemia when a lower virus titer was injected to the animals. In contrast, the Graffi MuLV induces 100% of myeloid leukemia when injected in three different strains of susceptible mice. This suggests that the target progenitor cell might be different for these two retroviruses. We were therefore interested to compare the LTR sequences of these two viruses in light of potential regulatory sequences recognized by specific DNA binding proteins. The results are summarized in Fig. 7. As shown, a large number of elements identified in Moloney and Friend MuLV U3 regions are also present in Cas-Br-E and Graffi U3 sequences. Indeed, Cas-Br-E LTR is highly homologous to GV-1.4 LTR (91.5%) but also to Moloney LTR (91%). The Moloney MuLV causes primarily T lymphomas, and sequences in its U3 region, in particular the core and LVb regions, have been shown to be the primary determinant for disease specificity [32]. Cas-Br-E and Graffi MuLV U3 sequences are very similar to those of Moloney in these regions although they do not induce T lymphomas in mice. However, knowing

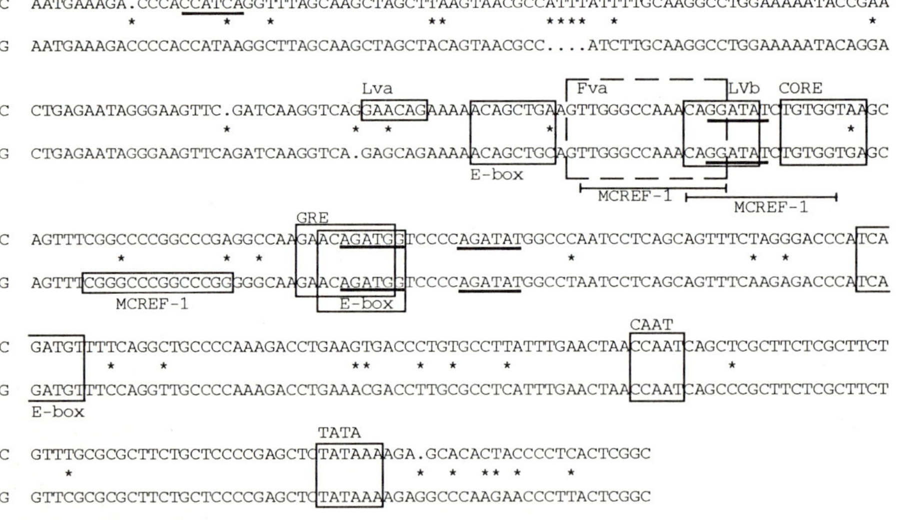

Fig. 7. Alignment of U3 LTR sequences of Cas-Br-E MuLV [16,20] and Graffi MuLV clone GV-1.4 [15]. The regulatory elements including the CAAT box and promotor (TATA) are boxed or underlined. Mismatch is represented by a *. LVa, LVb, core, GRE:binding sites identified in Moloney MuLV U3 region [32]; FVa:Binding site identified in Friend MuLV U3 region [32,35]; MCREF-1: Mammalian type C retrovirus enhancer factor-1 [35]; E-box:binding sites for basic helix-loop-helix transcription factors identified in U3 regions of Friend, Moloney, SL3-3 and AKV MuLVs [36]; GATA:binding sites for members of GATA protein family [34].

that a multitude of factors bind to U3 region and regulate transcription in a tissue-specific fashion, the few nucleotide differences between Cas-Br-E, GV-1.4 and Moloney U3 sequences could be sufficient to alter the very complex assembly of binding factors, thus changing the tissue specificity of the enhancer. For example, the sequence in the core region could be important, since a single nucleotide of the SL3-3 MuLV LTR was shown to be crucial [32,33]. Also, as another example, the GATA elements could be of importance as we have shown that members of this family are able to bind elements in Graffi and Cas-Br-E LTRs (Barat and Rassart, unpublished results), even if they do not match perfectly the GATA consensus element [34]. Alternatively, the disease determinant could be situated outside of the LTR, for example in *gag* or *env* coding regions. In this regard, the leukemogenic potential of Cas-Br-E was shown to be dispersed in the entire genome as any region of its genome could render the ampho4070 MuLV leukemogenic [7].

Additional studies including the construction of chimeric viruses between Cas-Br-E, Graffi and non-leukemogenic retroviruses, as well as site-directed mutagenesis, will be required to understand the molecular basis of their disease specificity. Moreover, the analysis of the various factors binding to Cas-Br-E and Graffi U3 regions will allow the identification of putative myeloid specific factors which may explain the targeting in myeloid progenitors.

References

1. Van Lohuizen M, Berns A (1990) Tumorigenesis by slow-transforming retroviruses - an update. Biochim Biophys Acta 1032:213-235
2. Kung HJ, Boerkoel C, Carter TH (1991) Retroviral mutagenesis of cellular oncogenes: a review with insights into mechanisms of insertional activation. Curr Top Microbiol Immunol 171:1-25
3. Chatis PA, Holland CA, Hartley JW, Rowe WP, Hopkins N (1983) Role of the 3'end of the genome in determining disease specificity of Friend and Moloney murine leukemia viruses. Proc Natl Acad Sci USA 80:4408-4411.
4. DesGroseillers L, Rassart E, Jolicoeur P (1983) Thymotropism of murine leukemia virus is conferred by its long terminal repeat. Proc Natl Acad Sci USA 80:4203-4207.
5. Li Y, Golemis E, Hartley JW, Hopkins N (1987) Disease specificity of nondefective Friend and Moloney murine leukemia viruses is controlled by a small number of nucleotides. J Virol 61:693-700
6. Holland CA, Hartley JW, Rowe WP, Hopkins N (1985) At least four viral genes contribute to the leukemogenicity of murine retrovirus MCF 247 in AKR mice. J Virol 53:158-165.
7. Jolicoeur P, Desgroseillers L (1985) Neurotropic Cas-Br-E murine leukemia virus harbors several determinants of leukemogenicity mapping in different regions of the genome. J Virol 56:639-643.
8. Oliff A, Signorelli K, Collins L (1984) The envelope gene and long terminal repeat sequences contribute to the pathogenic phenotype of helper-independent Friend viruses. J Virol 51: 788-794.
9. Gardner MB, (1978) Type C viruses of wild mice: characterization and natural history of amphotropic, ecotropic, and xenotropic MuLV. Curr Top Microbiol Immunol 79:215-239
10. Bryant ML, Scott JL, Pall BK, Estes JD, Gardner MB (1981) Immunopathology of natural and experimental lymphomas induced by wild mouse leukemia virus. Am J Pathol 104:272-282
11. Bergeron D, Poliquin L, Kozak CA, Rassart E (1991) Identification of a common viral integration region in Cas-Br-E murine leukemia virus-induced non-T-, non-B-cell lymphomas. J Virol 65:7-15
12. Bergeron D, Houde J, Poliquin L, Barbeau B, Rassart E (1993) Expression and DNA rearrangement of proto-oncogenes in Cas-Br-E-induced non-T, non-B-cell leukemias. Leukemia 7:954-962
13. Graffi A (1957) Chloroleukemia in mice. Ann NY Acad Sci 68:540-558

14. Graffi A, Fey F, Schramm T (1966) Experiments on the hematologic diversification of viral mouse leukemias. Natl Cancer Inst Monogr 22:21-31
15. Ru M, Shustik C, Rassart E (1993) Graffi murine leukemia virus: molecular cloning and characterization of the myeloid leukemia-inducing agent. J Virol 67:4722-4731
16. Rassart E, Nelbach L, Jolicoeur P (1986) Cas-Br-E murine leukemia virus: sequencing of the paralytogenic region of its genome and derivation of specific probes to study the origin and structure of its recombinant genomes in leukemic tissues. J Virol 60:910-919
17. Ott DE, Keller J, Sill K, Rein A (1992) Phenotypes of murine leukemia virus-induced tumors: influence of 3'viral coding sequences. J Virol 66:6107-6116
18. Ott DE, Keller J, Rein A (1994) 10A1 MuLV induces a murine leukemia that expresses hematopoietic stem cell markers by a mechanism that includes Fli-1 integration. Virol 205:563-568
19. Ben-David Y, Giddens E, Bernstein A (1990) Identification and mapping of a common proviral integration site Fli-1 in erythroleukemia cells induced by Friend. Proc Natl Acad Sci. USA 87:1332-1336
20. Bergeron D, Poliquin L, Houde J, Barbeau B, Rassart E (1992) Analysis of proviruses integrated in Fli-1 and Evi-1 regions in Cas-Br-E MuLV-induced non-T, non-B-cell leukemias. Virol 191:661-669
21. Morishita K, Parker DS, Mucenski ML, Jenkins NA, Copeland NG, Ihle JN (1988) Retroviral activation of a novel gene encoding a zinc finger protein in IL-3-dependent myeloid leukemia cell lines. Cell 54: 831-840
22. Dreyfus F, Sola B, Fichelson S, Varlet P, Charon M, Tambourin P, Wendling F, Gisselbrecht S (1990) Rearrangements of the Pim-1, c-myc and p53 Genes in Friend helper virus-induced mouse erythroleukemias. Leukemia. 4:590-594.
23. Selten G, Cuypers HT, Boelens W, Robanus-Maandag E, Verbeek J, Domen J, Van Beveren C, Berns A (1986) The primary structure of the putative oncogene Pim-1 shows extensive homology with protein Kinases. Cell 46:603-611
24. Wingett D, Reeves R, Magnuson NS (1991) Stability changes in Pim-1 proto-oncogene messenger RNA after mitogen stimulation of normal lymphocytes. J Immunol 147:3653-3659
25. Wingett D, Reeves R, Magnuson NS (1992) Characterization of the testes-specific Pim-1 transcript in rat. Nucleic Acids Res 20:3183-3189
26. Kreider BL, Orkin SH, Ihle JN (1993) Loss of erythropoietin responsivness in erythroid progenitors due to expression of the Evi-1 myeloid-transforming gene. Proc Natl Acad Sci USA 90:6454-6458
27. Pevny L, Simon MC, Robertson E, Klein WH, Tsai SF, D'Agati V, Orkin SH, Constantini F (1991) Erythroid differentiation in chimaeric mice blocked by a targerted mutation in the gene for transcription factor GATA-1. Nature 349:257-260
28. Rassart E, Paquette Y, Jolicoeur P (1988) Inability of Kaplan Radiation Leukemia Virus To Replicate on Mouse Fibroblasts Is Conferred by Its Long Terminal Repeat. J Virol 62:3840-3848
29. Adriaansen HJ, Soeting PWC, Wolvers-Tettero ILM, van Dongen JJM. (1991) Immunoglobulin and T-cell receptor gene rearrangements in acute non-lymphocyte leukemias. Analysis of 54 cases and a review of the literature. Leukemia 5:744-751.
30. di Cells PF, Cabone A, Lo Coco F, Guerrasio A, Diverio D, Rosso C, Lopez M, Saglio G, Foa R (1991) Identical utilization of T-cell receptor gene regions in B-lymphoid blast crisis of chronic myeloid leukemia and B-precursor acute lymphoblast leukemia. Leukemia 5:366-372
31. Saikevych IA, Kerrigan DP, McConnell TS, Head DR, Appelbaum FR, Willman CL (1991) Multiparameter analysis of acute mixed lineage leukemia: Correlation of a B/myeloid immunophenotype and immunoglobulin and T-cell receptor gene rearrangements with the presence of the Philadelphia chromosome translocation in acute leukemias with myeloid morphology. Leukemia 5:373-382
32. Speck NA, Renjifo B, Golemis E, Frederickson TN, Hartley JW, Hopkins N (1990) Mutation of the core and adjacent LVb elements of the Moloney murine leukemia virus enhancer alters the disease specificity. Genes Develop 4:233-242
33. Boral AL, Okenquist SA, Lenz J (1989) Identification of the SL3-3 enhancer core as a T-lymphoma cell-specific element. J Virol 63:76-84
34. Merika M, Orkin SH (1993) DNA-binding specificity of GATA family transcription factors. Mol Cell Biol 13:3999-4010
35. Manley NR, O'Connell M, Sun W, Speck NA, Hopkins NA (1993) Two factors that bind to highly conserved sequences in mammalian type C retroviral enhancers. J Virol 67:1967-1975
36. Nielson AL, Pallisgaard N, Pedersen FS, Jorgensen P (1994) Basic helix-loop-helix proteins in murine type C retrovirus transcriptional regulation. J Virol 68:5638-5647

Molecular Analysis of *Evi*1, a Zinc Finger Oncogene Involved in Myeloid Leukemia

M.C. Lopingco and A. S. Perkins
Department of Pathology, Yale University School of Medicine, P.O. Box 208023, New Haven, CT, 06520-8023

Introduction

Our molecular understanding of myeloid leukemogenesis has come largely from the identification and characterization of genes whose alteration, through deletion, translocation, mutation, or retroviral insertion contributes to this multistep process. In this pursuit, several mouse models have contributed significantly, including the BXH-2 [1] and AKXD23[2] recombinant inbred strains, as well as retroviral infection of a number of different strains. These studies have led to the identification of *Evi1*[3], *Nf1*[4], *Fli1*[5], *Spi1*[6], *Hox2.4*[7], and, more recently, *Meis1*[8] and *Hoxa9*[9] as genes involved in myeloid leukemogenesis. This review focuses on *Evi1*, a myeloid leukemia-specific site of retroviral insertion that encodes a zinc finger transcription factor.

Identification Of Evi1

Evi1 is a common site of retroviral insertion in murine myeloid leukemias, and has been found in tumors arising in the AKXD23 strain of mice[10], as well as in Cas-Br-E virus-induced tumors in NIH/Swiss[11] and their inbred derivative, NFS[10]. Tumors in Cas-Br-E-infected animals contain both ecotropic and MCF recombinant viruses. Interestingly, all of the proviral insertions at *Evi1* were ecotropic, whereas other sites of insertion, such as *Fli1*, have both ecotropic and MCFs [11]; the reason for this specificity is not known. While *Evi1* insertions are predominantly myeloid-specific, some were found as well in a small number of B and pre-B cell tumors [10]. Elucidation of the intron-exon structure of the 5' end of the gene revealed that the proviral insertions were in both transcriptional orientations, and were tightly clustered, occurring within a 2 kb stretch of DNA either upstream of exon one or between exon one and exon two. It is believed that the first coding exon is the third, since this is the first ATG on the mRNAs described to date, and is consistent with the size of observed protein products for the gene[12].

A distinct site of retroviral insertion was documented in myeloid tumors in mice, termed *Cb1/Fim3*[13], which by genomic mapping was localized near the *Evi1* locus. Proviral integrations at this site result in high expression of *Evi1*[14], yet analysis of the genomic organization of the *Cb1/Fim3/Evi1* region revealed that *Cb1/Fim3* is 90 kb upstream of the first exon of *Evi1*. Since there is no evidence of an exon in the vicinity of *Cbl/Fim3*, it is currently thought that these insertions act from a distance to activate *Evi1* transcription by an enhancer-insertion mechanism[15].

Both the retroviral insertions and the translocations (see below) at *Evi1* result in transcriptional activation of the gene, thus indicating that the gene acts as a dominantly acting oncogene in leukemia. Interestingly, most AMLs bearing *EVI1*-activating alterations are of the M0, M1, or M2 class, and are usually CD34 positive, suggesting an immature phenotype. Cell lines with such alterations are dependent on hematopoietic growth factors. These characteristics indicate that *Evi1* does not abrogate growth factor requirements [3] [16].

Involvement Of Evi1 In Human Disease

Rearrangements of *Evi1* have also been documented in human myelodysplasias and leukemias, indicating its involvement in human disease [16-22]. The gene is located on human chromosome 3q26 [23]. This is a region of nonrandom chromosomal rearrangement in a small subset of both myelodysplastic syndromes and myeloid leukemias [24-26]. There is a 3q21q26 myelodysplastic syndrome that is characterized by normal or elevated platelet counts, hyperplasia with dysplasia of megakaryocytes, multilineage involvement, short duration of the myelodysplastic stage, and a poor prognosis [26].

The involvement of *EVI1* rearrangements in MDS and AML has emerged from several studies. One study, involving 116 patients with human AML, eight expressed *EVI1*, and seven of these had cytogenetically detectable translocations or inversions of chromosome 3q26. In numerous cases, the site of rearrangement has been documented by pulsed field gel analysis, and these range from 13 kb 5' of the gene[16] to 400-kb region distal to the *EVI1* locus[20].

In cases with inv(3)(q21;q26) and t(3;3)(q21;q26), *EVI1* is located near the q26 breakpoint, while the ribophorin gene is located near the q21 breakpoint [21]. Ribophorin is a ubiquitously expressed gene that may provide enhancers for *EVI1* transcriptional activation. The size of the *EVI1* transcript in these cells, 6.0kb, is comparable to that seen in normal human kidney or ovary (normal sites of *EVI1* expression), suggesting that there is no significant alteration in the mRNA structure. The 3q26 breakpoint location relative to *EVI1* depended on the type of rearrangement: inv(3)(q21;q26) cases had breakpoints 3' of the gene, whereas in t(3;3)(q21;q26) cases, the breakpoints were 5' of the gene [21].

EVI1 is also involved in a t(3;21)(q26;q22), which is seen in the blast crisis phase of CML [27] and in myelodysplastic syndrome (MDS)-derived leukemia [28]. This is a reciprocal translocation and likely plays a causative role in these diseases [29, 30]. This rearrangement involves *AML1* on human chromosome 21q22, with the breakpoint between exons 6 and 7. *AML1* encodes the human homolog of the α subunit of the heterodimeric transcription factor PEBP2/CBF[31, 32], a factor that is involved in the transcriptional control of several myeloid-specific genes (see papers by Zhang et al, and Friedman, this volume). Interestingly, in between the t(3;21) breakpoint and *EVI1* (a distance of 470 kb), there are two other genes: *EAP*, which encodes the small ribosomal protein

L22[33], and *MDS*, a gene of unknown function that codes for an approximately 40kd protein [19](G. Nucifora, personal communication). All three chromosome 3 genes are in the same transcriptional orientation, directed away from the breakpoint site, and in the same orientation as *AML1*. Eight different chimeric RNA transcripts derived from this recombinant chromosome have been identified, and these join either exon five or six of *AML1* to one of the chromosome 3 genes [19]. In addition, *AML1/MDS/EVI1* transcripts have also been documented. In the latter, the translational reading frame is maintained across this tripartite chimera, and yields an 180 kd protein that reacts with both anti-AML1 and anti-EVI1 antisera [22]. The leukemogenic effect of these chimeric proteins is under investigation.

From the studies of *EVI1* in human myeloid disease, it is clear that it can contribute to leukemia by either increased level of expression or by alteration in protein structure.

Evi1 Protein Structure

cDNA cloning and analysis showed that *Evi1* encodes a 1030 residue protein with ten zinc finger motifs that are separated into two domains (Figure 1) [3]. There is also an acidic domain at the C-terminal end, although, as discussed below, *Evi1* does not appear to function as a transcriptional activator. A shorter isoform of *Evi1*, which migrates as 88kd, is produced via alternative splicing in both human [34] and mouse [35]. This form lacks zinc fingers six and seven, as well as 324 of the adjacent C-terminal amino acids. Since these fingers are important for DNA binding [36], it is likely that this isoform has different binding characteristics than the 145 kd isoform, although this has not been carefully examined.

EVI1 is localized to the nucleus [12, 37], and exhibits an interesting, dotted pattern of localization in cells that contain a provirally activated allele. A similar pattern of nuclear localization has been reported for the *PML* gene product in myeloid precursors [38, 39]. Murine EVI1 is highly phosphorylated at numerous sites, predominantly serine and threonine. The protein binds to wheat germ agglutinin, and can be eluted off with N-acetyl glucosamine (GlcNAc), suggesting the presence of O-linked GlcNAc residues, which is a feature seen on a number of other nuclear proteins [40]. The significance of these modifications is not clear.

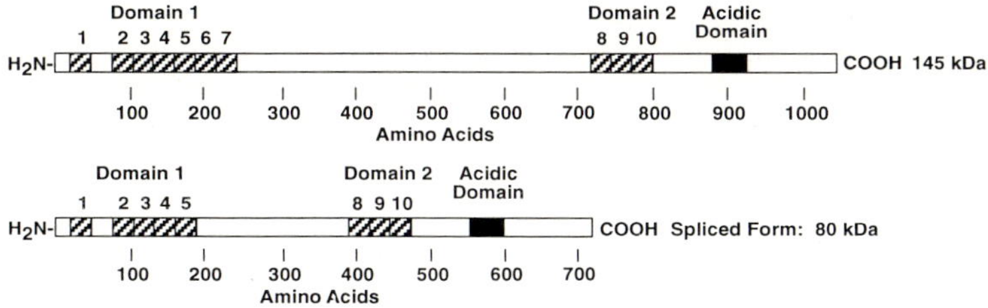

Figure 1. Structure of the 145kDa and 88kd isoforms of murine EVI1. Hatched areas denote zinc finger

Evi1 Expression

In Drosophila, zinc finger proteins play an important role in cell fate and body plan decisions. In the developing mouse, *Evi1* is expressed in a temporally and spatially restricted pattern of expression, suggesting a role for the gene in organogenesis[41]. Organs showing high level expression include the developing renal system and respiratory systems, where the gene is expressed in the epithelial component, and not in the mesenchymal cells. Expression of *Evi1* can be found in primary fetal kidney cells but not in numerous established kidney cell lines. Interestingly, *Evi1* expression in A704 human renal carcinoma cells could be induced with agents that raise cAMP levels, and blocked by the addition of TPA ; however, not all renal cell lines responded in a similar manner[42]. These data suggest that *Evi1* expression in the kidney may be modulated in vivo in response to various physiologic states.

In the developing limb buds, *Evi1* is expressed at 9-12 days postcoitum (dpc) in a non-graded pattern, with no significant expression seen in the apical ectodermal ridge. The developing heart also expresses *Evi1*; transcripts can be detected at 12.5 dpc in the endothelial cells that partition the truncus arteriosus, as well as the valve leaflets. At 14.5 dpc, when the truncus is fully separated into the two outflow channels, no expression can be seen in the heart [41].

The pattern of expression seen in the heart and limb buds, where the gene is on for brief periods of rapid growth and morphogenetic change, suggests a role for *Evi1* in cell migration or cell growth at a point after lineage commitment but prior to terminal differentiation. A similar pattern of expression, where *Evi1* is on at the midpoint of differentiation but off at later stages, is also seen in differentiating blood cells. Analysis of *EVI1* expression during G-CSF-induced differentiation of CD34 positive human myeloid precursors has revealed that while the gene is expressed at low levels early in myeloid differentiation, its expression increases dramatically at the promyelocyte stage; thereafter, its expression drops to undetectable levels. It is interesting to note that *EVI1* expression is downregulated as the cells become terminally differentiated and postmitotic. These data suggest a possible role for the gene in preventing terminal differentiation, a concept that has some experimental support [43].

Additional evidence of *EVI1* expression in hematopoietic cells, in the absence of retroviral insertion or chromosomal rearrangement, comes from a recent study that reported *EVI1* expression in a variety of myelodysplastic syndromes and leukemias that lacked cytogenetically evident rearrangements at *EVI1* [44]. It is not clear if a cytogenetically undetectable genetic lesion at *EVI1,* or a extragenic alteration, is causing *EVI1* overexpression in these instances, nor what role this expression plays in the transformed phenotype. It is possible that expression in these cells is stage-specific, and reflects the point in cellular maturation at which the cells arrested.

Biochemical Properties Of Evi1

The finding of ten zinc fingers in the protein argues that *Evi1* encodes a sequence-specific DNA binding protein that plays a role in RNA transcription, DNA recombination or replication, or some other aspect of nucleic acid regulation. By performing cycles of protein-DNA binding, selection, and amplification, it has been possible to identify consensus binding sites for both the first and second sets of zinc fingers: fingers 1-7 bind to TGACAAGATAA[36, 45], and the second set of fingers, numbers 8-10 bind to GAAGATGAG[46]. Studies on zinc finger structure and DNA binding indicate that each finger interacts with three nucleotides, and this interaction is mediated by key amino acids on the exposed face of an alpha helix that extends down the C-terminal half of each finger: the identities of these amino acids, to a certain extent, can determine the specificity of binding[47-49]. By such considerations, fingers 1-7 could interact with 21 base pairs, and fingers 8-10 with 9 bps. The latter appears to be the case. However, it is unlikely that fingers 1-7 bind 21 consecutive bps, for several reasons: this would require the protein wrap around two complete turns of the helix, and would likely encounter steric hindrance. In addition, the presence of uncharged amino acids such as serine at key positions in fingers 1 and 2 makes it unlikely that they interact strongly with DNA. Fingers 1 and 4 are His-X_4-His fingers instead of the more common His-X_3-His spacing, which allows more flexibility to the finger [50, 51], and yield a structure that makes no contact with DNA. It is argued that in TFIIIA, which has nine zinc fingers and the theoretical capacity to bind a 27 base site, some of the fingers make no contact with the DNA [49]. Indeed, DNA binding and amplification selections for optimal binding sites, and gel shift experiments, performed with triplets of fingers reveal that fingers 1-3, 2-4, and 3-5 do not bind DNA on their own, while triplets 4-6 and 5-7 are able to bind DNA, and can select the GATAA motif. The inclusion of finger 4 onto 5-7 yields increased selection for GACAA; fingers 2 and 3 appear to augment the specificity for GACAA[36].

The binding site for EVI1 that we identified - TGACAAGATAA - shows overlap with the binding site for the GATA family of transcription factors, which bind to the consensus motif (A/T)GATA(A/G) [52, 53]. This site was first identified as a common motif present in *cis*-acting elements of erythroid-specific genes, and through mutagenesis studies was found to be functionally important for erythroid gene transcription [54-57]. The overlap in binding sites for EVI1 and the GATA factors suggested the possibility that EVI1 may bind to GATA sites located in *cis* to erythroid-specific genes and influence their transcription. To investigate this possibility, we used competitive gel shift assay [58] to determined the affinity of binding of EVI1 for GATA sites located near various erythroid-specific genes. Remarkably, EVI1 did not exhibit significant affinity for any of the GATA sequences including those with very close similarity to the wildtype binding site. For instance, the chicken α^D globin gene has a tandemly repeated GATA site: GGATAAGATAAG, which has 83% identity to the EVI1 binding site, mismatching only at position four. This bound to EVI1 with 7% of the affinity with which the wildtype motif bound. None of the GATA sequences that have been functionally defined as playing a role in erythroid-specific gene expression that we tested showed any significant binding affinity for EVI1.

In light of the poor affinity of EVI1 for GATA sites, we wished to better define the points of contact points between EVI1 and DNA, which we did by performing methylation interference studies, using TGACAAGATAA as a substrate. These studies, summarized in Figure 2, reveal strong contacts between EVI1 and the G residues on both strands, as well as contacts with nearly all the bases of the binding site, including the initial T, but not the final A. The strong interaction with the G at position 4 on the bottom strand argues for the importance of that base in the motif. To further test the determinants of high affinity binding, we performed quantitative gel shift experiments with various competitor DNAs that contained single base differences from the TGACAAGATAA motif. These data show that changes at positions 1, 2, 3, 5, 6 or 8 preclude high affinity binding, whereas a change at position 10 is better tolerated, giving only 75% decrease in binding affinity (Figure 3). Interestingly, the longer binding motif identified by Delwel et al (GA(C/T)AAGA(T/C)AAGATAA) [36] binds with identical affinity as the shorter motif that we identified (TGACAAGATAA), which argues that the protein does not interact significantly with the final GATAA bases; these bases may have been selected by the binding of more than one protein molecule per DNA.

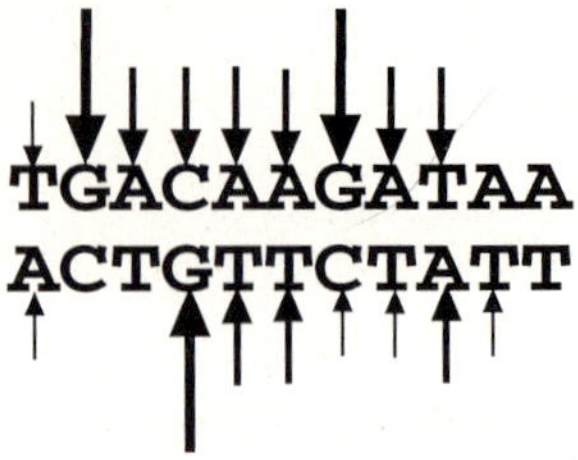

TGACAAGATAA
ACTGTTCTATT

Fig. 2. Summary of methylation interference studies. The points of contact between EVI1 and the DNA are indicated by the arrows, with the size of the arrow indicating the importance of the interaction to binding. The initial T is included, because by these studies, modification of either T or its base pair A led to slight but detectable decrease in binding.

Transcriptional Regulation by *Evi1*

Studies by ourselves and others show that EVI1, as expected, having multiple zinc fingers, binds to DNA with high affinity in a sequence-dependent manner. Sequence analysis of EVI1 also revealed the presence of an acidic domain [3], a feature found in numerous transcriptional activators [59]. To test if the effect of *Evi1* on transcription, we cotransfected cells with *Evi1* expression plasmids along with CAT reporters that contained concatemerized EVI1 binding sites upstream of the HSV thymidine kinase promoter. We found that *Evi1* dramatically repressed the activity of reporters that contained either the first or the second EVI1 binding site, and that this effect was most marked in cells where the reporters had high activity without *Evi1*, such as WEHI 3B cells. In other cells, such as L tk⁻ or NIH 3T3 cells, where the reporter had low activity, *Evi1* appeared to have no effect.

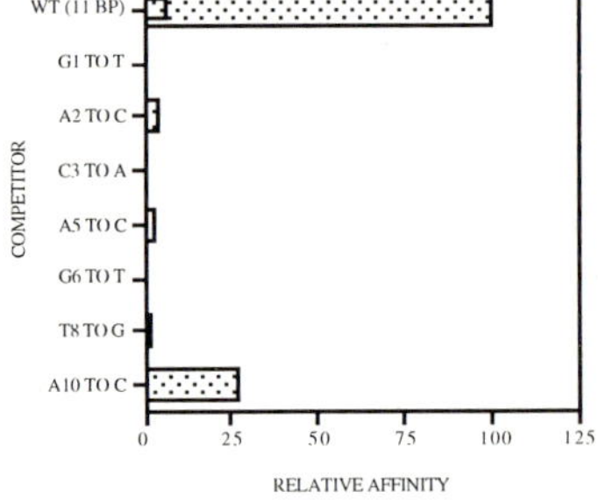

Fig. 3. Relative affinities of different oligonucleotides with variations to the GACAAGATAA EVI1 binding motif, as indicated. "WT" contains no change. Data were obtained by competitive gel shift performed in triplicate, at three different dilutions, and calculations were done as described by Glass et al, 1988. The 15 bp binding site described by Delwel et al, 1993, exhibited the same affinity as the GACAAGATAA oligo.

Thus, the ability of *Evi1* to cause transcriptional repression via its binding site appears to be a cell-specific phenomenon, and may depend on the array of transcription factors other than EVI1 that are present in the transfected cell. In no cotransfection experiment have we seen clear evidence of *Evi1*-mediated transactivation. This was surprising, given the presence of an acidic-rich domain in the C-terminal portion of the protein.

The ability of *Evi1* to repress transcription depended on the promoter/enhancer elements for the CAT reporter gene. The basal level of transcription observed with the TKCAT plasmid or with other reporter plasmids, such as those with the sequences of the albumin promoter to bp -35, was not repressed by *Evi1*, suggesting that *Evi1* does not interact with these sequences or with the basal transcriptional machinery. In addition, the transcription of more active reporters, such as ones having the SV40 early region promoter (which includes two AP1 sites and six Sp1 sites), with or without additional SV40 enhancer sequences, was not affected by *Evi1* in cotransfection studies using a variety of cells.

An 88 kd isoform of EVI1 is encoded by an alternatively spliced mRNA, which lacks zinc fingers 6, and 7 [35]. Since recent studies suggest these particular fingers are in large part responsible for the sequence-specific binding of EVI1 to the TGACAAGATAA sequence [36], it seems unlikely that this isoform would bind to the motif. As expected, in cotransfection studies, we found that expression of the 88 kd isoform of EVI1 had little effect on the activity of promoters that have the TGACAAGATAA motif.

Given that the first EVI1 binding site contains a perfect GATA binding motif, we considered it highly likely that GATA-1 would be able to bind to and activate reporters that contained the site. We have shown this is the case by cotransfection studies: CAT activity was 100-fold higher in transfections that contained GATA-1 expression plasmid and the first EVI1 binding site in the reporter, relative to control transfections without GATA-1 or without EVI1 binding sites on the CAT reporter. This indicated that GATA-1 can activate transcription via an EVI1 binding site. We were interested in determining whether *Evi1* could have any effect on the GATA-1-mediated transcriptional activation that we observed with reporters containing an EVI1 binding site. When the *Evi1* and GATA-1 expression plasmids were cotransfected with a CAT reporter containing EVI1 binding sites, there was a dramatic decrease in activity relative to transfections without the *Evi1* expression plasmid, indicating that in this context, *Evi1* can repress GATA-1-mediated transcriptional activation. These data are similar to those reported by Kreider et al[43].

The binding site for the second set of zinc fingers has been identified - GAAGATGAG [46] - and in cotransfection studies in WEHI-3B cells, we have shown that *Evi1* can repress transcription via this site: while the 88 kd isoform had a slight inhibitory effect on reporters bearing this motif, the 145 kd isoform inhibited significantly. This indicates that both via the first binding site and second binding site, *Evi1* can act as a transcriptional repressor, at least in the cells tested, which include WEHI 3B and NIH 3T3 cells. Whether this is true in all cell types, or on real target genes, is not known.

Evi1 Target Genes

The data on the binding specificity of EVI1 indicate that it is unlikely that the protein binds appreciably to GATA sites present on known erythroid genes with GATA sites. The identity of true target genes for EVI1 is not known. It is postulated that the protein acts on a defined number of target genes to alter transcription, most likely as a repressor, given the cotransfection data obtained to date. As a first step towards identifying EVI1 target genes, we performed cycles of binding selection using a plasmid-based mouse genomic library as a substrate[60]. The size of the DNA inserts in this library ranged from 2-7 kb, and thus may include, in *cis* to the selected EVI1 binding site, exon sequences for regulated genes. To isolate plasmids containing genomic fragments that bound to EVI1, we reacted purified DNA from the library with EVI1, and selected for DNA-protein complexes by filtration through nitrocellulose [61]. Selected fragments were amplified in *E.coli*, and subjected to another round of selection. With successive rounds of selection we were able to enrich for genomic fragments that bound to EVI1 with high affinity, as assessed by competitive gel shift. Sequence analysis of subfragments of the selected clones that bound to EVI1 revealed the presence of the TGACAAGATAA binding motif. The longer motif identified previously as an EVI1 binding site[36] was not identified in our selected fragments, suggesting that the shorter motif is more representative of the site used in vivo. This binding selection technique resulted in the isolation of approximately 150 different genomic fragments that are now being analyzed for the presence of exons.

Evi1 in Myelopoiesis and Leukemogenesis

The pattern of *Evi1* expression seen the heart, limb buds, and differentiating granulocytes suggest a possible role for the gene in preventing terminal differentiation, which is consistent with the immature phenotype in most AMLs bearing *Evi1* activating mutations. Direct support for this theory comes from the observation that ectopic expression of *Evi1* in primary bone marrow cells inhibits differentiation in response to erythropoietin [43]. Since differentiation along the erythroid lineage is dependent on GATA-1 activity[62], and since there is overlap between the GATA-1 and EVI1 binding sites, it is possible that loss of EPO responsiveness results from EVI1-induced repression of as yet unidentified GATA-1-responsive gene(s) harboring an EVI1 binding site.

Transforming oncogenes have been shown to raise activity of the transcription factor AP-1. It appears that the second set of zinc fingers raises AP-1 activity via the activation of c-*fos* [63]. While this finger region is necessary for c-*fos* activation, maximal activity requires the presence of fingers 5-7 of the first domain. Importantly, there are no binding motifs for the first or second domains of *Evi1* in the c-*fos* promoter, indicating that it is unlikely that this effect is direct. The significance of these findings to leukemogenesis is not clear.

Summary

Through chromosomal rearrangements and/or proviral insertions, a number of genes encoding nuclear transcription factors have been identified that play key roles in leukemogenesis. One of these is *Evi1*, which plays a role in both murine and human myeloid leukemia. The exact mechanism by which *Evi1* exerts its leukemogenic effect is not clear, but it may involve the inhibition of terminal differentiation, through the abnormal repression of genes necessary for cellular maturation. Our analysis of the DNA binding characteristics of EVI1 indicate a high degree of specificity, which likely indicates that the protein acts on a tightly defined number of targets in the cell. We are beginning to characterize candidate target genes located in the mouse genome near EVI1 binding sites with the expectation that these will yield insight into EVI1 function both in normal cells and in leukemogenesis.

References

1.Bedigian H, Johnson D, Jenkins N, Copeland N,Evans R (1984) Spontaneous and induced leukemias of myeloid origin in recombinant inbred BXH mice. J Virol 51:586-594

2.Mucenski ML, Taylor BA, Jenkins NA,Copeland NG (1986) AKXD recombinant inbred strains: Models for studying the molecular genetic basis of murine lymphomas. Mol Cell Biol 6:4236-4243

3.Morishita K, Parker DS, Mucenski ML, Jenkins NA, Copeland NG,Ihle JN (1988) Retroviral activation of a novel gene encoding a zinc finger protein in IL3-dependent myeloid leukemia cell lines. Cell 54:831-840

4.Buchberg AM, Bedigian HG, Jenkins NA,Copeland NG (1990) *Evi*-2, a common integration site, involved in murine myeloid leukemogenesis. Mol Cell Biol 10:4658-4666

5.Ben-David Y, Giddens EB,Bernstein A (1990) Identification and mapping of a common proviral integration site *Fli-1* in erythroleukemia cells induced by Friend murine leukemia virus. pnas 87:1332-1336

6.Moreau-Gachelin F, Tavitian A,Tambourin P (1988) *Spi-1* is a putative oncogene in virally induced murine erythroleukaemias. Nature 331:277-280

7.Aberdam D, Negreanu V, Sachs L,Blatt C (1991) The oncogenic potential of an activated hox-2.4 homeobox gene in mouse fibrolasts. Mol Cell Biol 11:554-557

8.Moskow J, Bullrich F, Huebner K,Buchberg A (1995) Personal communication.

9.Nakamura T,Copeland N (1995) personal communication

10.Mucenski ML, Taylor BA, Ihle JN, Hartley JW, Morse III HC, Jenkins NA,Copeland NG (1988) Identification of a common ecotropic viral integration site, Evi-1, in the DNA of AKXD murine myeloid tumors. Mol Cell Biol 8:301-308

11.Bergeron D, Poliquin L, Houde J, Barbeau B,Rassart E (1992) Analysis of proviruses integrated in *Fli-1* and *Evi-1* regions in Cas-Br-E MuLV-induced non-T-, non-B-cell leukemias. Virology 191:661-669

12.Matsugi T, Morishita K,Ihle JN (1990) Identification, nuclear localization, and DNA-binding activity of the zinc finger protein encoded by the *Evi*-1 myeloid transforming gene. Mol Cell Biol 10:1259-1264

13.Bordereaux D, Fichelson S, Sola B, Tambourin PE,Gisselbrecht S (1987) Frequent involvement of the *fim-3* region in Friend murine leukemia virus-induced mouse myeloblastic leukemias. J Virol 61:4043-4045

14.Bartholomew C, Morishita K, Askew D, Buchberg D, Jenkins NA, Copeland NG,Ihle JN (1989) Retroviral insertions in the CB-1/ *Fim-3* common site of integration active expression of the *Evi*-1 gene. Oncogene 4:529-534

15.Bartholomew C,Ihle J (1991) Retroviral insertions 90 kilobases proximal to the Evi-1 myeloid transforming gene activate transcription from the normal promoter. Mol Cell Biol 11:1820-1828

16.Morishita K, Parganas E, Willman CL, Whittaker MH, Drabkin H, Oval J, Taetle R, Valentine MB,Ihle JN (1992) Activation of EVI1 gene expression in human acute myelogenous leukemias by translocations spanning 300-400 kilobases on chromosome band 3q26. Proc Natl Acad Sci USA 89:3937-3941

17.Oval J, Smedsrud M,Taetle R (1992) Expression and regulation of the Evi-1 gene in the human factor-dependent leukemia cell line, UCSD/AML1. Leukemia 6:446-451

18.Fichelson S, Dreyfus F, Berger R, Melle J, Bastard C, Miclea J,Gisselbrecht S (1992) Evi-1 expression in leukemic patients with rearrangements of the 3Q25-Q28 chromosomal region. Leukemia 6:93-99

19.Nucifora G, Begy CR, Kobayashi H, Roulston D, Claxton d, Pedersen-Bjergaard J, Parganas E, Ihle JN,Rowley JD (1994) Consistent intergenic splicing and production of multiple transcripts between *AML1* at 21q22 and unrelated genes at 3q26 in (3;21)(q26;q22) translocations. Proc Nat'l Acad Sci USA 91:4004-4008

20.Levy ER, Parganas E, Morishita K, Fichelson S, James L, Oscier D, Gisselbrecht S, Ihle JN,Buckle VJ (1994) DNA rearrangements proximal to the EVI1 locus associated with the 3q21q26 syndrome. Blood 83:1348-1354

21.Suzukawa K, Parganas E, Gajjar A, Abe T, Takahashi S, Tani K, Asano S, Asou H, Kamada N, Yokota J, Morishita K,Ihle JN (1994) Identification of a breakpoint cluster region 3' of the ribophorin I gene at 3q21 associated with the transcriptional activation of the *EVI1* gene in acute myelogenous leukemias with inv(3)(q21q26). Blood 84:2681-2688

22.Mitani K, Ogawa S, Tanaka T, Miyoshi H, Kurokawa M, Mano H, Yazaki Y, Ohki M,Hirai H (1994) Generation of the *AML1 - EVI-1* fusion gene in the t(3;21)(126;q22) causes blastic crisis in chronic myelocytic leukemia. The EMBO Journal 13:504-510

23.Morishita K, Parganas E, Bartholomew C, Sacchi N, Valentine MC, Raimondi SC, Le Beau MM,Ihle JN (1990) The human *Evi*-1 gene is located on chromosome 3q24-q28 but is not rearranged in three cases of acute nonlymphocytic leukemias containing t(3;5) (q25;q34) translocations. Oncogene Research 5:221-231

24.Pintado T, Ferro M, San Roman C, Mayayo M,Larana J (1985) Clinical correlations of the 3q21;3q26 cytogenetic anomaly. A leukemic or myelodysplastic syndrome with preserved or increased platelet production and lack of response to cytotoxic drug therapy. Cancer 55:535-541

25.Bitter M, Neilly M, LeBeau M, Pearson M,Rowley J (1985) Rearrangements of chromosome 3 involving bands 3q21 and 3q26 are associated with normal or elevated platelet counts in acute non-lymphocytic leukemia. Blood 66:1362-1370

26.Bellomo M, Parlier V, Muhlematter D, Grob J,Beris P (1992) Three new cases of chromosome 3 rearrangements in bands q21 and q26 with abnormal thrombopoiesis bring further evidence to the existence of a 3q213q26 syndrome. Cancer Genet Cytogenet 59:138

27.Schneider N, Bowman W,Frenkel E (1991) Translocation (3;21)(q26;q22) in secondary leukemia. Ann Genet 34:256-263

28.Rubin C, Larson R, Anastasi J, Winter J, Thangavelu M, Vardiman J, Rowley J,LeBeau M (1990) t(3;21)(q26;q22): A recurring chromosomal abnormality in therapy-related myelodysplastic syndrome and acute myeloid leukemia. Blood 76:2594-2598

29.Coyle T,Najfeld V (1988) Tranlocation (3;21) in Philadelphia chromosome-positive chromic myelogenous leukemia prior to the onset of blast crisis. Am J Hematol 27:56-59

30.Rubin C, Larson R, Bitter M, Carrino J, LeBeau M, Diaz M,Rowley J (1987) Association of a chromosomal 3;21 translocation with the blast phase of chronic myelogenous leukemia. Blood 70:1338-1342

31.Kagoshima H, Shigesada K, Satake M, Ito Y, Miyoshi H, Ohki M, Pepling M,Gergen P (1993) The Runt domain identifies a new family of heteromeric trasncriptional regulators. Trends in Genetics 9:338-341

32.Miyoshi H, Shimizu K, Kozu T, Maseki N,Kaneko Y (1991) t(8;21) breakpoints on chromosome 21 in acute myeloid leukemia are clustered within a limited region of a single gene, *AML1*. Proc. Natl. Acad. Sci. USA 88:10431-10434

33.Nucifora G, Begy CR, Erickson P, Drabkin HA,Rowley JA (1993) The 3;21 translocation in myelodysplasia results in a fusion transcript between the *AML1* gene and the gene for EAP, a highly conserved protein associated with the Epstein-Barr virus small RNA EBER 1. Proc. Natl. Acad. Sci. USA 90:7784-7788

34.Morishita K, Parganas E, Douglass EC, Ihle JN (1990) Unique expression of the human *Evi*-1 gene in an endometrial carcinoma cell line: Sequence of cDNAs and structure of alternatively spliced transcripts. Oncogene 5:963-971

35.Bordereaux D, Fichelson S, Tambourin P,Gisselbrecht S (1990) Alternative splicing of the *Evi*-1 zinc finger gene generates mRNAs which differ by the number of zinc finger motifs. Oncogene 5:925-927

36.Delwel R, Funabiki T, Kreider B, Morishita K,Ihle J (1993) Four of the seven zinc fingers of the Evi-1 myeloid transforming gene are required for sequence-specific binding to GA(C/T)AAGA(T/C)AAGATAA. Mol Cell Biol 13:4291-4300

37.Khanna-Gupta A, Savinelli T, Berliner N,Perkins A (1995) Retroviral insertional activation of the Evi1 oncogene does not prevent G-CSF-induced maturation of the murine pluripotent myeloid cell line, 32Dcl3. Submitted

38.Dyck J, Maul G, Miller W, Chen J, Kakizuka A,Evans R (1994) A novel macromolecular structure is a target of the promyelocyte-receptor oncoprotein. Cell 76:333-343

39.Weis K, Rambaud S, Lavau C, Jansen J, Carvalho T, Carmo-Fonesca M, Lamond A,Dejean A (1994) Retinoic acid regulates aberrant nuclear localization of PML-RAR alpha in acute promyelocytic cells. Cell 76:345-356

40.Jackson SP,Tjian R (1988) O-glycosylation of eukaryotic transcription factors: Implications for mechanism of transcriptional regulation. Cell 55:125-133

41.Perkins AS, Mercer JA, Jenkins NA, Copeland NG (1991) Patterns of Evi-1 expression in embryonic and adult tissues suggest that Evi-1 plays an important role in mouse development. Development 111:479-487

42.Bartholomew C,Clark AM (1994) Induction of two alternatively spliced *Evi*-1 proto-oncogene transcripts by cAMP in kidney cells. Oncogene 9:939-942

43.Kreider B, Orkin S,Ihle J (1993) Loss of erythropoietin responsiveness in erythroid progenitors due to expression of the Evi-1 myeloid transforming gene. Proc Natl Acad Sci, USA 90:6454-6458

44.Russell M, List A, Greenberg P, Woodward S, Glinsmann B, Parganas E, Ihle J,Taetle R (1994) Expression of EVI1 in myelodysplastic syndromes and other hematologic malignancies without 3q26 translocations. Blood 84:1243-1248

45.Perkins AS, Fishel R, Jenkins NA,Copeland NG (1991) *Evi*-1, a murine zinc finger proto-oncogene, encodes a sequence-specific DNA-binding protein. Mol Cell Biol 11:2665-2674
46.Funabiki T, Kreider BL,Ihle JN (1994) The carboxyl domain of zinc fingers of the Evi-1 myeloid transforming gene binds a consensus sequence GAAGATGAG. Oncogene 9:1575-1581
47.Pavletich N,Pabo C (1993) Crystal structure of a five-finger GLI-DNA complex: new perspectives on zinc fingers. Science 261:1701-1707
48.Pavletich N,Pabo C (1991) Zinc finger-DNA recognition: crystal structure of a Zif268-DNA complex at 2.1 A,. Science 252:809-817
49.Klevit R (1991) Recognition of DNA by Cys2, His2 zinc fingers. Science 253:1367
50.Xu R, Horvath S,Klevit R (1991) ADR1a, a zinc finger peptide, exists in two folded conformations. Biochemistry 30:3365-3371
51.Kochoyan M, Havel TF, Nguyen DT, Dahl CE, Keutmann HT, Weiss MA (1991) Alternating zinc fingers in the human male associated protein ZFY: 2D NMR structure of an even finger and implications for "jumping-linker" DNA recognition. Biochemistry 30:3371-3386
52.Evans T, Reitman M,Felsenfeld G (1988) An erythroid-specific DNA-binding factor recognizes a regulatory sequence common to all chicken globin genes. Proc. Natl. Acad. Sci. USA 85:5976-5980
53.Wall L, deBoer E,Grosveld F (1988) The human β-globin gene 3' enhancer contains multiple binding sites for an erythroid-specific protein. Genes and Dev 2:1089-1100
54.Mignotte V, Eleouet F, Raich N,Romeo P (1989) Cis- and trans-acting elements involved in the regulation of the erythroid promoter of the human porphobilinogen deaminase gene. Proc Natl Acad Sci, USA 86:6548-6551
55.Reitman M,Felsenfeld G (1988) Mutational analysis of the chicken beta-globin enhancer reveals two positive-acting domains. Proc Natl Acad Sci USA 85:6267-6271
56.Plumb M, Frampton J, Wainwright H, Walker M, Macleod K, Goodwin G,Harrison P (1989) GATAAG: a *cis*-control region binding an erythroid-specific nuclear factor with a role in globin and non-globin gene expression. Nucleic Acids Research 17:73-93
57.Martin D, Tsai S,Orkin S (1989) Increased gamma-globin expression in a nondeletion HPFH mediated by an erythroid-specific DNA-binding factor. Nature 338:435-438
58.Glass C, Holloway J, Devary O,Rosenfeld M (1988) The thyroid hormone receptor binds with opposite transcriptional effects to a common sequence motif in thyroid hormone and estrogen response elements. Cell 54:313-323
59.Ptashne M (1988) How eukaryotic transcriptional activators work. Nature 335:683-689
60.Sompayrac L,Danna K (1990) Method to identify genomic targets of DNA binding proteins. Proc. Natl. Acad. Sci. USA 87:3274-3278
61.McEntee K, Weinstock G,Lehman I (1980) recA protein catalyzed assimilation. Stimulation by Escherichia coli single-stranded DNA-binding protein. Proc. Natl. Acad. Sci. USA 77:857-861
62.Pevny L, Simon MC, Robertson E, Klein WH, Tsai SF, D'Agati V, Orkin S,Costantini F (1991) Erythroid differentiation in chimaeric mice blocked by a targeted mutation in the gene for transcription factor GATA-1. Nature 349:257-260
63.Tanaka T, Nishida J, Mitani K, Ogawa S, Yazaki Y,Hirai H (1994) Evi-1 raises AP-1 activity and stimulates c-*fos* promoter transactivation with dependence on the second zinc finger domain. J Biol Chem 269:24020-24026

Activation of Stat-related DNA-binding Factors by Erythropoietin and the Spleen Focus-Forming Virus

T. Ohashi, M. Masuda and S. K. Ruscetti
Laboratory of Molecular Oncology, National Cancer Institute, Frederick, MD 21702-1201

Introduction

The proliferation and differentiation of normal erythroid cells is controlled by the hematopoietic growth factor erythropoietin (Epo), which interacts with its specific cell surface receptor (EpoR) to initiate a signal transduction pathway that is not fully understood (for reviews, see [1,2]). Erythroid cells infected with the Friend spleen focus-forming virus (SFFV), however, can proliferate and differentiate in the absence of Epo, and this results in uncontrolled growth of erythroid precursor cells and erythroleukemia (for review, see [3]). Although it is known that the unique envelope glycoprotein encoded by SFFV is responsible for its biological effects [4], the mechanism by which the viral protein activates erythroid cell growth in the absence of Epo is not known. The virus does not appear to be stimulating the production of Epo by erythroid cells, resulting in proliferation due to an autocrine mechanism [5]. A more likely possibility is that SFFV is activating the Epo signal transduction pathway. Studies have shown that only cells expressing a functional Epo receptor capable of generating a signal in response to Epo can be rendered factor-independent by SFFV [6,7]. Also, the SFFV envelope glycoprotein appears to be associated with the Epo receptor complex at the cell surface [8-10], putting it in a position where it could modulate the Epo signal transduction pathway.

The mechanism by which binding of Epo to the EpoR triggers the signal for erythroid cell growth and differentiation is unclear. Recent studies have shown that members of the cytokine receptor superfamily can signal through two major, functionally distinct pathways. One is the Ras signaling pathway, which involves a multistep cascade of intracellular events before gene regulation (for review, see [11,12]). There is some evidence that this pathway may be activated in erythroid cells since signal transducing molecules in the Ras pathway, such as Shc, Grb2 and Raf-1, have been shown to be tyrosine phosphorylated in response to Epo [13-15] and antisense oligonucleotides to Raf-1 mRNA can inhibit the growth of Epo-dependent erythroid cells [15]. The other pathway, called the Jak-Stat pathway, which was originally identified in interferon-stimulated cells, is a more rapid and direct signaling pathway that may have evolved for a quick response to changes in the extracellular environment (for reviews, see [16-18]). In this pathway, binding of a ligand to a receptor leads to activation of a member of the Jak tyrosine kinase family, which is then thought to phosphorylate latent transcription factors, called signal transducers and activators of transcription (Stat) proteins, in the cytoplasm. Tyrosine-phosphorylated Stat proteins form specific complexes which migrate into the nucleus and are thought to directly bind to the regulatory region of specific target genes. Four members of the Jak kinase family (Jak1, Jak2, Jak3, and Tyk2) and six different Stat proteins (Stat1 through 5 and IL-4 Stat) have been described to date, and various growth factors and cytokines have been shown to activate these proteins. It was recently shown by using a myeloid cell line expressing the transfected EpoR gene that Jak2 kinase is associated with a membrane proximal region of the EpoR and phosphorylated at its tyrosine residue in response to Epo stimulation [19,20]. Studies with a dominant-negative mutant of Jak2 kinase indicate that activation of Jak2 kinase is essential for growth stimulation induced by Epo [21]. However, it was not clear whether Epo stimulation resulted in activation of any Stat proteins in erythroid cells expressing the endogenous EpoR.

HCD-57 is an Epo-dependent erythroleukemic cell line established from a mouse infected with Friend murine leukemia virus [22]. Epo stimulation of HCD-57 cells induces tyrosine phosphorylation of various cellular substrates, including a very dominant protein with a molecular weight (97 kd) comparable to that of known Stat proteins (Fig. 1). When HCD-57 cells are infected with SFFV and their Epo-dependence is abrogated, the 97 kd protein as well as other proteins are constitutively phosphorylated (Fig. 1). These observations led us to investigate whether Epo can induce the activation of Stat proteins and to determine whether these proteins are activated in SFFV-infected cells in the absence of Epo.

Results

Epo Induces SIE and GRR Binding Factors in the Epo-dependent Erythroid Cell Line, HCD-57.

Electrophoretic mobility shift assays (EMSA) were carried out on nuclear extracts from HCD-57 cells before and after Epo stimulation using ^{32}P-labeled oligonucleotide probes corresponding to two sequences known to bind Stat proteins, the high-affinity sis-inducible element, or SIE of the c-*fos* gene [23], and the gamma response region, or GRR, of the Fcγ receptor factor 1 gene [24]. No DNA-binding activity was detected by these two probes before Epo stimulation of HCD-57 cells (Fig. 2A). However, when the cells were stimulated with Epo, two mobility shift signals were detected by the SIE probe and a slower migrating signal was detected by the GRR probe. Both the SIE and the GRR binding factors could be competed for by a 100-fold excess of the homologous probe, but not by an unrelated oligonucleotide.

We have shown that induction of DNA-binding factors by Epo is rapid, although the SIE binding factors have different kinetics of induction by Epo compared with the GRR-binding factors [25]. In both cases the DNA-binding activities can be detected first in the cytoplasm and then after a brief delay in the nucleus, suggesting that the activated factors are transported from the cytoplasm to the nucleus. High levels of both the SIE and GRR binding factors can be detected in the cytoplasm and nucleus

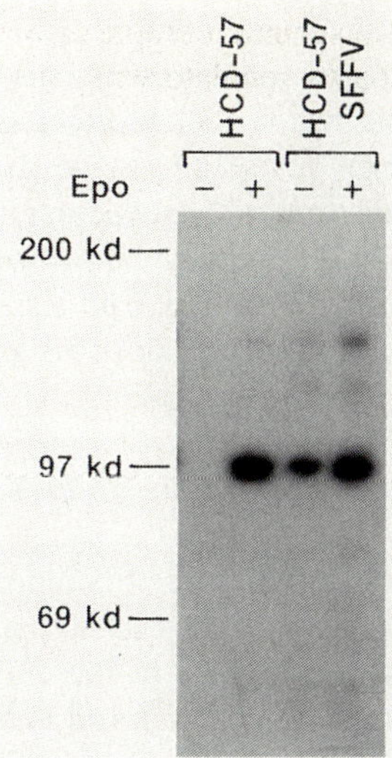

Fig. 1. Stimulation of Tyrosine Phosphorylation by Epo and SFFV. Total cell extracts were isolated from HCD-57 and SFFV$_P$-infected HCD-57 cells before and after stimulation with Epo (100 U/ml). The extracts were then precipitated with the anti-phosphotyrosine antibody 4G10 followed by immunoblotting with the same antibody.

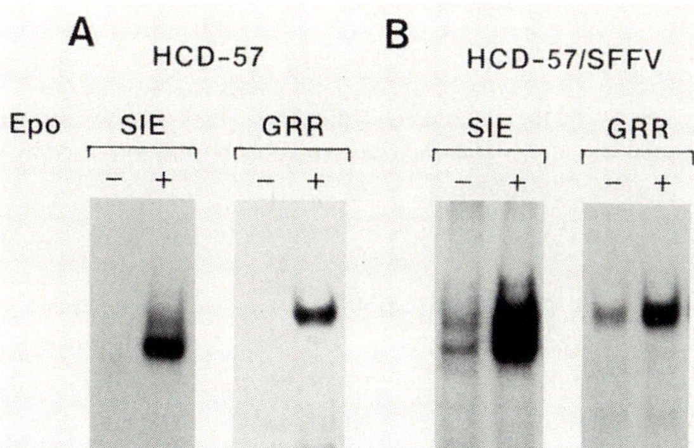

Fig. 2. Induction of DNA-binding Factors by Epo and SFFV. Nuclear extracts were prepared from HCD-57 cells and SFFV$_P$-infected HCD-57 cells before and after stimulation with Epo (100 U/ml). Extracts were then incubated with a ^{32}P-labeled SIE or GRR probe.

for up to 30 minutes after Epo stimulation. The activities are transient, however, and the levels decrease after 60 minutes and eventually disappear.

We've also shown that like activated Stat proteins, all three of the DNA-binding factors induced by Epo contain phosphotyrosine, because anti-phosphotyrosine antibodies affected complex formation by these factors, causing either a supershift or inhibition of complex formation [25]. Tyrosine phosphorylation appears to be crucial for the induction of DNA-binding factors by Epo since treatment of HCD-57 cells with the tyrosine kinase inhibitor genistein prior to Epo stimulation abolished the induction of the SIE and GRR binding factors.

DNA-binding Proteins Induced by Epo in HCD-57 Cells Are Related to Stat Proteins.

To determine whether the DNA-binding factors induced by Epo stimulation were related to Stat family proteins, we tested the reactivity of anti-Stat protein antibodies with these factors. Monoclonal antibodies to the amino terminal (Stat1N) or the carboxyl terminal region (Stat1C) of Stat1 caused either a supershift or loss of the signals corresponding to both of the SIE binding factors (Fig. 3A). Anti-Stat3 antibody inhibited the formation of only the SIE complex with a slower mobility, and caused a slight enhancement of the signal corresponding to the faster-migrating SIE complex. In contrast, anti-Stat1N antibody had no significant effect on the formation of the GRR complex, while both anti-Stat1C and anti-Stat3 antibodies showed

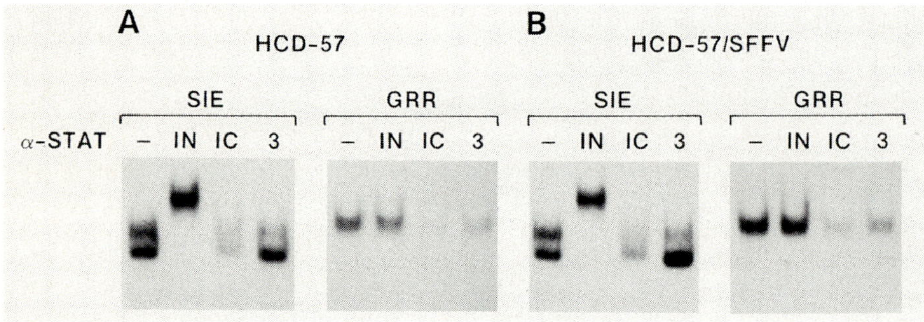

Fig. 3. The DNA-binding Proteins Induced by Epo and SFFV Are Related to Stat Proteins. Cytoplasmic extracts were prepared from HCD-57 cells or HCD-57 cells infected with SFFV$_P$ after stimulation with Epo (100 U/ml) for 15 minutes and then incubated without (-) or with various anti-Stat antibodies. The samples were then incubated with a ^{32}P-labeled SIE or GRR probe and subjected to 5% polyacrylamide gel electrophoresis.

inhibitory effects. We failed to detect an effect of anti-Stat2 antibodies on the complex formation by any of the SIE or GRR binding factors, and unrelated antibodies also failed to show any effects on the SIE or GRR complex formation. Consistent with these results, Stat1 and Stat3 mRNA can be detected in HCD-57 cells, and a 92-Kd tyrosine phosphorylated protein recognized by anti-Stat1 antibodies is induced after Epo treatment [25]. We are currently trying to determine if HCD-57 cells express other Stat proteins, such as Stat5, and whether they are involved in the Epo-induced DNA-binding complex formation.

Stat-related proteins are also activated by Epo in non-erythroid cell lines expressing the transfected EpoR gene (Ohashi et al., unpublished data). The mobilities of the SIE- and GRR-binding complexes in these cell lines as shown by EMSA, however, were often different from those induced in HCD-57 cells and did not always show the same reactivity with anti-Stat antibodies. This suggests that Epo induces similar but distinct Stat-related DNA-binding proteins in different Epo-responsive cell lines.

Stat-related DNA-binding Factors That Are Transiently Induced by Epo Are Constitutively Activated by SFFV.

After demonstrating that Epo induced SIE- and GRR-binding factors involving Stat-related proteins in HCD-57 cells, we examined whether SFFV, which abrogates Epo-dependence of HCD-57 cells, has any effects on the activation of DNA-binding factors. While HCD-57 cells express SIE- and GRR-binding factors only transiently after Epo stimulation, DNA-binding factors that recognize the same probes can be detected constitutively in SFFV-infected HCD-57 (HCD-57/SFFV) cells even in the absence of Epo (Fig. 2B). When HCD-57/SFFV cells were treated with Epo, an increase in all DNA-binding activities was observed. The DNA-protein complexes formed by the extracts from HCD-57/SFFV cells showed the same mobilities as those formed by extracts from Epo-stimulated HCD-57 cells (Fig. 2B) and showed similar reactivity to anti-Stat antibodies (Fig. 3B).

In order to determine if the DNA-binding factors activated by Epo and SFFV in erythroid cell lines can be induced in primary erythroid cells, we prepared erythroid cells from the spleens of uninfected or SFFV-infected NIH Swiss mice and examined the effects of Epo stimulation and SFFV infection on factor induction in these cells. When splenocytes of phenylhydrazine-treated mice, which contain greater than 90% erythroid cells, were treated with Epo, induction of the same SIE- and GRR-binding factors as those induced in HCD-57 cells was detected (Ohashi et al., unpublished data). In contrast, these factors were constitutively activated in splenic erythroid cells obtained from SFFV-infected mice, even in the absence of Epo. Therefore, the pattern of DNA-binding factor activation in primary erythroid cells was essentially the same as that shown by using HCD-57 cells.

Discussion

Our results indicate that Epo induces in erythroid cells the rapid and transient activation of several DNA-binding factors that contain components identical or related to known Stat proteins. We can detect similar DNA-binding factors in Epo-dependent erythroid cell lines and in primary erythroid cells from phenylhydrazine-treated mice. The similarity in the pattern of DNA-binding factor activation in primary erythroid cells and HCD-57 cells may indicate that this cell line retains a physiological signal transduction pathway for Epo and responsiveness to SFFV analogous to the biological events occurring in vivo. Epo-responsive, non-erythroid cell lines expressing the transfected EpoR gene exhibit distinct patterns of DNA-binding proteins in response to Epo. Although the DNA-binding factors induced by

Epo in HCD-57 cells react with antibodies to Stat1 and Stat3, it is possible that novel Stat proteins are involved in Epo-induced signal transduction. In order to further characterize the Epo-induced DNA-binding factors from HCD-57 cells, we are attempting to affinity purify them using agarose-conjugated probes. We are also searching for novel Stat-related genes expressed in HCD-57 cells.

SFFV infection of erythroid cells results in the constitutive activation of the same DNA-binding factors that are transiently induced by Epo in uninfected cells. Constitutively activated Stat-related proteins could be detected in SFFV-infected erythroid cell lines, as well as in erythroid cells from SFFV-infected mice. These virus-activated proteins might be triggering the signal that enables erythroid cells to proliferate autonomously without Epo. It is not known how SFFV activates Stat-related DNA-binding factors without Epo, but it is likely that it is activating these DNA-binding factors indirectly since neither the EpoR nor the SFFV envelope glycoprotein have an effector region, such as a tyrosine kinase domain. The tyrosine kinase Jak2 was recently shown to be associated with the cytoplasmic domain of the EpoR and activated after Epo stimulation [19,20]. It has now been proposed that Epo may bring together two molecules of the EpoR and the associated Jak2 kinases, allowing trans-phosphorylation and activation of the kinases [26-28] (Fig. 4). The Jak2 kinase would then phosphorylate on tyrosine both the EpoR and latent Stat proteins localized in the cytoplasm. In SFFV-infected cells, the viral protein is thought to interact with the EpoR complex at the cell surface [8,9]. The cell surface form of the SFFV envelope glycoprotein has been shown to oligomerize [31-33] and homodimers of the viral protein may bind to two molecules of the EpoR, bringing the two together in the absence of Epo. This may allow the constitutive activation of either Jak2 or another member of the Jak kinase family, leading to constitutive phosphorylation of Stat proteins. We are currently analyzing SFFV-infected HCD-57 cells to determine if Jak2 kinase or perhaps another Jak-related kinase is constitutively phosphorylated. Alternatively, SFFV interaction with the EpoR complex may interfere with the negative regulatory mechanism of the Jak-Stat pathway. Phosphorylation of the EpoR was recently shown to provide a docking site for hematopoietic cell phosphatase [29,30], which may lead to dephosphorylation of the Jak2 kinase and down-modulation of the signal. The SFFV envelope glycoprotein may block phosphorylation of the EpoR, so there is no docking site for the phosphatase, or it may block the phosphatase from binding to the docking site on the phosphorylated EpoR. We are currently carrying out studies to determine whether either of these events are occurring. Further studies using mutants of SFFV and the EpoR, as well as studies using erythroid cells from mice that are resistant to the biological effects of SFFV, will allow us to determine the various viral and cellular factors required for constitutive activation of the Epo signal transduction pathway by SFFV.

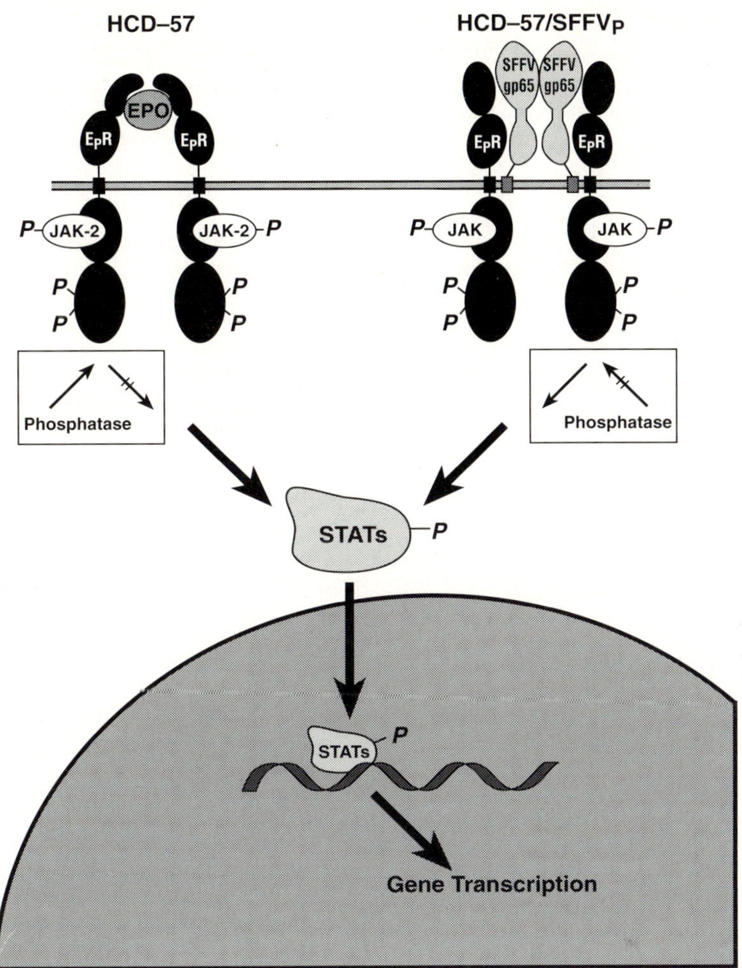

Fig. 4. Putative Model for the Activation of Stat-related Proteins by Epo and SFFV.

References

1. Youssoufian H, Longmore G, Neumann D, Yoshimura A, Lodish HF (1993) Structure, function, and activation of the erythropoietin receptor. Blood 81:2223-2236.
2. Wojchowski DM, He T-C (1993) Signal transduction in the erythropoietin receptor system. Stem Cells 11:381-392.
3. Ruscetti SK (1995) Erythroleukaemia induction by the Friend spleen focus-forming virus. In: Young N (ed) Bailliere's Clinical Haematology 8:225-247.
4. Wolff L, Ruscetti SK (1988) The spleen focus-forming virus (SFFV) envelope gene, when introduced into mice in the absence of other SFFV genes, induces acute erythroleukemia. J Virol 62:2158- .
5. Ruscetti SK, Ruscetti FW (1989) Apparent Epo-independence of erythroid cells infected with the polycythemia-inducing strain of Friend spleen focus-forming virus is not due to Epo production or change in number or affinity of Epo receptors. Leukemia 3:703-707.
6. D'Andrea AD, Yoshimura A, Youssoufian H et al (1991) The cytoplasmic region of the erythropoietin receptor contains nonoverlapping positive and negative growth-regulatory domains. Mol Cell Biol 11:1980-1987.
7. Yoshimura A, Zimmers T, Neumann D et al (1992) Mutations in the Trp-Ser-X-Trp-Ser motif of the erythropoietin receptor abolish processing, ligand binding, and activation of the receptor. J Biol Chem 267:11619-11625.
8. Casadevall N, Lacombe C, Muller O, Gisselbrecht S, Mayeux P (1991) Multimeric structure of the membrane erythropoietin receptor of murine erythroleukemia cells (Friend cells). J Biol Chem 266:16015.
9. Ferro FE Jr, Kozak SL, Hoatlin ME, Kabat D (1993) Cell surface site for mitogenic interaction of erythropoietin receptors with the membrane glycoprotein encoded by Friend erythroleukemia virus. J Biol Chem 268:5741.
10. Li J-P, Hu H-O, Niu Q-T, Fang C (1995) Cell surface activation of the erythropoietin receptor by Friend spleen focus-forming virus gp55. J Virol 69:1714-1719.
11. Feig LA (1993) The many roads that lead to Ras. Science 260:767-768.
12. Pawson T (1994) Regulation of the Ras signalling pathway by protein-tyrosine kinases. Biochem Soc Trans 22:455-460.
13. Cutler RL, Liu L, Damen JE, Krystal G (1993) Multiple cytokines induce the tyrosine phosphorylation of Shc and its association with Grb2 in hemopoietic cells. J Biol Chem 268:21463-21465.
14. Damen JE, Liu L, Cutler RL, Krystal G (1993) Erythropoietin stimulates the tyrosine phosphorylation of Shc and its association with Grb2 and a 145-Kd tyrosine phosphorylated protein. Blood 82:2296.
15. Carroll MP, Spivak JL, McMahon M, Weich N, Rapp UR, May WS (1991) Erythropoietin induces Raf-1 activation and Raf-1 is required for erythropoietin-mediated proliferation. J Biol Chem 266:14964.
16. Darnell JE Jr, Kerr IM, Stark GR (1994) JAK-STAT pathways and transcriptional activation in response to IFNs and other extracellular signaling proteins. Science 264:1415.
17. Ihle JN, Witthuhn BA, Quelle FW, Yamamoto K, Thierfelder WE, Kreider B, Silvennoinen O. (1994) Signaling by the cytokine receptor superfamily: JAKs and STATs. Trends Biochem Sci 19:222.
18. Taniguchi T (1995) Cytokine signaling through nonreceptor protein tyrosine kinases. Science 268:251-255.
19. Withuhn BA, Quelle FW, Silvennoinen O, Yi T, Tang B, Miura O, Ihle JN (1993) Jak2 associates with the erythropoietin receptor and is tyrosine phosphorylated and activated following stimulation with erythropoietin. Cell 74:227.
20. Miura O, Nakamura N, Quelle FW, Witthuhn BA, Ihle JN, Aoki N (1994) Erythropoietin induces association of the Jak2 protein tyrosine kinase with the erythropoietin receptor in vivo. Blood 84:1501.

21. Zhuang H, Patel SV, He T-C, Sonsteby SK, Niu Z, Wojchowski DM (1994) Inhibition of erythropoietin-induced mitogenesis by a kinase-deficient form of Jak2. J Biol Chem 269:21411.
22. Ruscetti SK, Janesch NJ, Chakraborti A, Sawyer ST, Hankins WD (1990) Friend spleen focus-forming virus induces factor independence in an erythropoietin-dependent erythroleukemia cell line. J Virol 63:1057.
23. Wagner BJ, Hayes TE, Hoban CJ, Cochran BH (1990) The SIF binding element confers sis/PDGF inducibility onto the c-fos promoter. EMBO J 9:4477.
24. Pearse RN, Feinman R, Ravetch JV (1991) Characterization of the promoter of the human gene encoding the high-affinity IgG receptor: Transcriptional induction by γ-interferon is mediated through common DNA response elements. Proc Natl Acad Sci USA 88:11305-11309.
25. Ohashi T, Masuda M, Ruscetti SK (1995) Induction of sequence-specific DNA-binding factors by erythropoietin and the spleen focus-forming virus. Blood 85:1454-1462.
26. Watowich SS, Yoshimura A, Longmore GD, Hilton DJ, Yoshimura Y, Lodish HF (1992) Homodimerization and constitutive activation of the erythropoietin receptor. Proc Natl Acad Sci USA 89:2140-2144.
27. Ohashi H, Maruyama K, Liu Y-C, Yoshimura A (1994) Ligand-induced activation of chimeric receptors between the erythropoietin receptor and receptor tyrosine kinases. Proc Natl Acad Sci USA 91:158-162.
28. Watowich SS, Hilton DJ, Lodish HF (1994) Activation and inhibition of erythropoietin receptor function: role of receptor dimerization. Mol Cell Biol 14:3535-3549.
29. Yi T, Zhang J, Miura O, Ihle JN (1995) Hematopoietic cell phosphatase associates with erythropoietin (Epo) receptor after Epo-induced receptor tyrosine phosphorylation: identification of potential binding sites. Blood 85:87-95.
30. Klingmuller U, Lorenz U, Cantley LC, Neel BG, Lodish HF (1995) Specific recruitment of SH-PTP1 to the erythropoietin receptor causes inactivation of JAK2 and termination of proliferative signals. Cell 80:729-738.
31. Gliniak BC, Kabat D (1989) Leukemogenic membrane glycoprotein encoded by Friend spleen focus-forming virus: transport to cell surfaces and shedding are controlled by disulfide-bonded dimerization and by cleavage of a hydrophobic membrane anchor. J Virol 63:3561-3568.
32. Kilpatrick DR, Srinivas RV, Compans RW (1989) The spleen focus-forming virus envelope glycoprotein is defective in oligomerization. J Biol Chem 264:10732-10737.
33. Yang Y, Tojo A, Watanabe N, Amanuma H (1990) Oligomerization of Friend spleen focus-forming virus (SFFV) env glycoproteins. Virology 177:312-316.

The Neurofibromatosis Type 1 (NF1) Tumor Suppressor Gene and Myeloid Leukemia

D. A. Largaespada[1], C. I. Brannan[2], J. D. Shaughnessy[1], N. A. Jenkins[1], and N. G. Copeland[1].
[1]Mammalian Genetics Laboratory, ABL-Basic Research Program, NCI-Frederick Cancer Research and Development Center, Frederick, MD 21702. [2]Department of Molecular Genetics and Microbiology and Center for Mammalian Genetics, University of Florida, Gainesville, FL 32610.

Abstract

Activating mutations of *RAS* genes are a common event in the genesis of a wide variety of human tumor types (for review see Barbacid 1990). It has also become increasingly clear that many tumors without *RAS* gene mutations have acquired genetic lesions that contribute to tumorigenesis by causing increased or prolonged activation of Ras. In this paper we review and describe studies using human and mouse cells on the association between *NF1* gene loss and the development of myeloid leukemia. The *NF1* gene product, neurofibromin, has GTPase activating protein (GAP) activity for Ras and thus, could negatively regulate Ras. Therefore, *NF1* gene loss in myeloid cells may contribute to leukemogenesis by activating Ras.

Ras Deregulation and Myeloid Leukemia

The RAS family genes encode 21 kilodalton (kd) proteins that bind to the guanine nucleotides, GDP and GTP (for review see Bourne et al. 1990). When Ras is bound to GTP it is biologically active and when bound to GDP it is biologically inactive. The interconversion of Ras between the GDP and the GTP bound state is tightly regulated by other proteins (for review see Boguski and McCormick 1993). The $p21^{ras}$ proteins have a weak intrinsic GTPase activity, but this activity is greatly accelerated by interaction with proteins that are called GTPase activating proteins or GAPs. Two well studied GAPs from mammals are $p120^{GAP}$ and the product of the *NF1* gene, neurofibromin. Proteins, called guanine nucleotide exchange factors (GNEFs), also regulate the rate of dissociation of GDP from Ras and the association of GTP to Ras. A large amount of data indicate that Ras proteins play a vital role in the transduction of signals received by cells. The targets of activated Ras, i.e. Ras-GTP, include phosphotidylinositol-3-OH kinase (PI-3 kinase) and the Raf family of serine/threonine kinases.

A recent review has pointed out that many of the consistent genetic lesions observed in human myeloid leukemias are predicted to affect the Ras signaling pathway (Sawyers and Denny 1994). Classic oncogenic point mutations in *NRAS* or *KRAS*, that cause constitutively high Ras-GTP levels, occur in roughly 10-50% of myelodysplastic syndromes, myeloproliferative syndromes, and acute myeloid leukemias (Janssen et al. 1987, Padua et al. 1988, Neubauer et al. 1991, Lubbert et al. 1992, Jacobs 1992). Additionally, Ras can be deregulated by indirect mutations. The adult form of chronic myelogenous leukemia (CML) is almost always associated with the generation of the BCR-ABL fusion gene. The Bcr-abl oncoprotein is a tyrosine kinase that activates Ras by binding to and apparently activating the Grb-2 protein, an adaptor protein that transmits signals from receptor tyrosine kinases to the Sos proteins, which are guanine nucleotide exchange factors for Ras (Pendergast et al. 1993, Puil et al. 1994). Some cases of chronic myelomonocytic leukemia (CMML) have a t(5;12) translocation that fuses an Ets family transcription factor to the platelet derived growth factor receptor β (PDGFRβ) (Golub et al. 1994). Ets HLH domain dimerization may constitutively activate the tyrosine kinase domain from PDGFRβ of this fusion protein, thus leading to high Ras-GTP levels.

NF1 Syndrome and Myeloid Leukemia

NF1 syndrome is a common autosomal dominant genetic disorder (1 in 3500) characterized by the development of abnormal areas of pigmentation on the skin called cafe-au-lait spots, benign subcutaneous neurofibromas composed primarily of fibroblasts and Schwann cells, and glial cell growths on the iris called Lisch nodules. NF1 patients also have a predisposition to certain malignancies including neurofibrosarcomas, astrocytoma, pheochromocytoma, and many types of juvenile myeloid leukemia, especially juvenile chronic myelogenous leukemia (JCML) and a related disorder called Monosomy 7 syndrome, in which part or all of one copy of chromosome 7 are missing (Bader and Miller, 1978, Hope and Mulvihill 1981, Bader 1986). The *NF1* gene was cloned in 1990 (for review see Gutmann and Collins, 1993) and encodes a large 2818 amino acid protein called neurofibromin with GTPase activating protein or GAP activity for N-, H-, K-, and R-ras. Many of the tumors that develop in NF1 patients show loss of heterozygosity (LOH) for markers within and near the NF1 gene with retention of the mutant allele, suggesting that the NF1 gene is a tumor suppressor gene. Because neurofibromin is a GAP, its loss in cells could lead to abnormally high levels of Ras-GTP in the cell, mimicking the effect of classic oncogenic point mutations in RAS. Indeed, neurofibrosarcoma cell lines that fail to express neurofibromin have very high Ras-GTP levels ten hours after serum stimulation compared to neurofibrosarcomas that do express neurofibromin.

Patients with NF1 syndrome have an increased risk of developing several juvenile forms of myeloid leukemia. Studies with affected NF1 patients have directly implicated *NF1* gene loss in myeloid leukemia and suggested that the consequence of such loss may be Ras activation. Leukemic cells from NF1 patients with JCML show LOH for markers within and near the *NF1* gene, but not the *P53* gene, with retention of the allele inherited from the parent with NF1 syndrome (Shannon et al. 1994). If NF1 gene loss is functionally equivalent to

Ras activation, then JCML cases in patients without NF1 should sometimes have *RAS* point mutations while JCML cases in patients without NF1 syndrome should never have *RAS* point mutations. This hypothesis was verified in a recent study by Kalra et al. (1994). The fact that mice with *Nf1* mutations are also predisposed to myeloid leukemia has opened up avenues for further study and for modeling the NF1 syndrome-associated myeloid diseases.

Studies Using Murine Models

An early indication that *Nf1* mutation in mice could lead to myeloid leukemia was provided by study of the BXH-2 recombinant inbred strain of mouse, which is highly prone to spontaneous myeloid leukemia with a retroviral etiology (Bedigian et al. 1984). BXH-2 mice develop lifelong viremia with a B-ecotropic retrovirus that is horizontally transmitted in these mice (Jenkins et al. 1982). Retroviruses can contribute to leukemia development by inserting near and activating proto-oncogenes or inserting within and inactivating tumor suppressor genes (for review see von Lohuizen and Berns 1990). We and others have attempted to identify the involved genes by searching for chromosomal sites that commonly have proviral insertions in BXH-2 tumors. One such site, called *Evi-2*, was found to be in a large intron of the *Nf1* gene (Buchberg et al. 1990, Cawthon et al. 1991). Proviral insertions at *Evi-2* often occurred in both alleles of the same clone (Fig. 1). This pattern of biallelic integration is rare for common viral integration sites that harbor dominantly acting oncogenes, but would be expected for a common viral integration site harboring a tumor suppressor gene.

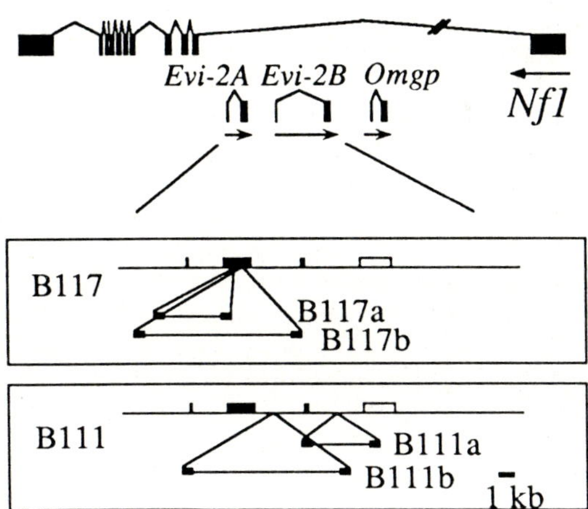

Fig. 1 Biallelic proviral integration at *Evi-2* in two BXH-2 myeloid leukemia cell lines. The top portion of the figure shows the relative size, orientation and location of exons from four genes (*Evi-2A*, *Evi-2B*, *Omgp* and *Nf1*) near the *Evi-2* locus. The position of proviral integrations in each allele of cell lines B117 and B111 are shown below.

To determine whether viral integration at *Evi-2* predisposes mice to myeloid disease by disrupting expression of *Nf1*, we have established about a dozen BXH-2 tumor cell lines (Table 1). The cell lines represent different stages of myeloid differentiation and all have multiple somatically acquired proviral integrations. These include cell lines with and without proviral integrations at *Evi-2*.

Table 1. Characteristics of BXH-2 leukemic cell lines.

		Cytochemistry[a]			Gene Rearrangements[b]				No. Proviruses[c]	
Line[d]	Growth	NSE	CAE	MPO	IgH	Igk	TcRb	Evi-2	Eco	MRV
B106E	Slow	+	-	-	+	-	-	-	4	0
B106L	Fast	+	-	-	+	-	-	+	2	0
B112	Fast	+	+	-	-	-	-	-	1	0
B113	Fast	+	-	-	-	-	-	-	3	0
B114	Slow	+	+	-	-	-	-	+	2	0
B117	Fast	+	+	-	-	-	-	+	6	1
B119	Fast	+	-	-	+	-	-	-	2	0
B132	Fast	+	-	-	-	-	-	-	7	1
B139	Fast	+	-	-	-	-	-	-	2	1
B140	Fast	-	-	-	-	-	-	-	2	0

[a] Cells were stained for a series of myeloid esterases including non-specific esterase (NSE), n-napthyl chloroacetate esterase (CAE), and myeloperoxidase (MPO). +, positive staining; -, negative staining.
[b] A plus (+) indicates that one or more DNA rearrangements were detected at the locus.
[c] Number of somatically acquired ecotropic (Eco) or MAIDS related (MRV) proviruses.
[d] B106E is an early passage and B106L is a late passage of the B106 cell line.

We have used the BXH-2 cell lines to determine whether proviral integration at *Evi-2* disrupts *Nf1* expression (Largaespada et al. 1995). Our results are consistent with the hypothesis that proviral disruption of *Nf1* expression is causally-associated with myeloid tumor development. In three BXH-2 cells lines that harbor *Evi-2* rearrangements only truncated *Nf1* transcripts are produced. These truncated transcripts appear to result from premature termination of *Nf1* transcripts within proviral sequences located at *Evi-2*. In addition, two of the cell lines expressing truncated *Nf1* transcripts produced no detectable neurofibromin (Largaespada et al. 1995).

It is likely that *Nf1* mutations alone are not sufficient to induce acute myeloid disease, such as that observed in BXH-2 mice. For example, deletions of one copy of chromosome 7, or 7q deletions, are frequently seen in myeloid leukemias that develop in patients with NF1 syndrome (Shannon et al. 1994). Human chromosome 7q seems to carry a tumor suppressor gene that is involved in myeloid tumor progression. The BXH-2 cell lines, produced in these studies, should provide useful reagents for identifying genes that cooperate with *NF1* to induce acute disease. Each BXH-2 cell lines carries, on average, 3.8 somatically acquired proviruses. It seems likely that some of these proviruses are affecting the expression of genes that cooperate with *Nf1* to induce acute disease. These cooperating genes will be identified by cloning the other somatically acquired proviruses from these cell lines, determining if they are located in common viral integration sites, and identifying nearby affected genes.

In addition to the studies described above, are results from Jacks et al. (1994) showing that a small percentage of very aged mice carrying one germline mutation at the *Nf1* gene developed myeloid leukemia with loss of the wild type allele. Taken together, these studies show that it will be possible to address the mechanisms by which *NF1* gene loss causes myeloid leukemia using mice as a model system.

Conclusions and Future Work

Studies reviewed here directly implicate *NF1* gene loss in the development of myeloid leukemia in both mouse and human. It seems likely, that the failure to express neurofibromin in myeloid cells disrupts regulation of Ras, leading to higher than normal levels of Ras-GTP in some circumstances. The Ras signal transduction pathway seems to be affected by many of the consistent genetic lesions found in several types of human myeloid leukemia. In particular, Ras activation seems to be associated with chronic myeloproliferative symptoms or myelodysplasia. Indeed, Bcr-abl, the Tel-PDGFRβ fusion protein, and *NF1* gene loss are all associated with chronic leukemia, that can then progress to acute leukemia. *RAS* point mutations are found in both chronic disease and many *de novo* cases of acute leukemia. However, since *RAS* point mutations are found in cases of chronic leukemia they probably cause chronic symptoms and are not associated directly with blast crisis or progression to acute leukemia.

We have begun to utilize mice, created using embryonic stem (ES) cell technology, that carry a germline null mutation in the *Nf1* gene to study the role of neurofibromin in myeloid cell development and disease. This null mutation, in the homozygous state, is lethal around mid-gestation (E11.5-E14.5), and causes a variety of phenotypes including generalized developmental delay, severe cardiac deformities, and overgrowth of certain sympathetic and parasympathetic ganglia (Brannan et al. 1994). In order to study *Nf1*$^{-/-}$ hematopoietic precursors we isolated fetal liver cells from *Nf1*$^{-/-}$ embryos prior to their death *in utero* , and studied them *in vitro* and *in vivo* by reconstituting lethally irradiated mice. We hope that such studies will shed light on the direct biological consequences of *Nf1* gene loss on myeloid cell development, response to growth factors, and neoplastic transformation. We also hope to test directly, whether *Nf1* gene loss leads to abnormal Ras activation in response to growth factors. As *Nf1* gene loss alone is likely to be insufficient for full transformation of myeloid cells, we hope to identify the oncogenes or tumor suppressor genes that collaborate to induce leukemia using proviral tagging. These goals will be most easily met using model murine systems that mimic, on a molecular level, the types of myeloid leukemia that occur in people.

Acknowledgments

This research was sponsored by the National Cancer Institute, DHHS, under contract with ABL. The contents of this publication do not necessarily reflect the views or policies of the Department of Health and Human Services, nor does mention of trade names, commercial products, or organizations imply endorsement by the U.S. Government. D. A. Largaespada is a Leukemia Society of America Fellow.

References

Bader, J. L. (1986). "Neurofibromatosis and cancer." Ann. N. Y. Acad. Sci. 486: 56-65.

Bader, J. L. and R. W. Miller (1978). "Neurofibromatosis and childhood leukemia." J. Pediat. 92: 925-929.

Barbacid, M. (1990). "Ras oncogenes: their role in neoplasia." Eur. J. Clin. Invest. 20: 225-235.

Basu, T. N., D. H. Gutmann, J. A. Fletcher, T. W. Glover, F. S. Collins and J. Downward (1992). "Aberrant regulation of ras proteins in malignant tumor cells from type 1 neurofibromatosis patients." Nature 356: 713-715.

Bedigian, H. G., D. A. Johnson, N. A. Jenkins, N. G. Copeland and R. Evans (1984). "Spontaneous and induced leukemias of myeloid origin in recombinant inbred BXH-2 mice." J. Virol. 51: 586-594.

Boguski, M. S. and F. McCormick (1993). "Proteins regulating ras and its relatives." Nature 366: 643-653.

Bourne, H. R., D. A. Sanders and F. McCormick (1990). "The GTPase superfamily: a conserved switch for diverse cell functions." Nature 348: 125-132.

Brannan, C. I., A. S. Perkins, K. S. Vogel, N. Ratner, M. L. Nordlund, S. W. Reid, A. M. Buchberg, N. A. Jenkins, L. F. Parada and N. G. Copeland (1994). "Targeted disruption of the neurofibromatosis type -1 gene leads to developmental abnormalities in heart and various neural crest-derived tissues." Genes Devel. 8: 1019-1029.

Buchberg, A. M., H. G. Bedigian, N. A. Jenkins and N. G. Copeland (1990). "Evi-2, a common integration site involved in murine myeloid leukemogenesis." Mol. Cell. Biol. 10: 4658-4666.

Cawthon, R. M., L. B. Andersen, A. M. Buchberg, G. Xu, P. O'Connel, D. Viskochil, R. B. Weiss, M. R. Wallace, D. A. Marchuk, M. Culver, et al. (1991). "cDNA sequence and genomic structure of EVI-2B, a gene lying within an intron of the neurofibromatosis type 1 gene." Genomics 9: 446-460.

DeClue, J. E., A. G. Papageorge, J. A. Fletcher, S. R. Diehl, N. Ratner, W. C. Vass and D. R. Lowy (1992). "Abnormal regulation of mammalian p21ras contributes to malignant tumor growth in von Recklinghausen (Type 1) neurofibromatosis." Cell 69: 265-273.

Emanuel, P. D., L. J. Bates, R. B. Castleberry, R. J. Gualtieri and K. S. Zuckerman (1991). "Selective hypersensitivity to granulocyte-macrophage colony-stimulating factor by juvenile chronic myeloid leukemia hematopoietic progenitors." Blood 77: 925-929.

Golub, T. R., G. F. Barker, M. Lovett and D. G. Gilliland (1994). "Fusion of PDGF receptor β to a novel ets-like gene, tel, in chronic myelomonocytic leukemia with t(5;12) chromosomal translocation." Cell 77: 307-316.

Gutmann, D. H. and F. S. Collins (1993). "The neurofibromatosis type 1 gene and its protein product, neurofibromin." Neuron 10: 335-343.

Hope, D. G. and J. J. Mulvihill (1981). "Malignancy in neurofibromatosis." Adv. Neurol. 29: 33-56.

Jacks, T., T. S. Shih, E. M. Schmitt, R. T. Bronson, A. Bernards and R. A. Weinberg (1994). "Tumor predisposition in mice heterozygous for a targeted mutation in Nf1." Nature Genet. 7: 353-361.

Jacobs, A. (1992). "Gene mutations in myelodysplasia." Leuk. Res. 16: 47-50.

Janssen, J. W. G., A. C. M. Steenvoorden, J. Lyons, B. Anger, J. U. Bohlke, J. L. Bos, H. Seliger and C. R. Bartram (1987). "RAS gene mutations in acute and chronic myelocytic leukemias, chronic myeloproliferative disorders, and myelodysplastic syndromes." PNAS 84: 9228-9232.

Jenkins, N. A., N. G. Copeland, B. A. Taylor, H. G. Bedigian and B. K. Lee (1982). "Ecotropic murine leukemia virus DNA content of normal and lymphomatous tissues of BXH-2 recombinant inbred mice." J. Virol. 42: 379-388.

Kalra, R., D. C. Paderanga, K. Olson and K. M. Shannon (1994). "Genetic analysis is consistent with the hypothesis that NF1 limits myeloid growth through p21ras." Blood 84: 3435-3439.

Largaespada, D. A., J. D. Shaughnessy, N. A. Jenkins and N. G. Copeland (1995). "Retroviral insertion at the Evi-2 locus in BXH-2 myeloid leukemia cell lines disrupts Nf1 expression without changes in steady state ras-GTP levels." J. Virol. 69: 5095-5102.

Lubbert, M., J. M. Jr., G. Kitchingman, F. McCormick, R. Mertelsmann, F. Herrmann and H. P. Koeffler (1992). "Prevalence of N-ras mutations in children with myelodysplastic syndromes and acute myeloid leukemia." Oncogene 7: 263-268.

Neubauer, A., K. Shannon and E. Liu (1991). "Mutation of the ras proto-oncogenes in childhood monosomy 7." Blood 77: 594-598.

Padua, R. A., G. Carter, D. Hughes, J. Gow, C. Farr, D. Oscier, F. McCormick and A. Jacobs (1988). "RAS mutations in myelodysplasia detected by amplification, oligonucleotide hybridization, and transformation." Leukemia 2: 503-510.

Pendergast, A. M., L. A. Quilliam, L. D. Cripe, C. H. Bassing, Z. Dai, N. Li, A. Batzer, K. M. Rabun, C. J. Der, J. Schlessinger, et al. (1993). "BCR-ABL-induced oncogenesis is mediated by direct interaction of the SH2 domain of the GRB-2 adaptor protein." Cell 75: 175-185.

Puil, L., J. Lui, G. Gish, G. Mbamalu, D. Bowtell, P. G. Pelicci, R. Arlinghaus and T. Pawson (1994). "Bcr-abl oncoproteins bind directly to activators of the ras signalling pathway." EMBO J. 13: 764-773.

Sawyers, C. and C. T. Denny (1994). "Chronic myelomonocytic leukemia: Tel-a-kinase what ets all about." Cell 77: 171-173.

Shannon, K. M., P. O'Connell, G. A. Martin, D. Paderanga, K. Olson, P. Dinndorf and F. McCormick (1994). "Loss of thenormal NF1 allele from the bone marrow of children with type 1 neurofibromatosis and malignant myeloid disorders." New Eng. J. Med. 330: 597-601.

von Lohuizen, M. and A. Berns (1990). "Tumorigenesis by slow transforming retroviruses--an update." Biochim. Biophys. Acta. 1032 (2-3): 213-235.

Molecular Aspects of Myeloid Leukemia. II. Human Studies

Rearrangement of the *AML1/CBFA2* Gene in Myeloid Leukemia with the 3;21 Translocation: Expression of Co-Existing Multiple Chimeric Genes with Similar Functions as Transcriptional Repressors, but with Opposite Tumorigenic Properties

C. Zent[1], N. Kim[2], S. Hiebert[3], D-E. Zhang[4], D. G. Tenen[4], J. D. Rowley[1,5], and G. Nucifora[1], Departments of [1]Medicine, of [2]Biochemistry, and of [5]Molecular Genetics and Cell Biology, University of Chicago, Chicago, IL, [3] Dept. Biochemistry, St. Jude Children's Research Hospital, Memphis, TN, [4]Hematology/Oncology Division, Harvard Medical School, Boston MA.

Abstract

Several recurring chromosomal translocations involve the *AML1* gene at 21q22 in myeloid leukemias resulting in fusion mRNAs and chimeric proteins between *AML1* and a gene on the partner chromosome. *AML1* corresponds to *CBFA2*, one of the DNA-binding subunits of the enhancer core binding factor CBF. Other CBF DNA-binding subunits are CBFA1 and CBFA3, also known as *AML3* and *AML2*. AML1, AML2 and AML3 are each characterized by a conserved domain at the amino end, the runt domain, that is necessary for DNA-binding and protein dimerization, and by a transactivation domain at the carboxyl end. *AML1* was first identified as the gene located at the breakpoint junction of the 8;21 translocation associated with acute myeloid leukemia. The t(8;21)(q22;q22) interrupts *AML1* after the runt homology domain, and fuses the 5' part of *AML1* to almost all of *ETO*, the partner gene on chromosome 8. AML1 is an activator of several myeloid promoters; however, the chimeric AML1/ETO is a strong repressor of some AML1-dependent promoters.

AML1 is also involved in the t(3;21)(q26;q22), that occurs in myeloid leukemias primarily following treatment with topoisomerase II inhibitors. We have studied five patients with a 3;21 translocation. In all cases, *AML1* is interrupted after the *runt* domain, and is translocated to chromosome band 3q26. As a result of the t(3;21), *AML1* is consistently fused to two separate genes located at 3q26. The two genes are *EAP*, which codes for the abundant ribosomal protein L22, and *MDS1*, which encodes a small polypeptide of unknown function. In one of our patients, a third gene *EVI1* is also involved. *EAP* is the closest to the breakpoint junction with *AML1*, and *EVI1* is the furthest away. The fusion of *EAP* to *AML1* is not in frame, and leads to a protein that is terminated shortly after the fusion junction by introduction of a stop codon. The fusion of *AML1* to *MDS1* is in frame, and adds 127 codons to the interrupted *AML1*. Thus, in the five cases that we studied, the 3;21 translocation results in expression of two coexisting chimeric mRNAs which contain the identical runt domain at the 5' region, but differ in the 3' region. In addition, the chimeric transcript *AML1/MDS1/EVI1* has also been detected in cells from one patient with the 3;21 translocation as well as in one of our patients.

Several genes necessary for myeloid lineage differentiation contain the target sequence for AML1 in their regulatory regions. One of them is the *CSF1R* gene. We have compared the normal AML1 to AML1/MDS1, AML1/EAP and AML1/MDS1/EVI1 as transcriptional regulators of the *CSF1R* promoter. Our results indicate that AML1 can activate the promoter, and that the chimeric proteins compete with the normal AML1 and repress expression from the *CSF1R* promoter. AML1/MDS1 and AML1/EAP affect cell growth and phenotype when expressed in rat fibroblasts. However, the pattern of tumor growth of cells expressing the different chimeric genes in nude mice is different. We show that when either fusion gene is expressed, the cells lose contact inhibition and form foci over the monolayer. In addition, cells expressing AML1/MDS1 grow larger tumors in nude mice, whereas cells expressing only AML1/EAP do not form tumors, and cells expressing both chimeric genes induce tumors of intermediate size. Thus, although both chimeric genes have similar effects in transactivation assays of the *CSF1R* promoter, they affect cell growth differently in culture and have opposite effects as tumor promoters in vivo. Because of the results obtained with cells expressing one or both genes, we conclude that *MDS1* seems to have tumorigenic properties, but that *AML1/EAP* seems to repress the oncogenic property of *AML1/MDS1*.

Introduction

AML1 is involved in a number of chromosomal translocations associated with leukemia, such as the t(8;21), identified in about 20% of AML-M2 patients [1], and the rarer t(3;21)(q26;q22), t(5;21)(q13;q22) and t(17;21)(q11;q22), all associated with myeloid leukemia [2,3]. *AML1* also called *CBFA2*, is the DNA-binding subunit of the heterodimeric transcription factor CBF, and encodes a protein with a novel DNA-binding domain at the amino end, and a transactivation domain at the carboxyl end. The DNA-binding region of *CBFA2* has homology to the DNA-binding domain of the *D. melanogaster* segmentation gene *runt*, and is often indicated as the runt homology domain [4]. *AML1* is a member of a growing family of related CBF subunits, and recently *AML2* or *CBFA3*, located at chromosome band 6p21, and *AML3* or *CBFA1*, located at chromosome band 1p34-pter, have been cloned and described [5,6]. Four isoforms of *AML1* have been reported all of which have different amino and carboxyl ends: two of them, of 250 and 259 amino acids (AML1A), terminate shortly after the runt homology domain, and do not have any transactivation region [1,7]. The remaining two, of 472 and 479 amino acids (AML1B), contain an extension at the carboxyl end consisting of the transactivation domain of the protein [5,8]. The runt homology domain is necessary not only to bind to DNA, but also to bind to CBFB. CBFB is the second subunit of CBF, and binding of the two subunits increases the affinity of the transcription factor for the DNA target [9]. *CBFB*, located at chromosome band 16q22, is involved in the t(16;16) and the inv(16), associated with almost all cases of AML-M4Eo [10]. In this translocation, *CBFB*, except for the last 17 codons, fuses to the 3' end of the myosin heavy chain gene *MYH11*, encoding the coiled coil motif of the myosin protein, to produce the chimeric gene *CBFB/MYH11* [10]. Thus, the two subunits of CBF are involved in the two most frequent translocations detected in AML. In all of the chromosomal translocations associated with myeloid leukemia involving 21q22 that have been studied at a molecular level, *AML1* is interrupted by the translocation breakpoint after the runt homology domain, and the 5' part of *AML1* is then fused to a gene on the other chromosome [11]. Recently, the translocation breakpoint of the t(12;21) has been cloned, and the genes involved at the breakpoint junction are *TEL* and *AML1* [12]. It is worth noting that in the t(12;21), associated with acute B-cell lymphoblastic leukemia, *AML1* is not truncated by the breakpoint, but it is fused in its entirety to the 5' end of *TEL* to produce the chimeric gene *TEL/AML1* [12]. Thus, truncation of *AML1* and substitution of the transactivation domain with part of a gene from another chromosome have only been observed so far in myeloid leukemia. The DNA target site of AML1, TGTGGT, is very common in eukaryotic promoter regions, and it has been identified in regulatory regions of several genes of the myeloid-specific lineage such as those of the myeloperoxidase, neutrophil-elastase, *GMCSF* [13] and *CFS1* receptor (*CSF1R*) genes [14], and in the enhancer of the T-cell receptors (TCR) β, γ and δ [15,16]. In addition, the target site of AML1 is also found in retroviral long terminal repeats and in viral promoter regions and it is also necessary for replication of the murine polyoma virus [17,18]. Recently, it was shown that AML1B can transactivate the TCR β [8,19] and *GMCSF* [13] promoters, whereas AML1A is a weak repressor of TCR β (8). In addition, it was shown that AML1/ETO and AML1/EVI1 are repressors of the AML1B-dependent TCR β (8,19).

The t(3;21), associated with therapy-related myeloid leukemia and with chronic myeloid leukemia in blast crisis [20,21] is a unique translocation in that at least two chimeric genes are produced with *AML1* in all of the patients whom we analyzed [2]. The two genes located at chromosome band 3q26 are *EAP*, which encodes the ribosomal protein L22 [22], and *MDS1*, whose product has unknown function. The fusion of *EAP* to *AML1* is not in frame, and translation is terminated shortly 17 codons after the fusion junction by a stop codon. Thus, *AML1/EAP* is similar to the normal short isoforms of *AML1* that have been described in all but the 17 distal codons. The fusion junction of *MDS1* to *AML1* occurs exactly at the same position as for *AML1/EAP* [2,22], however the part of *MDS1* that is added after the runt homology domain is in frame, and the predicted chimeric protein contains 127 codons fused to the runt homology domain of AML1 [2]. We studied five patients with myeloid leukemia and a t(3;21), and the two fusion genes *AML1/EAP* and *AML1/MDS1* have been detected at the same time in our patients, indicating that they co-exist in the cells with the translocation. Fusion genes between *AML1* and *EVI1* have also been described in cells with the t(3;21) [2,23]. However, only one of five t(3;21) patients whom we analyzed expressed the fusions with *EVI1* [2], indicating that they are not consistently expressed in the t(3;21). Fig. 1 shows a diagram of the fusions between *AML1* and the genes on 3q26 that we detected in our leukemia patients with a t(3;21).

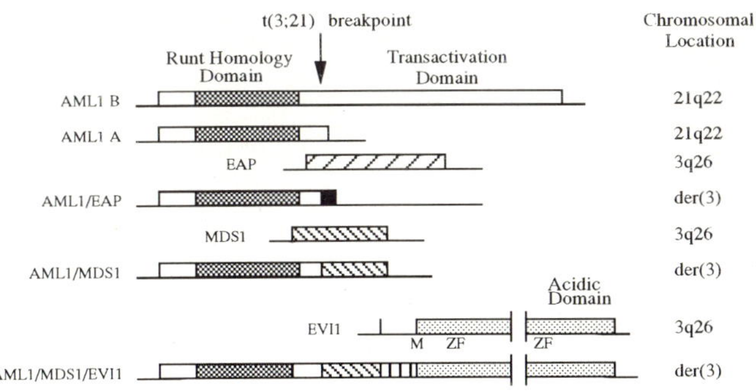

Fig. 1 Comparisons of the various normal and fusion *AML1* products and chromosomal location of the normal and fusion genes. Boxes indicate open reading frames, and thin lines indicate 5' or 3' untranslated regions. The breakpoint junction of the t(3;21) is indicated by a vertical arrow 3' of the runt homology domain (shaded box) and is located 9 codons before the end of *AML1A*. *AML1A* does not contain the transactivation domain. The fusion with *EAP* is not in frame, and a stop is introduced 17 codons after the *AML1/EAP* junction. The predicted translation of these 17 codons (indicated by a small black box) has no homology to either the AML1 or the EAP amino acid sequence. The fusion of *AML1* with *MDS1* is in frame, and adds 127 codons to *AML1*. The open reading frame of the normal *EVI1* starts in the third exon, where the first methionine (M) is located. Thus, the second exon of *EVI1*, indicated by a single line, is not normally translated although it is in frame. The fusion of *AML1* to *EVI1* includes *MDS1* as well, and joins *AML1* to the second exon of *EVI1* which is now translated, and is shown as a box with vertical stripes in the diagram. ZF indicates two groups of seven and three zinc fingers in EVI1. The cDNA of *EVI1* is not drawn to scale and is shown interrupted by vertical bars.

The three genes that are involved with *AML1* in the t(3;21) were mapped by fluorescence in situ hybridization (FISH) and pulsed field gel electrophoresis (PFGE). The combined results indicated that *EAP* was the closest gene to the breakpoint junction with *AML1*, and *EVI1* was the furthest away [2]. Fig. 2 shows a diagram of the germline configuration of the three genes.

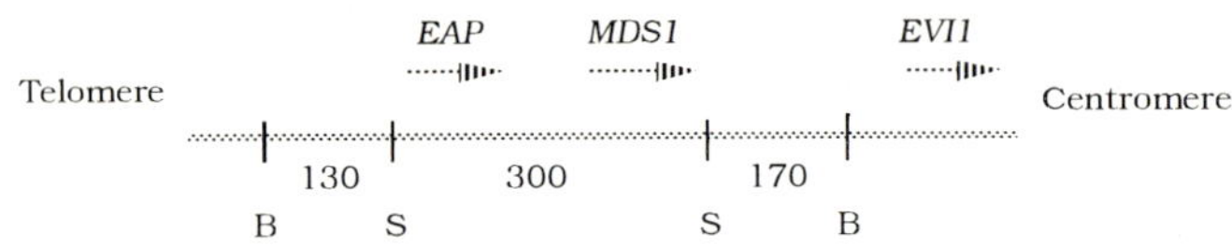

Fig. 2 Illustration of the germline configuration of the genes on chromosome 3 obtained by PFGE mapping and FISH analysis. The horizontal arrows indicate the direction of transcription of the genes. The short vertical bars indicate the restriction site for *BSS*HII (B) and *Sfi*I (S), and the numbers below the horizontal lines indicate the size of the restriction fragments in kb. In five patients with the t(3;21) whom we analyzed, the breakpoint and fusion with chromosome band 21q22 is telomeric of *EAP*, as determined by FISH analysis [2].

We have investigated the roles of the three fusion genes *AML1/EAP*, *AML1/MDS1* and *AML1/MDS1/EVI1* by comparing their effects in the activation of the luciferase gene cloned under the control of the AML1B-dependent *CSF1R* promoter. We have also evaluated the effects of *AML1/EAP* and *AML1/MDS1* on the growth characteristics and tumorigenicity of Rat 1A cells. Our results indicate that all three fusion products repress activation of the *CSF1R* promoter. However, whereas cells expressing either *AML1/EAP* or *AML1/MDS1* form foci in culture, only cells expressing *AML1/MDS1* form larger tumors in nude mice, whereas cells expressing *AML1/EAP* repress tumor growth.

Results

1. AML1/EAP, AML1/MDS1 and AML1/MDS1/EVI1 repress the AML1-dependent promoter of the *CSF1R* gene.

AML1B binding sites have been identified in the regulatory regions of several genes involved in myeloid cell growth and differentiation. One of them regulates the *CSF1R* gene [14]. To determine whether the chimeric genes *AML1/EAP*, *AML1/MDS1* and *AML1/MDS1/EVI1* affect the *CSF1R* promoter, and to compare their role as transcriptional activators to that of the normal AML1B, we measured the level of luciferase activity induced by the normal and chimeric genes in a system in which luciferase was cloned under the *CSF1R* promoter. Plasmids expressing the normal and chimeric *AML1* genes were transiently transfected into murine P19 cells with plasmid M-CSF-Luc as reporter. This plasmid has been described [14], and contains the luciferase gene under the control of the *CSF1R* promoter. To normalize the results, the transfecting DNA contained constant amounts of plasmid pCH110, that constitutively expresses β-galactosidase. Normalized representative results of the transactivation studies are shown in Fig. 3A. These results show that addition of *AML1B* consistently increased the level of the luciferase gene 3-5 fold over the background (Fig. 3A, bars 1 and 2), whereas addition of the chimeric genes did not affect the luciferase levels (Fig. 3A, bars 3, 4 and 5). Thus, similarly to what happens in the AML1/ETO fusion [8], the transactivation domain of AML1B cannot be substituted by the region of one of the t(3;21) partner genes, and each one of the three chimeric proteins that result from the t(3;21) is functionally inactive as transcriptional transactivator. In addition, when plasmids expressing fusion genes were added with *AML1B* at a 1:1 molar ratio, the activation of the reporter gene was reduced to the background value (Fig. 3A, bars 6, 7 and 8), indicating that although the chimeric proteins have lost the ability to transactivate the *CSF1R* promoter, they still maintain the capability to bind a factor (or DNA site) that is necessary for the transactivation. Titration of luciferase activation by the normal AML1B with increasing amount of a plasmid expressing one of the fusion genes confirmed these results, and showed that when any of the chimeric genes is in molar excess compared to the normal AML1B, the luciferase activity due to endogenous AML1B was also repressed. Representative titration results are shown in Fig. 3B.

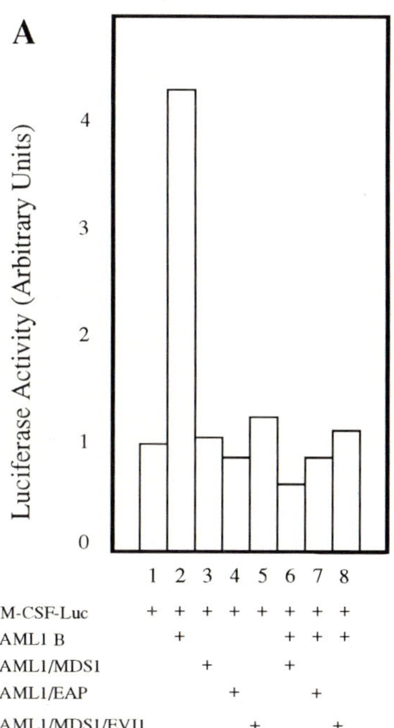

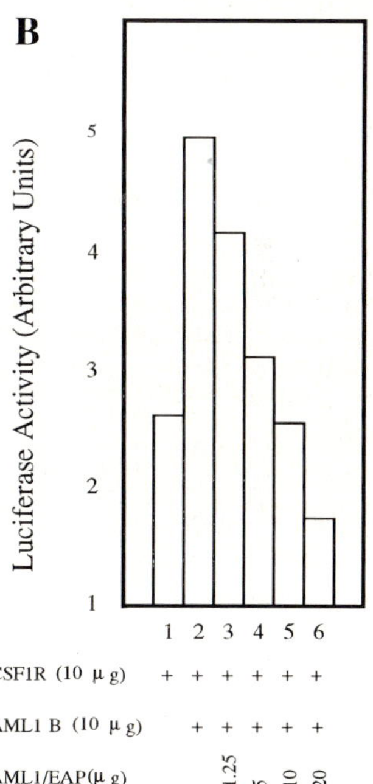

Fig. 3 A. The three fusion protein products of the t(3;21) fail to activate the *CSF1R* promoter but compete with the normal AML1B protein to repress the *CSF1R* promoter. 10 µg of each plasmid with 5 µg of pCH110, expressing β-galactosidase, was used to transfect $1x10^6$ murine P19 cells. Identical volumes of cell extracts were used to measure the luciferase activity. Luciferase activity readings were normalized to relative expression of β-galactosidase and plotted as shown. Addition of AML1B increased the luciferase activity 3-5 fold over the background (bars 1 and 2), whereas addition of fusion proteins failed to activate luciferase (bars 3, 4 and 5) or they repressed activation by AML1B (6, 7 and 8). B. Titration of luciferase activity by AML1/EAP. Increasing amounts of plasmid expressing *AML1/EAP* were added as indicated to a constant amount of plasmid expressing *AML1B*. Total repression of luciferase was observed at a molar ratio of 1:1. At higher molar ratio of AML1/EAP:AML1B, endogenous AML1B activity was also repressed.

2. AML1/EAP and AML1/MDS1 transform Rat1A cells in culture.

AML1/EAP and *AML1/MDS1* are consistently detected in our patients with myeloid leukemia and the t(3;21) [2], and by analogy with other proto-oncogenes that are activated by chromosomal translocations and that have been shown to be directly involved in transformation, it is possible that *AML1/EAP* and *AML1/MDS1* contribute to the development or the progress of the disease. To address the question of their role in cell transformation, we stably transfected Rat1A cells with plasmids expressing one or both chimeric cDNAs, and analyzed the transfected cells to detect any change in morphology or in growth pattern compared to the parental cells or to cells expressing the vector plasmid only. Although the cells expressing the chimeric genes did not show any variation in the size of colonies grown in soft agar, they clearly lost contact inhibition, and acquired the ability to form foci expanding over the monolayer. This is illustrated in Fig. 4, in which cells transfected with the vector plasmid only (Fig. 4, panel 1) or cells transfected with *AML1/EAP* (Fig. 4, panel 2) or *AML1/MDS1* (Fig. 4, panel 3) are shown. In this assay, cells were transfected with 10 µg of plasmid and were allowed to grow without selection. After two weeks, foci extending over the monolayer were clearly observed in cells transfected with one of the fusion genes (Fig. 4, panels 2 and 3), but were not detected in cells transfected with the vector only (Fig. 4, panel 1).

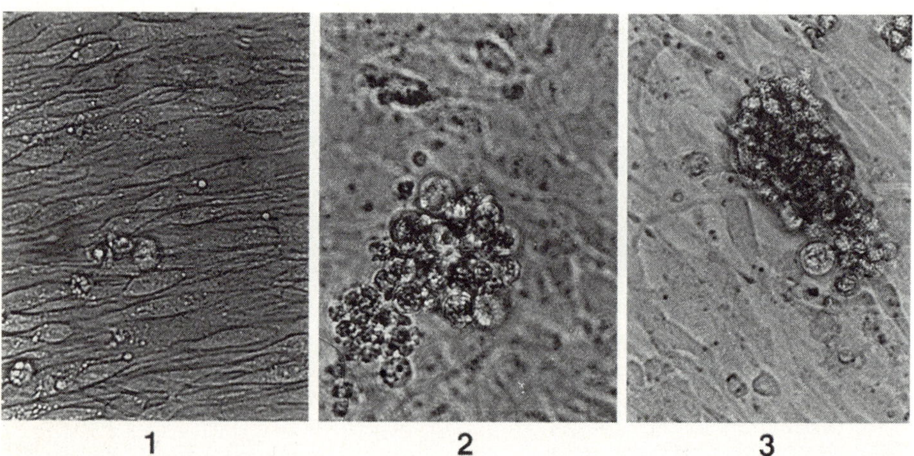

Fig. 4. *AML1/EAP* and *AML1/MDS1* transform Rat1A cells in culture. Rat1A fibroblasts were transfected separately with 10 µg of each plasmid DNA by the calcium phosphate

precipitate method. The cells were allowed to grow undisturbed without antibiotic selection for several days after reaching confluency. Cells transfected with the vector plasmid DNA stopped growing at confluency (panel 1). However, cells that had integrated either *AML1/EAP* (panel 2) or *AML1/MDS1* (panel 3) did not stop growing after confluency, and formed large foci expanding over the cell monolayer.

Several clumps of cells detached from the foci and remained in suspension. By using the trypan blue exclusion assay, we determined that 90 to 95% of the cells in suspension were viable. These cells were transferred to flasks to allow growth in suspension, and their growth pattern was observed for about three weeks. After this time, the cells were discarded. Cells expressing *AML1/MDS1* grew in suspension with a doubling time of approximately 4-5 days (data not shown). Suspension cells expressing *AML1/EAP* remained viable for all the time they were observed, although they did not increase in number significantly (data not shown). Interestingly, once the cells had detached from the plate, they continued to grow in suspension and did not adhere to the plate surface any longer.

3. In vivo tumorigenicity of transfected Rat1A cells.

Rat1A cells produce slow-growing tumors in nude mice. To determine whether *AML1/EAP* or *AML1/MDS1* chimeric genes had any effect on the growth of tumors in nude mice, we compared the size of tumors produced in nude mice by Rat1A cells either untransfected or stably transfected with one or both chimeric genes. To avoid effects due to the integration site of the transfected DNA in the Rat1A genome, we used a cell population obtained by combining equal numbers of cells grown from 6-8 different transfected clones for each type of plasmid DNA. The single clones as well as the mixtures of clones obtained from them were analyzed and were found to have similar growth characteristics (data not shown). In this assay, we used 8 week old female nude mice as recipients. The mice were divided into five groups of four animals each, and injected with 0.4 and 0.8 x 10^6 cells on either side as follows: three groups of four mice each received cells transfected either with one of the chimeric genes or with both chimeric genes. Another group of four mice received cells transfected with the vector plasmid only, and the last group of four mice received untransfected Rat1A cells as controls. One of the four mice injected with cells expressing both chimeric genes died unexpectedly and without a known cause within a week of injection. Palpable swellings at the site of the injection were first detected in mice which had received the *AML1/MDS1* cells 26 days after the injections, and in the other animals by day 30. However, the swellings in animals with *AML1/EAP* cells did not grow with time. On day 40, the animals were sacrificed, and the tumors were dissected from the skin and subcutaneous tissue and weighed. Rat1A untransfected cells and cells containing the cloning vector only produced small tumors of similar size, whereas the tumors from the *AML1/MDS1* cells were overall significantly larger. Surprisingly, the cells that contained the *AML1/EAP* chimeric gene did not produce any tumor at all, but the swellings consisted of enlarged subcutaneous lymph nodes with features of acute and chronic inflammatory changes, suggesting a sustained immune response. Tumors derived from cells expressing both chimeric genes were significantly smaller than the background. Microscopic examination of the tumors revealed the presence of undifferentiated fibrosarcomas in all resected tissue except that derived from the *AML1/EAP* cells. All tumors had a high mitotic index (> 10 mitoses per high power field), were locally invasive and were histologically indistinguishable on light microscope. Comparison of the tumors is shown in Fig. 5. To demonstrate that the tumors derived from the injected cells which were hygromycin resistant, part of all types of tumors, except those derived from untransfected cells, were minced and grown in vitro in medium containing hygromycin. Cell growth was obtained for all tumors, except for the cells harvested from the *AML1/EAP* growth. There was no difference in phenotype between the original clones injected into the mice and the cells grown from the tumors, thus confirming that the tumors derived from the injected cells.

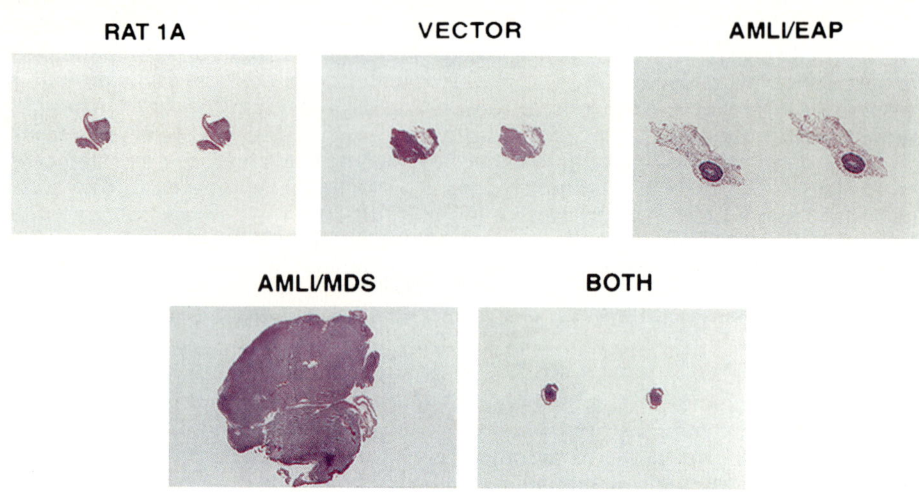

Fig. 5. Tumor growth in nude mice. Rat1A cells untransfected (top left), stably transfected with the vector plasmid (top middle) or with constructs expressing *AML1/EAP* (top right), *AML1/MDS1* (bottom left) or both chimeric genes (bottom right), were injected into 8 weeks old female nude mice. The animals were sacrificed 40 days later, and the sites of the injections were analyzed. No tumors were detected at the site of injection with cells expressing *AML1/EAP*, but enlarged lymph nodes were observed.

Discussion

In this report we have shown that the normal *AML1B* gene, corresponding to the DNA-binding subunit of the transcription factor CBF, is a transactivator of the promoter of the *CSF1R* gene in murine P19 cells. We also show that the chimeric genes *AML1/EAP*, *AML1/MDS1* and *AML1/MDS1/EVI1*, isolated from a myeloid leukemia patient with a t(3;21), fail to activate the *CSF1R* promoter in P19 cells, and that they compete with the normal *AML1B* to repress transactivation of the *CSF1R* promoter when co-expressed in P19 cells with *AML1B*. The molecular basis for the repression is not known, but it probably involves competition and higher binding affinity of the chimeric proteins for one or more factors, perhaps CBFB, which are rate limiting. We also show that Rat1A cells expressing either *AML1/EAP* or *AML1/MDS1* acquire a partially transformed phenotype in that they loose contact growth inhibition and form foci when grown in culture, although they are unable to form large colonies in soft agar. Although Rat1A cells expressing either *AML1/EAP* or *AML1/MDS1* acquire similar characteristics in in vitro assays, our preliminary in vivo results indicate that they have opposite tumor promoting effects in nude mice. Whereas *AML1/MDS1* is a strong tumor promoter, and cells expressing this chimeric gene will form significantly larger tumors in mice than those produced by the untransfected Rat1A cells, cells that express *AML1/EAP* consistently failed to grow as tumors in nude mice, and the site of injection of these cells showed sustained immune response, suggesting that expression of *AML1/EAP* could in fact repress in vivo growth of Rat1A cells. Because Rat1A cells expressing both chimeric genes formed tumors of much smaller size than the *AML1/MDS1* cells, our results suggest that the effect of *AML1/EAP* as tumor repressor is dominant over that of *AML1/MDS1* as tumor promoter. The mechanisms by which the two chimeric genes regulate tumor growth in nude mice are not known, but they clearly depend on the fusion partner with *AML1*. EAP was first isolated for binding to the small nuclear RNAs associated with the Epstein-Barr virus (EBV) [24] and was later identified as the component L22 of the large ribosomal subunit [25]. EAP has been detected in the nucleus and in the cytoplasm as well. However, because the fusion between *AML1* and *EAP* is not in frame [22,26], presumably only the short stretch of amino acids corresponding to the 17 codons 3' of the fusion junction are added to the truncated AML1. These amino acids are not related to EAP and their physiological function is not known. To determine their role, we are currently comparing the short AML1A lacking the transactivation domain to AML1/EAP in in vitro and in vivo assays. With regards to *MDS1*, this is a gene with a complex pattern of expression that can either be selectively expressed in normal human pancreas and kidney as a small protein of 169 amino acids, or that can be spliced to *EVI1* to yield a protein much larger than MDS1 or EVI1 by themselves (G.N., unpublished results). The analysis of the normal genes and of the two chimeric genes will be necessary to ultimately understand their role in leukemia.

References

1. Miyoshi H, Shimizu K, Kozu T, Maseki N, Kaneko Y and Ohki M. (1991) The t(8;21) breakpoints on chromosome 21 in acute myeloid leukemia are clustered within a limited region of a novel gene, *AML1*. Proc Natl Acad Sci USA: 88 10431-10435
2. Nucifora G, Begy CR, Kobayashi H, Claxton D, Pedersen-Bjergaard J, Parganas E, Ihle J, and Rowley JD (1994) Consistent intergenic splicing and production of multiple transcripts between *AML1* at 21q22 and unrelated genes at 3q26 in the (3;21)(q26;q22) translocation. Proc Natl Acad Sci USA 91:4004-4008.
3. Roulston D, Nucifora G, Dietz-Band J, LeBeau MM and Rowley JD [1993] Detection of rare 21q22 translocation breakpoints within the *AML1* gene in myeloid neoplasms by fluorescence in situ hybridization. Blood 82 (Suppl. 1): 532a
4. Erickson P, Gao J, Chang K-S, Look T, Whisenant E, Raimon S, Lasher R, Trujillo J, Rowley JD and Drabkin H (1992) Identification of breakpoints in t(8;21) AML and isolation of a fusion transcript with similarity to *Drosophila* segmentation gene *runt*. Blood 80:1825-1831
5. Levanon D, Negreanu V, Bernstein Y, Bar-Am I, Avivi L and Groner Y. AML1, AML2 and AML3, the human members of the *runt* domain gene-family: cDNA structure, expression and chromosomal localization. (1994) Genomics 23:425-430
6. Wijmenga C, Speck N A, Dracopoli N C, Lewis A F, Hofker M H and Collins F S (1995) Identification of a new murine *runt* domain-containing gene, *Cbfa3*, and localization of the human homolog, *CBFA3*, to chromosome 1p35-ter. Genomics 26:611-614
7. Nucifora G, Birn D, Espinosa R, LeBeau M, Erickson P, Roulston D, McKeithan TW, Drabkin H and Rowley JD (1993) Involvement of the *AML1* gene in the t(3;21) in therapy-related leukemia and in chronic myeloid leukemia in blast crisis. Blood 81:2728-2733
8. Meyers S, Lenny N and Hiebert S. (1995) The t(8;21) fusion protein interferes with AML1B-dependent transcriptional activation. Mol Cell Biol 15:1974-1982
9. Kagoshima H, Shigesada K, Satake M, Ito Y, Miyoshi M, Ohki M, Pepling M and Gergen P. (1993) The Runt domain identifies a new family of heterodimeric transcriptional activators. Trends Genet. 9:338-341

10. Liu P, Tarle SA, Hajra A, Claxton DF, Marlton P, Freedman M, Siciliano MJ and Collins FS (1993) Fusion between transcription factor CBFβ/PEBP2β and α myosin heavy chain in acute myeloid leukemia. Science 261:1041-1044
11. Nucifora G and Rowley JDR. *AML1* and the 8;21 and 3;21 translocations in acute and chronic myeloid leukemia. Blood, in press
12. Golub TR, Barker GF, Bohlander SK, Hiebert SW, Ward DC, Bray-Ward P, Morgan E, Raimondi SC, Rowley JD and Gilliland DG. Fusion of the *TEL* gene on 12p13 to the *AML1* gene on 21q22 in acute lymphoblastic leukemia. Proc Natl Acad Sci USA, in press.
13. Frank R, Zhang J, Hiebert S, Meyers S and Nimer S (1994) AML1B but not the AML1/ETO fusion protein can transactivate the GM-CSF promoter. Blood 84(Suppl. 1):229a
14. Zhang D-R, Fujioka K-I, Hetherington CJ, Shapiro LH, Look AT and Tenen DG. Identification of a region which directs monocytic activity of the CSF-1 (M-CFS) receptor promoter and binds PEBP2/CBF (AML1). Mol Cell Biol, in press.
15. Gottschalk LR and Leiden JM (1990) Identification and functional characterization of the human T-cell receptor β gene transcriptional enhancer: common nuclear proteins interact with the transcriptional regulatory elements of the T-cell receptor α and β genes. Mol Cell Biol 10:5486-5495
16. Kappes DJ, Browne CP and Tonegawa S (1991) Identification of a T-cell-specific enhancer at the locus encoding T-cell antigen receptor δ chain. Proc Natl Acad Sci USA 88:2204-2208
17. Speck NA, Renjifo B, Golemis E, Fredrickson TN, Hartley JW and Hopkins N (1990) Mutation of the core of adjacent LVb elements of the Moloney murine leukemia virus enhancer alters disease specificity. Genes Dev 4:233-242
18. Wang S and Speck NA (1992) Purification of core-binding factor, a protein that binds the conserved core site in murine leukemia virus enhancers. Mol Cell Biol 12:89-102
19. Tanaka T, Mitani K, Kurokawa M, Ogawa S, Tanaka K, Nishida J, Yazaki Y and Hirai H (1995) Dual functions of the AML1/EVI1 chimeric protein in the mechnism of leukemogenesis in t(3;21) leukemia. Mol Cell Biol 15:2383-2392
20. Rubin CM, Larson RA, Anastasi J, Winter JN, Thangavelu M, Vardiman J, Rowley JD and LeBeau MM (1990) t(3;21)(q26;q22): a recurring chromosomal abnormality in therapy-related myelodysplastic syndrome and acute myeloid leukemia. Blood 76:2594-2599
21. Rubin CM, Larson RA, Bitter MA, Carrino JJ, LeBeau MM, Diaz MO and Rowley JD (1987) Association of a chromosomal translocation with the blast phase of chronic myelogenous leukemia. Blood 70:1338-1344
22. Nucifora G, Begy CR, Erickson P, Drabkin H A and Rowley JD. (1993) The 3;21 translocation in myelodysplasia results in a fusion transcript between the *AML1* gene and the gene for EAP, a highly conserved protein associated with the Epstein-Barr virus small RNA EBER-1. Proc Natl Acad Sci USA 90: 7784-7788
23. Mitani K, Ogawa S, Tanaka T, Miyoshi H, Kurokawa MK, Mano H, Yazaki Y, Ohki M and Hirai H. (1994) Generation of the *AML1/EVI1* fusion gene in the t(3;21)(q26;q22) causes blastic crisis in chronic myelocytic leukemia. EMBO J. 13:504-510
24. Toczyski DPW and Steitz JA (1991) EAP, a highly conserved cellular protein associated with Epstein-Barr virus small RNAs (EBERs) EMBO J. 10:459-466
25. Toczyski D, Matera AG, Ward DC and Steitz JA The Epstein-Barr virus (EBV) small RNA EBER1 binds and relocalizes ribosomal protein L22 in EBV-infected human B lymphocytes (1994) Proc Natl Acad Sci USA 91:3463-3467
26. Sacchi N, Nisson P, Watkins P, Faustinella F, Wijsman J and Hagemeijer A. (1994) AML1 fusion transcripts in t(3;21) positive leukemia: Evidence of molecular heterogeneity and usage of splicing sites frequently involved in the generation of normal AML1 transcripts. Genes Chromosomes Cancer 11:226-231

Aknowledgments

This work was supported by National Institutes of Health grants CA 42557 (J.D.R.) and CA 67189 (G.N.), and by Department of Energy grant DE-FG02-86ER60408 (J.D.R.). G.N. is Special Fellow of the Leukemia Society of America.

Transcriptional Regulation by the t(8;21) Fusion Protein, AML-1/ETO

S. W. Hiebert[1], J. R. Downing[2], N. Lenny[1], and S. Meyers[1].
Departments of [1]Tumor Cell Biology and [2]Pathology, St. Jude Children's Research Hospital, 332 N. Lauderdale, Memphis, TN 38105.

Introduction

Certain chromosomal translocations are characteristically associated with acute myeloid leukemia (AML), suggesting that alteration of specific genes at the translocation breakpoints contribute to the genesis or maintenance of leukemia. The t(8;21) is the second most frequent chromosomal abnormality associated with AML, occurring in 12-15% of cases (Downing et al. 1993). The t(8;21) breakpoint was cloned and the disrupted gene on chromosome 21 was termed *AML1* (Miyoshi et al. 1991). Cytogenetic evidence from t(8;21)-containing leukemia cells with complex translocations indicates that the der(8) chromosome is conserved in each case; tightly linking this chromosomal aberration to the pathogenesis of AML (Rowley 1982). In addition, *AML1* is interrupted in the less commonly observed (3;21) and (12;21) translocations found in chronic myeloid leukemia and acute lymphocytic leukemia, respectively (Nucifora et al. 1993; Golub et al. 1995), suggesting that AML-1 may be a principal regulatory protein.

The *AML1* cDNA encodes a 250 amino acid protein (Miyoshi et al. 1991), whose predicted sequence suggests little about its normal cellular role. The sequence does contain a region of 118 amino acids that is 69% identical to the Drosophila *runt* protein sequence. *Runt* is a pair-rule gene that regulates pattern development and segmentation in the fly (Ingham and Gergen, 1988).
We have found that AML-1 regulates transcription through the "enhancer core" motif and that the AML-1/ETO fusion protein interferes with this regulatory function, suggesting a mechanism for leukemogenesis.

Studies of AML-1/ETO Function

Identification of the AML-1 DNA-Binding Site

To examine whether AML-1 could be a site specific DNA binding protein, we created a glutathione S-transferase-AML-1 fusion protein for use in DNA site selection experiments (Fig. 1). After seven successive rounds of DNA binding/PCR amplification, we cloned the DNA that bound to our fusion protein. Sequence analysis of these selected oligomers led us to identify the AML-1 consensus binding site as TGT/cGGT (Meyers et al. 1993). This site, which is known as the enhancer core, has been implicated in the tissue specific regulation of a large number of genes including myeloperoxidase, IL-3, GM-CSF, lck, and the T-cell receptor (TCR) enhancer (Meyers et al. 1993). AML-1 sites are also required for expression of the viral enhancers of polyoma virus and the Moloney murine leukemia virus (MoMLV) long terminal repeat (Redondo et al. 1991; Speck et al. 1990).

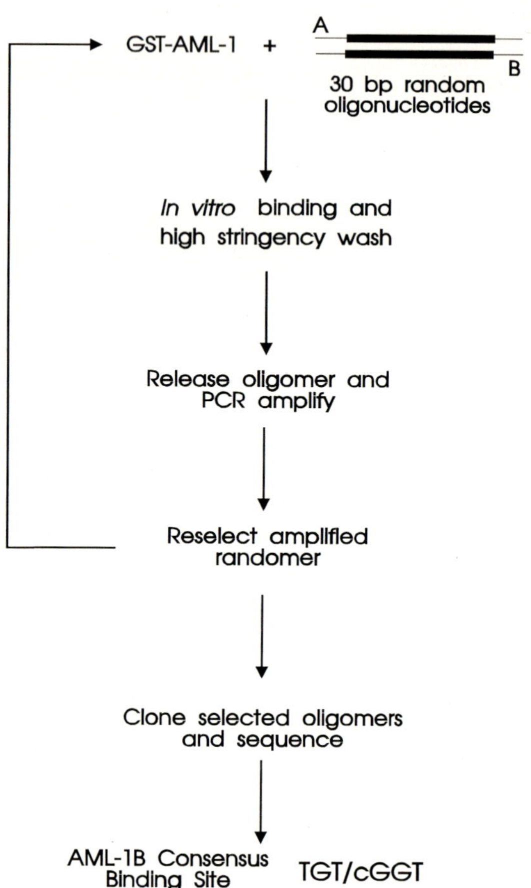

Fig. 1. Schematic diagram of procedures used to identify the AML-1 DNA binding site.

When we probed whole cell extracts with oligonucleotides containing the AML-1 binding site in electrophoretic mobility shift assays (EMSA), we identified two major AML-1-containing DNA-protein complexes, which we termed A and B (Meyers et al. 1993). By synthesizing AML-1 *in vitro* and adding it to cell lysates, we determined that AML-1 binds DNA in association with another protein, which was subsequently shown to be core binding factor β (CBFβ) (Meyers et al. 1993; Meyers et al. 1995; Wang et al. 1993). CBFβ is not a DNA binding protein, but is thought to modulate AML-1 activity

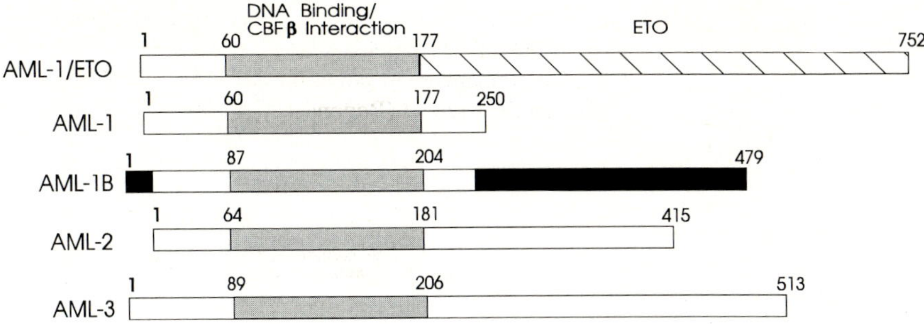

Fig. 2. Schematic diagram of the AML gene family and the t(8;21) fusion protein.

(Ogawa et al. 1993a). In gel mobility shift assays using whole cell extracts, DNA-protein complexes with slower mobilities than *in vitro* produced AML-1 interacted with anti-AML-1 serum, suggesting the presence of larger forms of AML-1 (Meyers et al. 1993). For this reason, we screened a human B-cell cDNA library and isolated AML-1B (Meyers et al. 1995) (Fig. 2), which contains both N- and C-terminal extensions to the original AML-1 cDNA. When AML-1 or AML-1B was synthesized *in vitro* and added to cell extracts or to *in vitro*-produced CBFβ for use in gel mobility shift assays, the A and B complexes were reconstituted, suggesting that these proteins constitute the endogenous DNA binding activities.

Having defined AML-1 and AML-1B as sequence-specific DNA binding proteins, we have begun to analyze their ability to activate transcription through consensus binding sites, and to examine the effect of the t(8;21) fusion protein on AML-1-dependent transcription. We co-transfected AML-1 or AML-1B with a reporter gene under the control of the TCRβ enhancer (Gottschalk and Leiden, 1990). This test gene contains two potential AML-1 binding sites which are required for tissue specific expression of the TCRβ enhancer (Gottschalk and Leiden, 1990; Prosser et al. 1992). The human cervical carcinoma cell line C33A was used for these experiments because it expresses relatively low amounts of endogenous AML-1 DNA binding activities and excess CBFβ (as judged by reconstitution assays (Meyers et al. 1995)). As an internal control for transfection efficiency, a plasmid expressing a secreted form of alkaline phosphatase was also included in each sample. AML-1 failed to activate the TCRβ enhancer when compared to the vector control. However, AML-1B could activate the enhancer, indicating that the additional C-terminal sequences of AML-1B are required for transcriptional activation (Meyers et al. 1995). By contrast, AML-1/ETO, like AML-1, failed to activate the test gene. Thus, one consequence of the t(8;21) is the loss of the C-terminal sequences present in AML-1B that are both unique to AML-1B and are required for transcriptional activation.

Because the t(8;21) affects only one AML-1 allele, the AML-1/ETO fusion protein must act dominantly to promote leukemogenesis. Accordingly, we performed mixing experiments to determine if the fusion protein could affect AML-1B-dependent transcription. Transient co-transfection of AML-1 and AML-1B with the TCRβ-CAT reporter plasmid indicated that, at high plasmid concentrations, the transcriptionally inactive AML-1 could compete with AML-1B and thus partially inhibit transactivation (Meyers et al. 1995). By contrast, AML-1/ETO could prevent AML-1B-dependent transactivation even at sub-stoichiometric DNA concentrations, suggesting that ETO possesses a repressive function (Meyers et al. 1995).

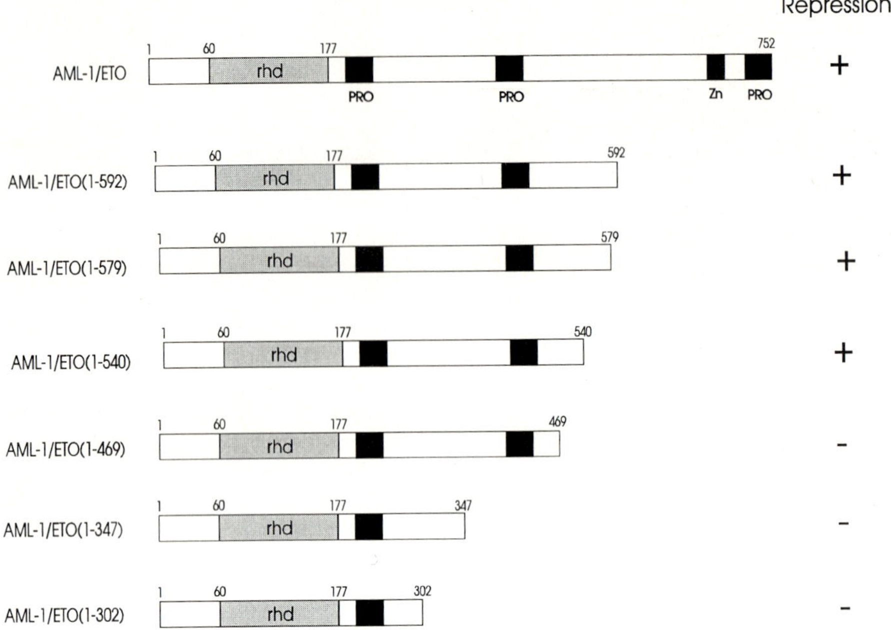

Fig. 3. Requirement of ETO sequences for Transcriptional Interference

Analysis of the AML-1/ETO Sequences Required to Suppress AML-1B Transactivation

Every t(8;21) case analyzed to date retains the same AML-1 and ETO sequences, suggesting that both the *runt* homology domain and ETO are required for leukemogenesis. To confirm that ETO sequences are indeed required for AML-1/ETO to interfere with AML-1B-dependent transcription, and to begin to define the sequences required, we constructed a series of progressive C-terminal AML-1/ETO deletion mutants (Fig. 3). Using these mutants, we showed that the predicted metal binding "zinc finger" motifs of AML-1/ETO are not required for repression, and that residues 1-540 of the 752 amino acid fusion protein comprise the minimal sequence necessary for transcriptional interference (Lenny et al. 1995). These results confirm that ETO sequences are required for interference with AML-1-mediated transactivation.

Conclusions

The (8;21) translocation causes the transcriptional activation domain of AML-1B to be replaced by ETO, creating a dominant interfering protein that is specific for AML-1B-dependent transcription. However, if simply blocking AML-1B activity potentiates leukemogenesis, then both AML-1B alleles should be inactivated in these tumors. Since this is not the case, at least two possibilities remain. First, the DNA binding domains of AML-1B, AML-2 and AML-3 are virtually identical, and AML-1 and murine AML-3 bind the same site (Meyers et al. 1993; Ogawa et al. 1993b; Levanon et al. 1994). Given that more than one family member can be expressed in hematopoietic cells, the t(8;21) fusion protein might also affect transactivation mediated by these other family members. Indeed, AML-1/ETO can interfere with transcription mediated by the murine homologue of AML-3 (S.W.H., unpublished data). Second, the ETO gene product might also regulate gene expression. In fact, our data suggest that ETO does harbor a transcriptional repressor function. Given that the majority of ETO is retained in the translocation, it is possible that ETO is also a DNA binding protein. Support for this belief comes from at least one case of AML with the t(3;21), in which AML-1 was linked to Evi-I, a DNA binding protein that represses GATA-1 transcriptional activation (Kreider et al. 1993).

CBFβ is a component of the murine PEBP2αB transcription factor complex, and we have demonstration that human CBFβ binds AML-1B (Meyers et al. 1995). CBFβ is involved in the inv(16) chromosomal rearrangement, the most common abnormality associated with acute myeloid leukemias, accounting for 15-18% of cases. The inversion forms a fusion gene that contains the majority of CBFβ coding sequences fused in frame with a portion of the smooth muscle myosin heavy chain coding region (Liu et al. 1993). Thus with CBFβ's involvement in the inv(16) and AML-1's involvement in the t(8;21) (Downing et al. 1993), some 30-35% of the *de novo* cases of AML with chromosomal abnormalities (approximately 3,000 cases per year in the U.S. alone) involve the AML-1 transcription factor complex (AML-1 or CBFβ). Because CBFβ does not bind DNA on its own, the inv(16) fusion protein likely targets AML-1 to exert its transforming effects suggesting that AML-1B is a key regulator of transcription. Extrapolating from our results with AML-1/ETO, we would hypothesize that the inv(16) fusion protein also inhibits AML-1B-dependent transcription.

The cumulative evidence suggests that AML-1 is an important regulator of transcription during hematopoiesis. The involvement of the *AML1* locus in both the (8;21) and (3;21) translocations, the role of the CBFβ subunit in inv(16), the conservation of the DNA binding domain of AML-1 from humans to *Drosophila*, and the importance of the AML-1 binding site for tissue specific transcriptional regulation, all point toward *AML1* as an important regulatory gene. Future studies promise to yield valuable information concerning the role of AML-1B in development and myelopoiesis.

Acknowledgements

This work was supported by NIH grant RO1-64140 and ACS Grant CB-39 to SWH, by Cancer Center support CORE grant 5 P30 CA 21765, and by the American Lebanese and Syrian Associated Charities of St. Jude Children's Research Hospital.

References

Downing JR, Head DR, Curcio-Brint AM, Hulshof MG, Motroni TA, Raimondi SC, Carroll AJ, Drabkin HA, Willman C, Theil KS, Civin CI, and Erickson P (1993) An AML1/ETO fusion transcript is consistently detected by RNA-based polymerase chain reaction in acute myelogenous leukemia containing the (8;21) (q22;q22) translocation Blood 81: 2860

Golub TR, Barker GF, Bohlander SK, Hiebert SW, Ward DC, Bray-Ward P, Raimondi SC, Rowley JD, and Gilliland DG (1995) Fusion of the *Tel* gene on 12p13 to the *AML1* gene on 21q22 in acute lymphoblastic leukemia Proc Natl Acad Sci U S A in press

Gottschalk LR and Leiden JM (1990) Identification and functional characterization of the human T-cell receptor B gene transcriptional enhancer: common nuclear proteins interact with the transcriptional regulatory elements of the T-cell receptor α and β genes Mol Cell Biol 10: 5486-5495

Ingham PW and Gergen JP (1988) Interactions between the pair-rule genes *runt*, *hairy*, *even-skipped* and *fushi tarazu* and the establishment of periodic pattern in the *Drosophila* embryo Devel 104: 51-60

Kreider BL, Orkin SH, and Ihle JN (1993) Loss of erythropoietin responsiveness in erythroid progenitors due to expression of the Evi-I myeloid-transforming gene Proc Natl Acad Sci USA 90: 6454-6458

Lenny N, Meyers S, and Hiebert SW (1995) Functional domains of the t(8;21) fusion protein Submitted

Levanon D, Negreanu V, Bernstein Y, Bar-Am I, Avivi L, Groner Y (1994) *AML1*, *AML2*, and *AML3*, the human members of the runt domain gene-family Genomics 23: 425-432

Liu, P, Tarle SA, Hajra A, Claxton DF, Marlton P, Freedman M, Siciliano MJ, and Collins FS (1993) Fusion between transcription factor CBF beta/PEBP2 beta and a myosin heavy chain in acute myeloid leukemia Science 261: 1041-1044

Meyers S, Downing JR, and Hiebert SW (1993) Identification of AML-1 and the t(8;21) translocation protein (AML-1/ETO) as sequence-specific DNA-binding proteins: the runt homology domain is required for DNA binding and protein-protein interactions Mol Cell Biol 13: 6336-6345

Meyers S, Lenny N, and Hiebert SW (1995) The t(8;21) fusion protein interferes with AML-1B-dependent transcriptional activation Mol Cell Biol 15: 1974-1982

Miyoshi H, Shimizu K, Kozu T, Maseki N, Kaneko Y, and Ohki M (1991) t(8;21) breakpoints on chromosome 21 in acute myeloid leukemia are clustered within a limited region of a single gene, AML1 Proc Natl Acad Sci U S A 88: 10431-10434

Nucifora G, Birn DJ, Espinosa RIII, Erickson P, LeBeau MM, Roulston D, McKeithan TW, Drabkin H, and Rowley JD (1993) Involvement of the AML1 gene in the t(3;21) in therapy-related leukemia and in chronic myeloid leukemia in blast crisis Blood 81: 2728-2734

Ogawa E, Inuzuka M, Maruyama M, Satake M, Naito-Fujimoto M, Ito Y, and Shigesada K (1993a) Molecular cloning and characterization of PEBP2 beta, the heterodimeric partner of a novel Drosophila runt-related DNA binding protein PEBP2 alpha Virol 194: 314-331

Ogawa E, Maruyama M, Kagoshima H, Inuzuka M, Lu J, Satake M, Shigesada K, and Ito Y (1993b) PEPB2/PEA2 represents a family of transcription factors homologous to the products of the Drosophila runt gene and the human AML1 gene Proc Natl Acad Sci USA 90: 6859-6863

Prosser HM, Wotton D, Gegonne A, Ghysdael J, Wang S, Speck NA, and Owen MJ (1992) A phorbol ester response element within human T-cell receptor beta chain enhancer Proc Natl Acad Sci U S A 89: 9934-9938

Redondo JM, Pfohl JL, and Krangel MS (1991) Identification of an essential site for transcriptional activation within the human T-cell receptor delta enhancer Mol Cell Biol 11: 5671-5680

Rowley JD (1982) Identification of the constant chromosome regions involved in human hematologic malignant disease Science 216: 749-750

Speck NA, Renjifo B, Golemis E, Fredrickson TN, Hartley JW, and Hopkins N (1990) Mutation of the core or adjacent LVb elements of the Moloney murine leukemia virus enhancer alters disease specificity Genes Dev 4: 233-242

Wang S, Wang Q, Crute BE, Melnikova IN, Keller SR, and Speck NA (1993) Cloning and characterization of subunits of the T-cell receptor and murine leukemia virus enhancer core-binding factor EMBO J 13: 3324-3339

The Mixed Lineage Leukemia (MLL) Protein Involved in 11q23 Translocations Contains a Domain that Binds Cruciform DNA and Scaffold Attachment Region (SAR) DNA

P. L. Broeker, A. Harden, J. D. Rowley, and N. Zeleznik-Le
Section of Hematology & Oncology, Department of Medicine, University of Chicago, Chicago IL 60637

Abstract

Translocations involving chromosome band 11q23, found in acute lymphoid and myeloid leukemias, disrupt the *MLL* gene. This gene encodes a putative transcription factor with regions of homology to several other proteins including the zinc fingers and other domains of the *Drosophila* trithorax gene product, and the "AT-hook" DNA-binding motif of high mobility group proteins. We have previously demonstrated that MLL contains transcriptional activation and repression domains using a GAL4 fusion protein system (21). The repression domain, which is capable of repressing transcription 3-5-fold, is located centromeric to the breakpoint region of *MLL*. The activation domain, located telomeric to the breakpoint region, activated transcription from a variety of promoters including ones containing only basal promoter elements. The level of activation was very high, ranging from 10-fold to more than 300-fold, depending on the promoter and cell line used for transient transfection.

In translocations involving *MLL*, the protein produced from the der(11) chromosome which contains the critical junction for leukemogenesis includes the AT-hook domain and the repression domain. We assessed the DNA binding capability of the MLL AT-hook domain using bacterially expressed and purified AT-hook protein. In a gel mobility shift assay, the MLL AT-hook domain could bind cruciform DNA, recognizing structure rather than sequence of the target DNA. This binding could be specifically competed with Hoechst 33258 dye and with distamycin. In a nitrocellulose protein-DNA binding assay, the MLL AT-hook domain could bind to AT-rich SARs, but not to non-SAR DNA fragments. The role that the AT-hook binding to DNA may play *in vivo* is unclear, but it is likely that DNA binding could affect downstream gene regulation. The AT-hook domain retained on the der(11) would potentially recognize a different DNA target than the one normally recognized by the intact MLL protein. Furthermore, loss of an activation domain while retaining a repression domain on the der(11) chromosome could alter the expression of various downstream target genes, suggesting potential mechanisms of action for MLL in leukemia.

Introduction

Chromosomal translocations involving chromosomal band 11q23 with as many as 30 different chromosomal regions have been observed in 5-10% of patients with acute lymphoblastic leukemia (ALL), and acute myeloid leukemia (AML) [17]. The most frequent translocations observed are the t(4;11) and t(11;19) in ALL, and the t(9;11), and t(6;11) in AML [23,33]. We and others have previously cloned the *MLL* gene (for mixed-lineage leukemia or myeloid-lymphoid leukemia; also called *ALL-1, HRX*, and *Htrx*) at 11q23 [38]. We have demonstrated that almost all of the chromosomal breaks occur in an 8.3 kb BamHI fragment of *MLL*, designated the breakpoint cluster region (BCR) [40]. In addition 5-15% of patients treated for a prior neoplasm with DNA topoisomerase II (topo II) inhibitors, especially the epipodophyllotoxins, develop therapy related acute myeloid leukemia (t-AML) [15,21]. We showed that all t-AML patients who had a rearranged *MLL* gene had balanced 11q23 translocations and a history of treatment with topo II inhibitors [15].

At present, 10 different chromosomal partners which rearrange with *MLL* have been cloned [1,8,16,32,38,39,42,44]. The fusion gene on the der (11) chromosome, which has been proposed to contain the critical junction for leukemogenic transformation because it is conserved in complex translocations, consists of 5' *MLL* sequences and 3' sequences from the partner chromosome [37]. Moreover, in about 25% of patients studied, the translocated telomeric region of *MLL* is deleted [41]. Thus, the der(11) fusion protein is most likely responsible for leukemogenesis in patients with *MLL* rearrangements. The different partners provide quite variable protein domains, but all fusion proteins appear in frame.

The *MLL* gene spans approximately 100 kb and contains at least 21 small exons which code for a protein of predicted molecular mass of 430 kd [43]. The cellular function of the MLL protein is unknown, but homologies to other proteins may provide some clues. MLL shares some regions of homology with the *Drosophila* trithorax (trx) protein including a zinc-finger domain, and the COOH-terminal region of the protein [16,27,43]. Trithorax is also >400 kd and is involved in maintaining the proper spatial pattern of expression of homeotic genes of the Bithorax and Antennapedia complexes by interacting with cis regulatory elements in their promoters [7,25,27]. Other MLL motifs defined by homology are three AT hooks, which are homologous to HMGI(Y) chromatin binding proteins [43], and a motif shared with mammalian DNA methyltransferase [26]. Therefore, by homology to domains of other proteins, it seems likely that MLL is a DNA binding protein. We have previously reported that MLL contains an extended activation domain in the C-terminal part of the protein, and a repression domain in the N-terminal part of the protein [46]. Furthermore, the AT-hook region of MLL is capable of binding cruciform DNA, recognizing the structure rather than the sequence of the target DNA [46].

In this paper, we demonstrate that the MLL AT-hooks are capable of binding to scaffold attachment region (SAR) DNA, but not to non-SAR DNA from the same gene region. We also demonstrate that distamycin and Hoechst 33258 can compete for MLL AT-hook binding to cruciform DNA. Therefore, in translocations involving *MLL*, the SAR DNA binding and cruciform DNA binding domain and a repression domain would remain on the der(11) chromosome, and a very strong activation domain would either be lost [24,41] or translocated to the der(other) chromosome. The splitting of these functional domains is likely to contribute to leukemogenesis in cells containing *MLL* translocations.

Materials and Methods

DNA clones and proteins.

DNA encoding MLL amino acids 142-400 (AT-hook), or as controls, MLL amino acids 1101-1400, MLL amino acids 2772-3114, or Egr-1 amino acids 281-304 (Egr-1 repression domain) [12], were subcloned into pGEX-KT [18]. MLL glutathione S transferase (GST) fusion proteins were expressed in bacterial cells, purified, and quantitated by polyacrylamide gel electrophoresis as previously described [46]. The histone H1 protein was obtained from Sigma (St. Louis, MO). DNA fragments from the the interleukin-2 3' untranslated region (UTR) [10] (kindly provided by Raymond Reeves, U Wash.), and the *IFNA2* gene flanking regions ([4], Broeker et al., manuscript in preparation) were previously characterized as SARs or non-SARs using functional SAR mapping assays [9,30,45]. The SAR and non-SAR DNA inserts were purified on agarose gels, radiolabeled by filling in with Klenow fragment of DNA polymerase, and used as a probe for the SAR-protein binding assay.

SAR-protein binding assay.

One ug of each GST-fusion protein was transferred onto nitrocellulose membrane using a Dot blot apparatus. Dot blots were first pre-incubated at room temperature in a buffer containing10 mM Tris (ph 7.5), 50 mM NaCl, 2 mM EDTA, and 5% non-fat dry milk [13]. Blots were then incubated with ^{32}P labelled SAR and non-SAR DNAs in this same buffer and temperature but with 0.5% non-fat dry milk and competitor DNA (E. coli). Blots were then washed in the binding buffer several times [13]. DNA/protein hybridization signals were analyzed by autoradiography.

Cruciform DNA binding assays.

Artificial cruciform DNAs were prepared as described [2,3,46]. Cruciform DNA consists of oligonucleotides 1-4 [3]. Gel mobility shift assays were performed as described [2,46]. Hoechst 33258 and distamycin were purchased from Sigma (St. Louis, MO). Amounts of distamycin and Hoechst 33258 ranging from 50ng to 1ug were added as competitors at the beginning of the reactions. For the repression domain binding reactions, 1ug of the competitors were added.

Results

The AT-hook domain was first described in the HMGI(Y) group of proteins [35]. The AT-hooks were demonstrated to bind to the AT-rich DNA present in the 3' untranslated region (UTR) of the *IL2* gene [34]. This same region of the IL2 3' UTR has also been shown to be a SAR [10]. MLL contains three AT-hooks clustered in a single domain near the amino terminus of the protein (Fig. 1). Although these AT hooks contain the same core amino acid sequence as the HMGI(Y) AT hooks, flanking amino acids are very different but tend to be lysine, arginine or proline (Fig. 2).

The MLL AT hook domain contains three AT hooks of similar but not identical sequence to the HMGI(Y) AT hooks, and thus we wished to test whether it could also bind to the AT-rich DNA present in the IL2 3' UTR. We found that the MLL AT hook domain bound strongly to the IL2 3' UTR SAR using a nitrocellulose binding assay (Fig. 3A). Histone H1 protein was used as a positive control because it has previously been shown that histone H1 protein binds cooperatively to SAR DNA [22]. We also tested binding of the MLL repression

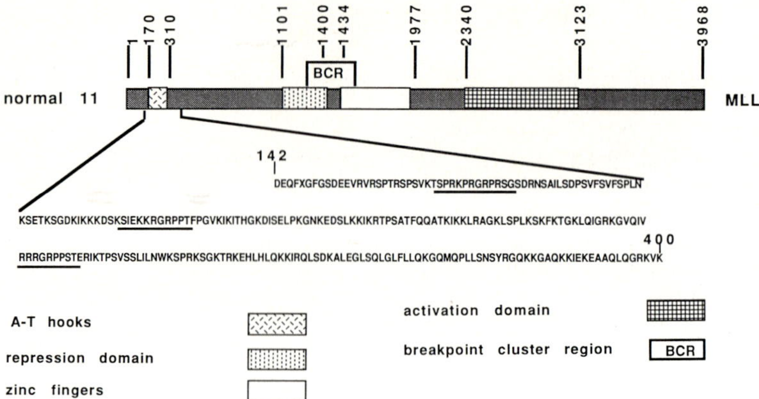

Fig.1. Schematic representation of the normal MLL cDNA on chromosome 11. The figure is drawn to scale and the position of pertinent MLL amino acids are indicated by numbers. The AT hook amino acids subcloned and expressed as protein for binding studies are indicated and the AT hooks underlined. The locations of the 0.8kb breakpoint cluster region (BCR), the zinc finger region, and the transcriptional activation and repression domains are also shown.

Fig. 2. Comparison of HMGI(Y) and MLL AT hook amino acid sequence. Each protein contains three AT hooks separated by a variable number of amino acids. The sequences of the AT hooks are denoted by the single-letter amino acid code, and their positions in the protein denoted by the amino acid numbers. Amino acid identity is indicated by a vertical line.

domain (amino acids 1101-1400) to the IL2 3' UTR SAR. This domain has previously been shown to bind to cruciform DNA as well as to duplex DNA [46]. We found that the MLL repression domain also bound to the IL2 3' UTR SAR (Fig. 3A). In contrast, the repression domain of another protein, the early growth response gene Egr-1, did not bind to the IL-2 3' UTR SAR (Fig. 3A). We then wished to test whether this was a general observation; could the MLL AT hooks bind all SARs or just a subset of SARs. Furthermore, we wished to determine whether the MLL AT hook binding was specific for SAR DNA or whether the AT hooks were also capable of binding to non-SAR DNA. Therefore, we tested the known SAR and non-SAR DNA fragments from the *IFNA2* gene region [4] for binding to the MLL AT hook domain. Both SAR (SAR2) DNA fragments from the *IFNA2* gene showed strong binding to the MLL AT hook domain (Fig. 3B) whereas non-SAR fragments (*IFNA2* coding region, *IFNA2* flanking region 1.0 kb BglII/BamHI) did not bind (Fig. 3B). None of the SAR and non-SAR DNA fragments bound to the MLL activation domain or to the Egr-1 repression domain (Fig. 3B, and data not shown). Neither of these proteins was expected to bind to DNA, and thus they represented negative controls. All SAR and non-SAR DNA fragments bound to histone H1 protein and to the MLL repression domain (Fig. 3B). These protein species showed a non-specific pattern of binding.

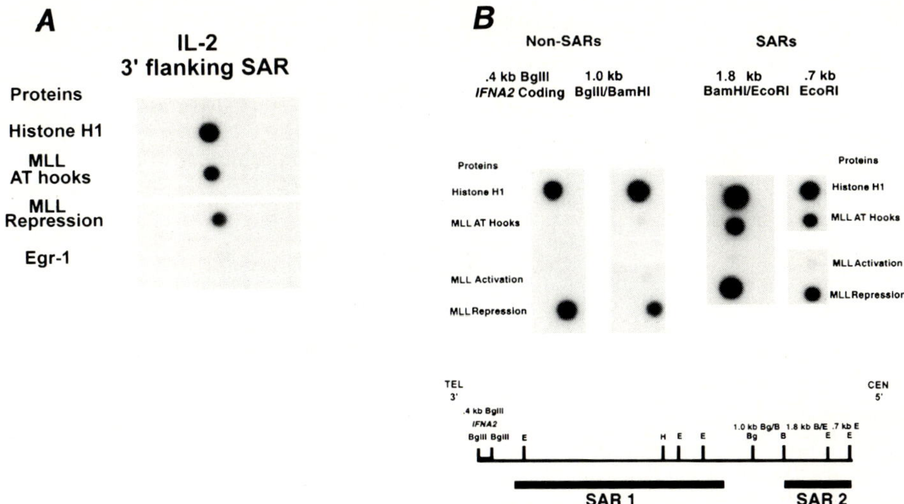

Fig. 3. The MLL AT hook domain binds to SAR DNA but not to non-SAR DNA. Proteins were expressed in bacteria, purified, spot blotted onto nitrocullulose membrane, and hybridized with radiolabeled SAR or non-SAR DNA fragments. Proteins assayed were MLL AT hook domain (amino acids 142-400), MLL repression domain (amino acids 1101-1400), MLL activation domain (amino acids 2772-3114), Egr-1 repression domain (amino acids 281-304), and histone H1 protein. (A) Binding to the IL2 3' UTR SAR. (B) Binding to interferon alpha 2 SAR and non-SAR DNA fragments. Left panel shows the binding results of the *IFNA2* non-SAR DNA regions (*IFNA2* coding region=.4 kb BglII, and 1.0 kb BglII/BamHI 5' upstream region). Right panel shows the binding results of the *IFNA2* 5' upstream SARs (1.8 kb BamHI/EcoRI, and .7 kb EcoRI). The map at the bottom of the figure (top black line) shows the *IFNA2* gene (left and thicker black line), and the two 5' upstream SARs (SAR1, and SAR2 = thick black bars below the map). Note that only SAR2 was tested for protein binding.

We have previously demonstrated that the MLL AT hook region, as well as the repression domain, could bind cruciform DNA using a gel mobility shift assay and synthetic cruciform DNA structures [46]. The AT hook domain binding was specific, because retarded bands were competed with an excess of nonradiolabeled competitor cruciform DNA [46]. Surprisingly, the binding of the MLL repression domain to cruciform DNA probes was also partially competed with a 100-fold molar excess of nonradiolabeled cruciform DNA [46]. The MLL AT-hook domain was recognizing structure rather than sequence of the DNA, because duplex DNAs that together contained all of the same DNA sequence did not bind the MLL AT hooks [46]. In contrast, the MLL repression domain was able to bind to duplex DNA, however, this binding was not competed with a 100-fold molar excess of nonradiolabeled duplex DNA [46]. Therefore, the AT-hook domain of MLL can specifically recognize and bind to a cruciform DNA structure. It has previously been demonstrated that the dye Hoechst 33258 and the drug distamycin, which both have a similar three-dimensional structure as the HMGI(Y) AT hooks, compete for binding of the HMGI(Y) AT hooks to AT-rich DNA [14,35]. We wished to

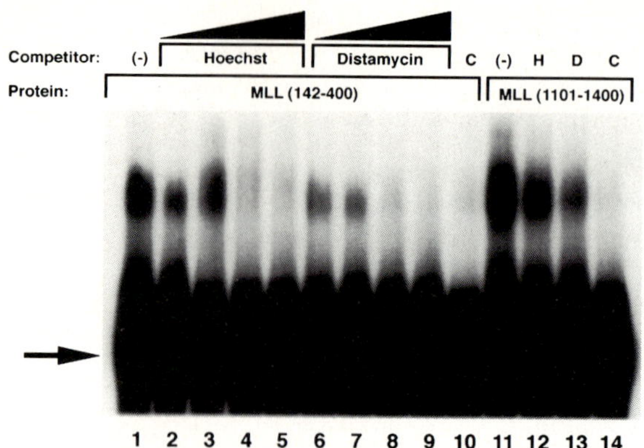

Fig. 4. Cruciform DNA binding activity of the MLL AT hook domain is specifically competed with Hoechst 33258 or distamycin. Proteins were expressed in bacteria, purified, and tested by gel mobility shift analysis for their ability to bind radiolabeled cruciform DNA probes. Protein-DNA complexes were formed in the presence of increasing amounts of Hoechst 33258 (lanes 2-5, 12) or distamycin (lanes 6-9, 13), or a 100-fold molar excess of cold competitor DNA (lanes 10, 14), or without competitor (lanes 1, 11). Free probe is indicated by an arrow. Proteins assayed were MLL AT hook domain (amino acids 142-400) (lanes 1-10) and MLL repression domain (amino acids 1101-1400) (lanes 11-14). Proteins were incubated with radiolabeled cruciform probe.

determine whether the MLL AT hook binding to cruciform DNA was similarly competed with Hoechst 33258 or distamycin. For competition, amounts of distamycin or Hoechst 33258 ranging from 50ng to 1ug were added at the beginning of the incubation period. Both distamycin and Hoechst 33258 were able to compete the binding of the MLL AT hook domain to cruciform DNA in a manner similar to the nonradiolabeled competitor cruciform DNA (Fig. 4, lanes 1-10). In contrast, even 1ug of distamycin or Hoechst 33258 were unable to compete for MLL repression domain binding to cruciform DNA (Fig. 4, lanes 12,13).

Discussion

We are studying the potential cellular functions of MLL which is involved in chromosomal translocations that result in acute lymphoid and acute myeloid leukemias [38]. One particularly intriguing aspect of the MLL protein that we have been studying is the ability of its AT-hook domain to bind cruciform DNA in a structure-dependent, sequence-independent manner. It had been proposed that the AT-hook region of MLL would bind DNA because of its homology to the AT-hooks of the HMG-I(Y) proteins [35]. We found that this region of MLL could bind to cruciform DNA [46] as has been shown for the HMG box of the HMG1/2 proteins [2,3]. This binding is specific for DNA structure rather than DNA sequence because binding was not observed to duplex DNAs of the same sequence [46]. Two types of natural cruciform structures occur: Holliday junctions formed during recombination [19], and the structures formed when inverted repeats are extruded from superhelical DNAs. These structures may be associated with initiation of mammalian DNA replication [11] and are also commonly found in the vicinity of structural genes, so that they might play a role in gene expression [2].

As has been suggested for HMG1 protein binding to DNA [31], the AT-hook region of MLL may also recognize sharp bends in DNA that are not necessarily part of a cruciform structure. It is unclear what role the AT-hook binding to DNA may play *in vivo*, but it is conceivable that it may recognize cruciform and/or bent DNA and thereby affect gene regulation.

In this investigation we have concentrated on several aspects of MLL AT hook binding to DNA fragments including cruciform DNA and SAR DNA interactions. Our data demonstrate specific binding of the MLL AT hook domain to cruciform DNA. Furthermore, this binding is specifically competed with compounds that form a similar three dimensional structure as the HMGI(Y) AT hook domain (Hoechst 33258 and distamycin). In contrast to the specific binding of the MLL AT hooks, the MLL repression domain seems to bind DNA fairly nonspecifically, and this binding is not competed by distamycin or Hoechst 33258.

Scaffold attachment regions (SARs) are defined as regions of binding between the DNA axis and non-histone "scaffold" proteins. SARs are AT-rich regions, whose DNA is bent, and they may be protein-protected regions in the genome [13]. They map to specific non-random positions, i.e. near gene regulatory elements, and appear to organize interphase and mitotic chromatin into a series of loop domains [13]. Reeves et al. demonstrated that the HMGI(Y) AT hooks bind the IL2 3' UTR SAR [10,34]. We have similarly studied binding of the MLL AT hook domain to SARs. Although the MLL AT hooks have the same core amino acid sequence as the HMGI(Y) AT hooks, they differ significantly in sequence outside of the core region. We have demonstrated that the MLL AT hook domain is capable of binding to the IL2 3' UTR SAR. Furthermore, the MLL AT hook domain binds other SARs, but not non-SAR regions from the same gene.

It may be that the MLL AT-hooks recognize bent DNA in general, or a particular subset of bent DNA structures, rather than only those present in cruciform DNA structures. Bent DNA would be found at sequences containing tracts of oligo(dA)-oligo(dT) spaced periodically along the DNA [17], and it has been found at origins of replication [5,17], scaffold associated regions (SARs) [45], and promoters [17]. It has also been demonstrated that retroviral integration into chromosomal DNA occurs preferentially at intrinsically bent DNA [29] and that bent DNA is present at chromosomal sites of illegitimate recombination events [28]. Furthermore, bent DNA is a preferential substrate for DNA topoisomerase II enzymes [20] and DNA nicking enzymes including topoisomerase I [6]. Because of the ability of the MLL AT-hook region to bind to cruciform DNA and SAR DNA, it will be important to determine the extent of MLL AT-hook binding to various bent DNAs, as well as the ability of the MLL protein itself to cause DNA to bend. It is possible that MLL could play an important role in illegitimate recombination which is the dominant mechanism of recombination in mammalian somatic cells. Illegitimate recombination is responsible for most genome rearrangements such as translocations, deletions and nonhomologous integrations [36]. The mechanism of illegitimate recombination and the enzymes or regulatory proteins involved are poorly understood. It is interesting that *MLL* is involved in a variety of translocations that result in acute leukemia. Furthermore, we have shown that patients with therapy-related AML with balanced translocations of 11q23 had received prior treatment with DNA topoisomerase II inhibitors and had rearrangements involving *MLL* [15]. Perhaps the DNA binding of MLL may give a clue to the mechanism of these translocations and to the normal function of the MLL protein in cells.

References

1. Bernard OA, Mauchauffe M, Mecucci C, Vandenberghe H, Berger R (1994) A novel gene, *AF-1p*, fused to *HRX* in t(1;11)(p32;q23), is not related to *AF-4, AF-9* nor *ENL*. Oncogene 9: 1039-1045
2. Bianchi ME (1988) Interaction of a protein from rat liver nuclei with cruciform DNA. EMBO J 7:843-849.
3. Bianchi ME, Beltrame M, Paonessa G (1989) Specific recognition of cruciform DNA by nuclear protein HMG1. Science 243:1056-1059.
4. Broeker PL (1993) The relationship of chromatin structure to breakpoints in translocations and deletions in human cancer, specifically gliomas and leukemias. Thesis, University of Chicago, Illinois
5. Caddle MS, Lussier RH, Heintz NH (1990) Intramolecular DNA triplexes, bent DNA and DNA unwinding elements in the initiation region of an amplified dihydrofolate reductase replicon. J Mol Biol 211:19-33.
6. Caserta M, Amadei A, Di Mauro E, Camilloni G (1989) In vitro preferential topoisomerization of bent DNA. Nucl Acids Res 17:8463-8474.
7. Castelli-Gair JE, Garcia-Bellido A (1990) Interactions of Polycomb and trithorax with cis regulatory regions of Ultrabithorax during the development of Drosophila melanogaster. EMBO J 9:4267-4275.
8. Chaplin T, Ayton P, Bernard O, Vaskar S, Della Valle V, Hillion J, Gregorini A, Lillington D, Berger R, Young BD (1995) A novel class of zinc finger/leucine zipper genes identified from the molecular cloning of the t(10;11) translocation in acute leukaemia. Blood 85: 1435-1441
9. Cockerill PN, Garrard WT (1987) Chromosomal loop anchorage of the kappa immunoglobulin gene occurs next to the enhancer in areion containing topoisomerase II sites. Cell 44: 273-282
10. Elton TS (1986) Purification and characterization of the high mobility group nonhistone chromatin proteins. Thesis, Washington State University, Washington.
11. Frisque RJ (1983) Nucleotide sequence of the region encompassing the JC virus origin of DNA replication. J Virol 46:170-176.
12. Gashler AL, Swaminathan S, Sukhatme VP (1993) Novel repression module, extensive activation domain, and bipartite nuclear localization signal defined in the immediate early transcription factor Egr-1. Mol Cell Biol 13:4556-4571.
13. Gasser SM, Amati BB, Cardenas ME, Hofmann JF -X (1989) Studies on scaffold attachment sites and their relation to genome function. In: Jeon KW, Friedlander M (eds) International Review of Cytology, Acad Press, Inc (vol 119)
14. Geierstanger BH, Volkman BF, Kremer W, Wemmer DE (1994) Short peptide fragments derived from HMG-I/Y proteins bind specifically to the minor groove of DNA. Biochem 33:5347-5355
15. Gill Super HJ, McCabe NR, Thirman MJ, Larson RA, Le Beau M M, Pedersen-Bjergaard J, Preben P, Diaz M, Rowley JD (1993) Rearrangements of the *MLL* gene in therapy-related acute myeloid leukemia in patients previously treated with agents targeting DNA-topoisomerase II. Blood 82: 3705-3711
16. Gu Y, Nakamura T, Alder H, Prasad R, Canaani O, Cimino G, Croce CM, Canaani E (1992) The t(4;11) chromosome translocation of human acute leukemias fuses the ALL-1 gene, related to Drosophila trithorax, to the *AF-4* gene. Cell 71: 701-708
17. Hagerman PJ (1990) Sequence-directed curvature of DNA. Annu Rev Biochem 59:755-781.
18. Hakes DJ, Dixon JE (1992) New vectors for high level expression of recombinant proteins in bacteria. Anal Biochem 202:293-298.
19. Holliday R (1964) A mechanism for gene conversion in fungi. Genet Res 5:282-304.
20. Howard MT, Lee MP, Hsieh TS, Griffith JD (1991) Drosophila topoisomerase II-DNA interactions are affected by DNA structure. J Mol Biol 217:53-62.
21. Hunger SP, Tkachuk DC, Amylon MD, Link, MP, Carroll AJ, Welborn JL, Willman CL, Cleary ML (1993) *HRX* involvement in de novo and secondary leukemias with diverse chromosome 11q23 abnormalities. Blood 81: 3197-3203

22. Izaurralde E, Kas E, Laemmli UK (1989) Highly preferential nucleation of histone H1 assembly on scaffold-associated regions. J Mol Biol 210:573-585.
23. Kaneko Y, Shikano T, Maseki N (1988) Clinical characteristics of infant acute leukemia with or without 11q23 translocations. Leukemia 2: 672-676
24. Kobayashi H, Espinosa III R, Thirman MJ, Gill HJ, Fernald AA, Diaz MO, Le Beau MM, Rowley JD (1993) Heterogeneity of breakpoints of 11q23 rearrangements in hematologic malignancies identified with fluorescence in situ hybridization. Blood 82:547-551.
25. Kuzin B, Tillib S, Sedkov Y, Mizrokhi L, Mazo A (1994) The Drosophila trithorax gene encodes a chromosomal protein and directly regulates the region-specific homeotic gene fork head. Genes Dev 8:2478-2490
26. Ma Q, Alder H, Nelson K, Chatterjee D, Gu Y, Nakamura T, Canaani E, Croce C, Siracusa L, Buchberg AM (1993) Analysis of the murine *ALL-1* gene reveals conserved domains with human *ALL-1* and identifies a motif shared with DNA methyltransferases. Proc Natl Acad Sci USA 90: 6350-6354
27. Mazo AM, Huang D-H, Mozer BA, Dawid IB (1990) The trithorax gene, a trans-acting regulator of the bithorax complex in Drosophila, encodes a protein with zinc-binding domains. Proc Natl Acad Sci USA 87:2112-2116.
28. Milot E, Belmaaza A, Wallenburg JC, Gusew N, Bradley WEC, Chartrand P (1992) Chromosomal illegitimate recombination in mammalian cells is associated with intrinsically bent DNA elements. EMBO J 11:5063-5070.
29. Milot E, Belmaaza A, Rassart E, Chartrand P (1994) Association of a host DNA structure with retroviral integration sites in chromosomal DNA. Virol 201:408-412.
30. Mirkovitch J, Mirault M E, Laemmli UK (1984) Organization of the higher order chromatin loop: specific DNA attachment sites on nuclear scaffold. Cell 39: 223-232
31. Onate SA, Prendergast P, Wagner JP, Nissen M, Reeves R, Pettijohn DE, Edwards DP (1994) The DNA-bending protein HMG-1 enhances progesterone receptor binding to its target DNA sequences. Mol Cell Biol 14:3376-3391.
32. Prasad R, Leshkowitz D, Gu Y, Alder H, Nakamura T, Saito H, Huebner K, Berger R, Croce CM, Canaani E (1994) Leucine-zipper dimerization motif encoded by the AF17 gene fused to *ALL-1* (*MLL*) in acute leukemia. Proc Natl Acad Sci USA 91: 8107-8111
33. Raimondi SC (1993) Current status of cytogenetic research in childhood acute lymphoblastic leukemia. Blood 81: 2237-2251
34. Reeves R, Elton TS, Nissen MS, Lehn D, Johnson KR (1987) Posttranscriptional gene regulation and specific binding of the nonhistone protein HMG-I by the 3' untranslated region of bovine interleukin 2 cDNA. Proc Natl Acad Sci USA 84:6531-6535.
35. Reeves R, Nissen MS (1990) The AT-DNA-binding domain of mammalian high mobility group I chromosomal proteins. J Biol Chem 265:8573-8582
36. Roth DB, Wilson JH (1988) Illegitimate recombination in mammalian cells. In Kucherlapati R, Smith GR (eds), Genetic Recombination, American Society for Microbiology, Washington DC, pp621-653.
37. Rowley JD (1992) The der (11) chromosome contains the critical breakpoint junction in the 4;11, 9;11, and 11;19 translocations in acute leukemia. Genes Chromosom Cancer 5:264-266
38. Rowley JD (1993) Rearrangments involving chromosomal band 11q23 in acute leukemia. In Rabbitts TH (ed) Seminars in Cancer Biology, Acad. Press, London, pp 377-385 (vol 4)
39. Schichman S, Caligiuri M, Gu Y, Strout M, Canaani E, Bloomfield CD, Croce CM (1994) *ALL-1* partial duplication in acute leukemia. Proc Natl Acad Sci USA 91: 6236-6239
40. Thirman MJ, Gill HJ, Burnett RC, Mbangkollo D, Mccabe NR, Kobayashi H, Ziemin-Van der poel S, Kaneko Y, Morgan R, Sandberg AA, Chaganti R S K, Larson RA, Lebeau MM, Diaz M O, Rowley JD (1993) Rearrangement of the *MLL* gene in acute lymphoblastic and acute myeloid leukemias with 11q23 chromosomal translocations. N Engl J Med 329: 909-914
41. Thirman MJ, Mbangkollo D, Kobayashi H, McCabe NR, Gill HJ, Rowley JD, Diaz MO (1993) Molecular analysis of 3' deletions of the *MLL* gene in 11q23 translocations reveals that the zinc finger domains of *MLL* are often deleted. Proc of the American Assoc for Cancer Research 34, 425

42. Thirman MJ, Levitan DA, Kobayashi H, Simon C, Rowley JD (1994) Cloning of *ELL*, a gene that fuses to *MLL* in a t(11;19)(q23;p13.1) in acute myeloid leukemia. Proc Natl Acad Sci USA 91: 12110-12114
43. Tkachuk D, Kogler S, Cleary M (1992) Involvement of a homolog of Drosophila trithorax by 11q23 chromosomal translocations in acute leukemias. Cell 71:691-700
44. Tse W, Zhu W, Chen HS, Cohen A (1995) A novel gene, *AF1q*, fused to *MLL* in t(1;11)(q21;q23), is specifically expressed in leukemic and immature hematopoietic cells. Blood 85: 650-656
45. von Kries JP, Phi-Van L, Diekmann S, Stratling W (1990) A non-curved chicken lysozyme 5' matrix attachment site is 3' followed by a strongly curved DNA sequence. Nuc Acids Res 18: 3881-3885.
46. Zeleznik-Le NJ, Harden A, Rowley JD (1994) 11q23 translocations split the "AT-hook" cruciform DNA-binding region and the transcriptional repression domain from the activation domain of the mixed-lineage leukemia (*MLL*) gene. Proc Natl Acad Sci USA 91: 10610-10614

Acknowledgments

We wish to thank Vikas Sukhatme, and Jack Dixon for plasmids. This research was supported in part by grants from the National Institutes of Health (CA42557 to J.D.R.), from the Department of Energy (DE-FG02-86ER60408 to J.D.R.), and from the American Cancer Society, Illinois Division (95-42 to N.J.Z.-L.).

Pathogenetic Role of the PML/RARα Fusion Protein in Acute Promyelocytic Leukemia

F. Grignani and P. G. Pelicci
Istituto di Medicina Interna e Scienze Oncologiche, Perugia University.
Policlinico Monteluce, 06100 Perugia, Italy

Introduction

Acute promyelocytic leukemia (APL) is characterized by a differentiation block at the promyelocytic stage of myeloid differentiation [1]. The differentiation block is the most distinctive biological feature of this leukemia since APL blasts have a modest proliferative rate [2] and the release of the block and terminal differentiation obtained with retinoic acid (RA) correlates in vivo with complete hematological remission in the great majority of APL patients [3-5].

Clinical response to RA is accompanied, in APL blasts, by rearrangements of the retinoic acid receptor α gene (RARα) as a result of a translocation of chromosomes 15 and 17 [t(15;17)] [6], that involves the PML (for Promyelocyte [7]) and RARα genes [8-12]. The resulting PML/RARα fusion gene encodes a PML/RARα fusion protein [13-16] (fig. 1). Since RARα is involved in the regulation of myeloid differentiation [17,18], the alteration in the RA signalling pathway caused by PML/RARα could contribute to the differentiation block. Paradoxically, the activation of the same signalling pathway by RA induces terminal differentiation, indicating APL as the first example of a leukemia that can be treated by specifically targeting therapy to the transforming protein.

Expression of the t(15;17) has never been reported in any other neoplasia. Interestingly, gene transfer experiments have shown that the expression of the PML/RARα fusion protein induces cell death in most lymphoid and non-hematopoietic cell lines (our unpublished results). This suggests that, *in vivo,* a negative selection may take place against non myeloid cells developing a 15;17 translocation.

Rare variant translocations in APL involve 17 with chromosomes other than 15 [6]. In the 11;17 chromosome translocation the chromosome 17 breakpoint on RARα is the same as in t(15;17), whereas in chromosome 11 it is located in a zinc finger gene named PLZF [19]. In the t(5;17) the RARα gene is fused to the NPM gene, which encodes a nucleolar phosphoprotein involved in RNA processing [20]. In both cases, the recombination leads to the formation of fusion proteins (PLZF/RARα and NMP/RARα).

The PML/RARα protein and its role in the pathogenesis of APL

The PML/RARα protein retains functional domains of the RARα and PML proteins (Fig.1). Since the chromosome 17 break of the 15;17 translocation always occurs in RARα intron 2, the RARα component of the PML/RARα protein is the same in all cases of APL and corresponds to regions B to F [21,22], which include the DNA binding, ligand binding, dimerization and nuclear localization properties of RARα [23].

The chromosome 15 breakpoint of the t(15;17) has been mapped in three regions of the PML locus: intron 6 (breakpoint cluster region 1; bcr1, 45%), intron 3 (bcr3 45%) (fig.1), exon 6 (bcr2 10%) [21,24]. However, there is a PML region retained in all PML/RARα fusion genes. This region includes the putative DNA binding domain and the dimerization interface (Fig.1). The DNA binding region includes three cysteine-histidine rich regions, referred to as RING finger, B1 and B2 box [25-27]. The dimerization interface is the coiled-coil portion of the α-helix, and contains four clusters of heptad repeats with hydrophobic aminoacids at first, fourth and eighth positions [28-29]. The region of PML encoding variable portions of the C-terminal region of the α-helix, the serine/proline rich region, including the CKII phosphorylation site, and the variable C-termini are contained in the reciprocal RARα/PML gene [30].

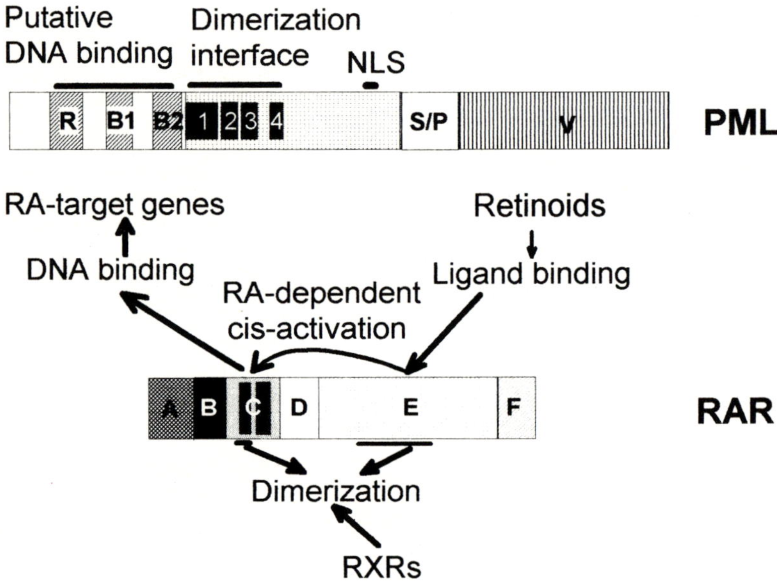

Fig.1: Normal PML and RARα functions retained in PML/RARα.
NLS, nuclear localization signal; R: Ring finger; B1, B2: B-boxes; 1,2,3,4: cluster of heptad repeats of hydrophobic aminoacids: V:variable COOH termini. A,B,C,D,E,F: RARα modular domains (23).

I. Biological Effects of PML/RARα on Differentiation and Growth of Hemopoietic Precursors.

The expression of the PML/RARα protein and the outburst of APL coincide, suggesting that PML/RARα is responsible for the transformed phenotype in APL. Experimental evidence of a pathogenetic role of PML/RARα has been obtained in an in vitro model system. Hematopoietic precursor cell lines can be used as a model for hematopoietic differentiation because they can be induced to mature upon exposure to various physiological and synthetic molecules. The phenotypic changes induced by the expression of PML/RARα in these cells recapitulate critical features of the promyelocytic leukemia phenotype [31].

Effects of PML/RARα on differentiation. Vitamin D3 (VD) or VD and TGFβ1 (TGF) treatment induces terminal monocytic differentiation of the U937 promonocytic cell line [32]. U937 cells that express the PML/RARα protein fail to terminally differentiate, as shown by both surface marker analysis and functional tests (proliferation and phagocytosis). Differentiation is blocked when PML/RARα expression levels are higher than those of the normal RARα protein as occurs in APL blasts [31], suggesting a dominant negative mechanism. The block of differentiation is not restricted to the effects of VD in myelomonocytic differentiation. Expression of PML/RARα inhibits erythroid differentiation induced by hemin in K562 cells [33].

The effect that PML/RARα exerts on differentiation is reversed when the cells expressing the fusion protein are exposed to high RA concentrations, comparable to the peak plasma levels achieved during RA therapy of APL patients [34]. The percentage of cells that enter the differentiation program is higher in the cell population where PML/RARα is expressed [31]. This effect is not due simply to a loss of PML/RARα-mediated maturation block, but also to a direct positive effect since cells that without PML/RARα do not differentiate, now enter the maturation cascade. Thus, a high concentration of RA converts the activity exerted by the fusion protein from a RA-independent inhibition to a ligand dependent stimulation of differentiation [31]. Exposure to RA of the APL derived NB4 cell line or of U937 cells expressing PML/RARα releases the block of VD-induced differentiation. Therefore, the RA-dependent differentiation activity dominates over the block of vitamin D3 induced differentiation. The maturation level achieved by these cells is more advanced than that obtained with VD or RA alone, indicating that, once the activity of PML/RARα is reversed, the two inducers have a synergistic effect [35].

The PML/RARα protein, in summary, could alone be responsible for two major features of the APL phenotype: block of differentiation and sensitivity to RA.

Effects of PML/RARα on cell growth. The effects of PML/RARα are not restricted to cell differentiation. U937 cells expressing the PML/RARα protein do not undergo programmed cell death in conditions of serum deprivation that induce apoptosis in control cells [31]. In addition, in the GM-CSF dependent cell line TF-1 survival in the absence of GM-CSF growth factor is prolonged by the expression of PML/RARα [36]. Whereas control cells become irreversibly committed to apoptosis after a short GM-CSF deprivation, PML/RARα expressing cells can gain competence to GM-CSF driven proliferation after as long as 20 days of growth factor deprivation. Thus, PML/RARα promotes cell survival by inhibiting the commitment of cells to enter the genetic program of cell death. By prolonging cell survival, PML/RARα could support the expansion of a population of leukemic cells with a low proliferative index [2]. This effect, combined with the differentiation block, provides a cellular mechanism to account for the oncogenic potential of PML/RARα.

II. Mechanisms of PML/RARα activity

PML/RARα is a multifunctional protein retaining activities of the normal PML and RARa proteins, including dimerization functions, DNA binding, retinoid binding transactivating functions. Experimental evidence (Fig.2) showed that the fusion of the PML and RARα components is required for the biological activity of PML/RARα. Overexpression of RARα or PML does not to block U937 cells differentiation or increase RA response (manuscript in preparation). In addition, the biological activity of the fusion protein is not mimicked by the overexpression of either one of the PML or RARα truncated components of PML/RARα (Fig.2). Even independent coexpression of the PML and RARα moieties of the fusion protein, outside the context of the fusion, does not produce a clear phenotypic change (manuscript in preparation).

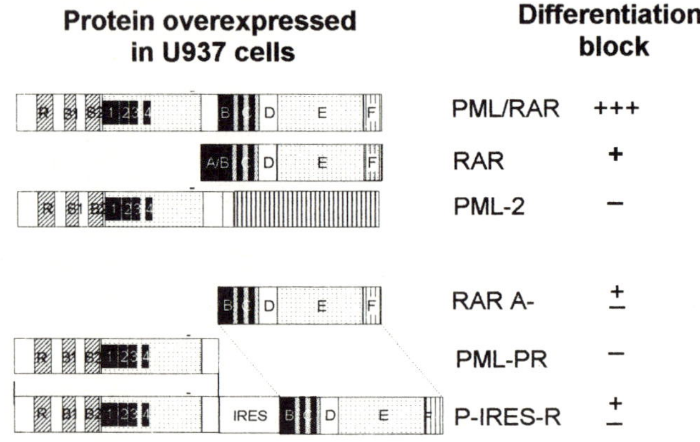

Fig.2: Differentiation potential with vitamin D3 of U937 cells overexpressing the indicated protein. IRES: internal ribosomal entry site. See figure 1 for the other symbols.

These observations indicate that the fusion confers to the PML/RARα protein novel features and modifies the normal PML and RARα activities. These new characteristics are responsible for the oncogenic effects of PML/RARα (see below and fig. 3).

Homo and heterodimerization of the PML/RARα protein. Macromolecular nuclear complexes. The ability of the PML/RARα protein to dimerize with various cellular proteins is currently considered one important mechanism by which the fusion protein exerts its leukemogenetic effect. PML/RARα can form homodimers, heterodimers with PML and heterodimers with RXR [16,29,37-40]. PML/RARα homodimers bind DNA in the absence of RXR and with a different affinity or specificity than RAR/RXR heterodimers [29]. Since PML/RARα expression is high in APL [24] dimerization of PML/RARα with RXR could interfere directly with the normal RARα molecular pathway and also influence indirectly the activity of other nuclear receptors like, TR and DR, which physiologically dimerize with RXR [41,42]. Accordingly, an excess of PML/RARα prevents both VDR binding to VDRE *in vitro*, and activation of a reporter gene by VDR [29]. PML/RARα homodimerizes and heterodimerizes with PML through the common heptad cluster dimerization domain of the coiled-coil region [29]. These interaction can conceivably alter PML function by competing with the normal protein for the same molecular targets and generating PML/RARα–PML heterodimers with a different biochemical functions than PML homodimers. In addition, the nuclear localization of PML/RARα homodimers and PML/RARα-PML heterodimer in APL cells is different from that of PML homodimers (see below).

HPLC analysis of RA binding proteins in APL blasts and in the APL cell line NB4 have demonstrated the presence of APL-specific high molecular weight nuclear protein complexes of 600 and 1200 kd (the apparent molecular weight of PML/RARα is 110 kd) [43]. These complexes are probably generated through the multiple dimerization properties of PML/RARα and contain RARα, PML, PML/RARα, RXR (our unpublished results), and may contain other nuclear proteins.

DNA binding, retinoid binding, transactivating activity. PML/RARα bind retinoids and function as RA-inducible transcription factors [13-16,43]. PML/RARα and RARα have different transactivating properties. These effects can be either stimulatory and inhibitory, depending on the promoter and cell type used, either in presence or absence of RA. However, PML/RARα is consistently more active than RARα. The high variability of the results depending on the experimental setting makes it difficult to correlate these data to the mechanisms of leukemogenesis.

PML/RARα binds DNA through the RARα DNA binding domain. PML/RARα homodimers bind responsive elements with different efficiency and specificity than the physiological heterodimers RARα-RXR, whereas PML/RARα-RXR complexes partially overlap to the RAR-RXR dimers in terms of binding activity to RAREs [29]. These observation suggest that i) the fusion protein could regulate a different set of genes than wild type RARα protein, and ii) that the two proteins have a different activity on the genes regulated by RARα. It remains to be established how this may alter differentiation, since the genetic program regulated by RARα is unknown.

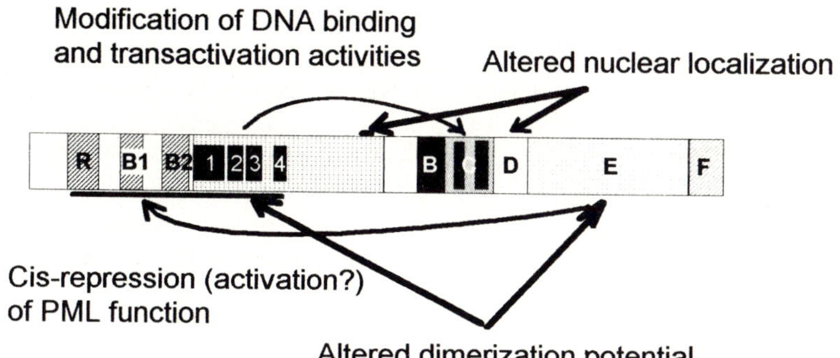

Fig. 3. Altered PML or RARα funtions in PML/RARα. See figure 1 for symbols.

Alteration of subcellular localization. PML/RARα may act as a dominant negative on both the PML or RARα molecular pathways. PML has been regarded as a growth regulator and may play a role in maturation and activation of specific cell types. PML protein shows preferential expression in differentiated post mitotic cells. Endothelial and epithelial cells and activated tissue macrophages, are the cell types with the highest amounts of PML protein. PML protein expression is also increased during differentiation of the monoblastic cell line U937 or after its activation by interferon-γ treatment [40,44]. When overexpressed in rat embryo fibroblasts transformed by Ha-ras and p53 or Ha-ras and c-myc, PML reduces the malignant behaviour of the cells. In addition, overexpression of PML in the APL cell line NB4 suppress its clonogenicity and tumorigenicity in nude mice, suggesting that overexpression of PML may antagonize the activity of PML/RARα. The fusion with RARα may disrupt the growth controlling activity of the normal PML protein, since PML/RARα has no effect on transformation of REF by activated oncogenes [45].

A dominant negative action of PML/RARa on PML may have a morphological counterpart in the altered intracellular distribution of the PML protein in APL blasts. Normal PML protein has both a cytoplasmic and nuclear distribution, with isoforms exclusively or predominantly localized in the cytoplasm [16,29,37-40,44]. The nuclear component of PML appears to be prevalent and has a typical distribution in discrete nuclear subdomains, referred to as POD (PML Oncogenic Domain) or PML nuclear bodies, giving the characteristic speckled pattern in immunohystochemical and immunofluorescence [37-40].

PML colocalizes in the nuclear bodies with other proteins defined by their speckled intranuclear distribution and reactivity with monoclonal antibodies raised against nuclear antigens or with sera from patients affected by autoimmune diseases. These proteins include the autoantigen SP100, identified in patients with biliary cirrhosis, and two undefined proteins of 55 and 65 kd [37-40]. The interaction among the components of the POD domain appears to be very stringent since overexpression of one of the proteins causes enlargement of the subnuclear domain and increased immunological labelling of the other antigens. However, coimmunoprecipitation studies have failed to show a physical interaction between PML and SP100 [40].

The intracellular localization of PML/RARα differs from that of PML and RARα. PML/RARα has a cytoplasmic fraction variably described as diffuse or perinuclear [16,29,37-40,44]. This fraction is difficult to detect because of its fine distribution as opposed to the nuclear fraction which is localized in defined domains. The nuclear component has micropunctated nuclear pattern, with hundreds of small dots, distinguished from the speckled nuclear pattern of PML and from the finely dispersed pattern of RARα [45]. Anti-PML antibodies reveal that APL cells, which express all three proteins, display the PML/RARα-like micropunctated nuclear pattern, indicating that the fusion protein localization dominates over the two wild type proteins and PML colocalize with PML/RARα; the PML protein is, accordingly, displaced from the nuclear body structure. Immunoelectromicroscopic studies have shown that the nuclear body structure is disrupted in APL cells and that the PML/RARα and the PML proteins are localized in small particles with poor structural organization, tightly bound to chromatin. In addition, at least part of the SP100 protein and of the RXR cellular pool are redistributed to the small PML/RARα containing particles. This observation suggests that one of PML/RARα mechanisms of action could be the displacement and sequestration of PML and RXR outside their normal intranuclear location. Interestingly, the treatment of APL blasts with RA converts the micropunctated nuclear pattern to the speckled, PML-like, pattern [16,29,37-40]. Alternatively, PML displacement could be the detectable component of a more complex alteration of protein-protein interactions among multiple partners.

In addition to its effects on the PML pathway PML/RARα could also affect the activity of RARα and other steroid receptors. In the presence of low near-physiological concentration of RA [34] in vitro, PML/RARα acts as a transcription repressor of certain RA-target genes [13-16]. Furthermore, PML/RARα may also function as a VD signalling antagonist by acting directly on VD-target genes or, indirectly, by sequestering RXR a cofactor essential for D3 receptor activity. As both RA and VD are implicated in myeloid differentiation [46], these effects may contribute to the inhibition of differentiation by PML/RARα. This view has received support from the ultrastructural observations on the PML/RARα-dependent RXR displacement in APL cells [39]. However, PML/RARα seems to be able to inhibit myeloid differentiation induced by TPA (our unpublished results), and erythroid differentiation induced by heme K562 cells [33]. These facts suggest that the activity of PML/RARα is not limited to RXR sequestration and may be directed to the regulation of fundamental differentiation master genes. In summary, it is likely that PML/RARα act as a multifunctional protein on several different pathways leading to cell maturation.

III. Mechanism of RA sensitivity of APL blasts.

APL undergoes disease remission upon RA treatment. This correlates with the induction of terminal differentiation of APL blasts [47,48]. APL blasts differentiate in the presence of T-RA *in vitro* and the clinical response can be predicted by *in vitro* differentiation tests [3,49].

The presence of the PML/RARα protein seems to be strictly necessary for RA sensitivity of leukemic blasts. First, there is a strict correlation between the expression of the PML/RARα fusion transcript and response to treatment [50]. Second, PML/RARα expression increases the sensitivity to RA in myeloid precursors *in vitro* [31]. Third, clones of the APL NB4 cell line become T-RA-resistant when they lose PML/RARα expression [51]. Taken together these evidences indicate an active role for PML/RARα in causing the RA sensitivity of APL. The fusion protein retains the RARα DNA and retinoid binding domains and, consequently, it could directly influence the RARα-dependent pathway that controls terminal myeloid differentiation. In addition, PML/RARα possesses the ability to act on DNA as an homodimer, independently from the availability of RXR [29 and our unpublished results]. In transactivation experiments PML/RARα can overstimulates the expression of RARα target genes when RA is present. It should be also kept in mind that the interaction between RA and the RA binding domain of PML/RARα could cis-activate the PML portion of the fusion protein [52]. RA may also reconstruct molecular interactions with diverse protein partners within nuclear bodies. RA causes a redistribution of PML/RARα, PML, RXR and probably other proteins within the nucleus. The PML/RARα containing microgranules of APL blasts are reconverted into the physiological nuclear body structure [37-40]. The pathogenesis of APL could depend on the disruption of multiple molecular partnership within nuclear bodies, due to PML/RARα. RA would release the differentiation block by modifying the PML/RARα molecule, restoring a physiological distribution of important functional proteins and reconstituting the nuclear bodies.

Acknowledgements

This research was supported by grants from the "Associazione italiana per la ricerca sul cancro (A.I.R.C.), the Italian Council for Research (ACRO project) and the European Community (Biomed and Biotech).

References

1. Bennet, J. M., Catovski, D., Daniel, M. T., Flandrin, G., Galton, D. A. G., Gralnick, H. R., and Sultan, C. (1976). Proposals for the classification of the acute leukemias. British J. of Haematol. 33:451-458
2. Raza A, Yousuf N, Abbas A, Umerani A, Mehdi A, Bokhari SKJ, Sheikh Y, Qadir K, Freeman J, Masterson M, Miller MA, Lampkin B, Browman G, Bennett J, Goldberg J, Grunwald H, Larson R, Vogler R, Preisler HD (1992). High expression of transforming growth factor-β prolonged cell cycle times and a unique clustering of S-phase cells in patients with acute promyelocytic leukemia. Blood 79:1037-1048
3. Huang M, Yu-Chen Y, Shu-Rong C, Chai J, Lin Z, Long J, Wang Z (1988). Use of all-trans retinoic acid in the treatment of acute promyelocytic leukemia. Blood 72:567-572
4. Castaigne S, Chomienne C, Daniel MT, Berger R, Fenaux P, Degos L (1990). All-trans retinoic acid as a differentiating therapy for acute promyelocytic leukemia I. Clinical results. Blood 76:1704-1709
5. Warrell RP Jr, Frankel SR, Miller WH Jr, Scheinberg DA, Itri LM, Hittelman WN, Vyas R, Andreeff M, Tafuri A, Jakubowski A, Gabrilove J, Gordon MS, Dmitrovsky E (1991). Differentiation therapy for acute promyelocytic leukemia with tretinoin (all-trans-retinoic acid). N Engl J Med 324:1385-1392
6. Mitelman F: Catalog of chromosome aberrations in cancer. Third edition. Alan R. Liss, Inc., New York, 1988.
7. Goddard AD, Borrow J, Freemont P, Solomon E (1991). Characterization of a zinc finger gene disrupted by the t(15;17) in acute promyelocytic leukemia. Science 254:1371-1374.
8. de Thè H, Chomienne C, Lanotte M, Degos L, Dejean A. (1990). The t(l5;17) translocation of acute promyelocytic leukaemia fuses the retinoic acid receptor α gene to a novel transcribed locus. Nature 347:558-561,.
9. Borrow J, Goddard AD, Sheer D, Solomon E (1990). Molecular analysis of acute promyelocytic leukemia breakpoint cluster region on chromosome 17. Science 249:1577-1580
10. Lemmons RS, Eilender D, Waldmann RA, Rebentisch M, Frej AK, Ledbetter DM, Willmann C, McConnell T, O'Connell P (1990). Cloning and characterization of the t(15;17) translocation breakpoint region in acute promyelocytic leukemia. Genes Chromosom Cancer 2:79-87
11. Longo L, Pandolfi PP, Biondi A, Rambaldi A, Mencarelli A, Lo Coco F, Diverio D, Pegoraro L, Avanzi G, Tabilio A, Zangrilli D, Alcalay M, Donti E, Grignani F, Pelicci PG (1990). Rearrangements and aberrant expression of the retinoic acid receptor a gene in acute promyelocytic leukemia. J Exp Med 172:1571-1575
12. Alcalay M, Zangrilli D, Pandolfi PP, Longo L, Mencarelli A, Giacomucci A, Rocchi M, Biondi A, Rambaldi A, Lo Coco F, Diverio D, Donti E, Grignani F, Pelicci PG (1991). Translocation breakpoint of acute promyelocytic leukemia lies within the retinoic acid receptor α locus. Proc Natl Acad Sci U S A 88:1977-1981
13. Pandolfi PP, Grignani F, Alcalay M, Mencarelli A, Biondi A, Lo Coco F, Grignani F, Pelicci PG (1991). Structure and origin of the acute promyelocytic leukemia myl/RARα cDNA and characterization of its retinoid-binding and transactivation properties. Oncogene 6:1285-1292

14. de Thè H, Lavau C, Marehio A, Chomienne C, Degos L, Dejean A (1991). The PML-RARα fusion mRNA generated by the t(l5;17) translocation in acute promyelocytic leukemia encodes a functionally altered RAR. Cell 66:675-684.
15. Kakizuka A, Miller WH Jr, Umesono K, Warrel RP Jr, Frankel SR, Murty VVVS, Dmitrovsky E, Evans RM (1991). Chromosomal translocation t(15;17) in human acute promyelocytic leukemia fuses RARα with a novel putative transcription factor, PML. Cell 66:663-674
16. Kastner P, Perez A, Lutz Y, Rochette-Egly C, Gaub MP, Durand B, Lanotte M, Berger R, Chambon P (1992). Structure, localization and transcriptional properties of two classes of retinoic acid receptor a fusion proteins in acute promyelocytic leukemia (APL): structural similarities with a new family of oncoproteins. EMBO J 11:629-642
17. Collins SJ, Robertson K, Mueller (1990). Retinoic acid induced granulocytic differentiation of HL60 myeloid leukemia cells is mediated directly through the retinoic acid receptor (RAR-α). Mol Cell Biol 10:2154-2161
18. Robertson KA, Emami B, Collins SJ (1989). Retinoic acid-resistant HL-60R cells harbor a point mutation in the retinoic acid receptor ligand binding domain that confers dominant negative activity. Blood 80:1885-1889.
19. Chen Z, Brand NJ, Chen A, Chen SJ, Tong JH, Wang ZY, Waxman S, Zelent A (1993). Fusion between a novel *Kruppel*-like finger gene and the retinoic acid receptor-α locus due to a variant t(ll;l7) translocation associated with acute promyelocytic leukemia. EMBO J 12:1161-1167
20. Redner RL, Rush EA, Faas S, Rudert WA, Corey SJ (1994). The t(5;17) translocation in acute promyelocytic leukemia generates a nucleophosmin-RARα fusion transcript. Blood 84:1486 Supplement 1, Thirty-Sixth Annual Meeting of the American Society of Hematology
21. Diverio D, LoCoco F, D'Adamo F, Biondi A, Fagioli M, Grignani Fr, Rambaldi A, Rossi V, Avvisati G, Petti MC, Testi AM, Liso V, Specchia G, Fioritoni G, Recchia A, Frassoni F, Ciolli S, Pelicci PG (1992). Identification of DNA rearrangements at the RARa locus in all patients with acute promyelocytic leukemia (APL) and mapping of APL breakpoints within the RARα second intron. Blood 79:3331-3336.
22. Chen SJ, Zhu YJ, Tong JH, Dong S, Huang W, Chen Y, Xiang WM, Zhang L, Song Li X, Qian GQ, Wang ZY, Chen Z, Larsen CJ, Berger R (1991). Rearrangements in the second intron of the RARα gene are present in a large majority of patients with acute promyelocytic leukemia and are used as molecular marker for retinoic acid induced leukemic cell differentiation. Blood 78:2696-2701
23. Evans RM (1988). The steroid and thyroid hormone receptor superfamily. Science 240:889-895
25. Freemont PS, Hanson IM, Trowsdale J (1991). A novel cysteine-rich sequence motif. Cell 64: 483-484.
27. Lovering R, Hanson IM, Borden KLB, Martin S, O'Reilly NJ, Evan GI, Rahman D, Pappin DJC, Trowsdale, Freemont P (1993). Identification and preliminary characterization of a protein motif related to the zinc finger. Proc Natl Acad Sci U.S.A. 90: 2112-2116
28. Forman BM, Samuels HH (1990). Interaction among a subfamily of nuclear hormone receptors: The regulatory Zipper model. Mol Endocr 4:1293-1301.
29. Perez A, Kastner P, Sethi S, Lutz Y, Reibel C, Chambon P (1993). PMLRAR homodimers: distinct DNA binding properties and heterodimeric interaction with RXR. EMBO J 12:3171-3182
30. Alcalay M, Zangrilli D, Fagioli M, Pandolfi PP, Mencarelli A, Lo Coco F, Biondi A, Grignani F, Pelicci PG (1992). Expression pattern of the RARα-PML fusion gene in acute promyelocytice leukemia. Proc Natl Acad Sci USA 89:4840-4844
31. Grignani Fr, Ferrucci PF, Testa U, Talamo G, Fagioli M, Alcalay M, Mencarelli A, Grignani F, Peschle C, Nicoletti I, Pelicci PG (1993). The acute promyelocytic leukaemia specific PML\RARα fusion protein inhibits differentiation and promotes survival of myeloid precursor cells. Cell 74: 423-431.
32. Testa U, Masciulli R, Tritarelli E, Pustorino R, Mariani G, Martucci R, Barberi T, Camagna A, Valtieri M, Peschle C (1993). Transforming growth factor-β potentiates vitamin D3 induced terminal monocytic differentiation of human leukemic cell lines. J Immunol 150:2418-2430
33. Grignani Fr, Testa U, Fagioli M, Barberi T, Masciulli R, Mariani G, Peschle C, Pelicci PG (1995). Promyelocytic leukemia-specific PML-retinoic acid α receptor fusion protein interferes with erythroid differentiation of human erythroleukemia K562 cells. Cancer Research, in press.

34. Muindi, J., Frankel, S.R., Miller, W.H., Jakuboski, A., Scheinberg, D.A., Young C.W., Dmitrovsky, E. and Warrel, R.P (1992). Continous treatment with All-trans retinoic acid causes progressive reduction in plasma drug concentrations: implication for relapse and retinoid "resistance" in patients with acute promyelocytic leukemia. Blood 79:299-303
37. Daniel MT, Koken M, Romagne O, Barbey S, Bazarbachi A, Stadler M, Guillemin MC, Degos L, Chomienne C, de Thé H (1993). PML protein expression in hematopoietic and acute promyelocytic leukemia cells. Blood, 82: 1858-1867.
38. Weis K, Rambaud S, Lavau C, Jansen J, Carcalho T, Carmo-Fonseca M, Lamond A, Dejean A (1994). Retinoic acid regulates aberrant nuclear localization of PML-RARα in acute promyelocytic leukemia cells. Cell 76:345-358
39. Dyck JA, Maul GG (1994). A novel macromolecular structure is a target of the promyelocyte-retinoic acid receptor oncoprotein. Cell 76:333-343
40. Koken MHM, Puvio-Dutilleul F, Guillemin MC, Viron A, Cruz-Linares G, Stuurman N, de Jong L, Szostecki C, Calvo F, Chomienne C, Degos L, Puvion E, de Thè H (1994). The t(15;17) translocation alters a nuclear body in a retinoic acid-reversible fashion. The EMBO J. 13:1073-1083
41. Yu VC, Delsert C, Andersen B, Holloway JM, Devary OV, Naar AM, Kim SY, Boutin JM, Glass CK, Rosenfeld MG (1991). RXRß: a coregualtor that enhances binding of retinoic acid, thyroid hormone, and vitamin D receptors to their cognate response elements. Cell 67:1251-1266
42. Kliewer SA, Umesono K, Mangelsdorf DJ, Evans RM (1991). Retinoid X receptor interacts with nuclear receptors in retinoic acid, thyroid hormone and vitamin D_3 signalling. Nature 355:446-449
43. Nervi C, Poindexter CE, Grignani F, Pandolfi PP, Lo Coco F, Avvisati G, Pelicci PG, Jetten AM (1992). Characterization of the PML/RARa chimeric product of the acute promyelocytic leukemia-specific t(15;17) translocation. Cancer research 52:3687-3692
44. Flenghi L, Fagioli M, Tomassoni L, Pileri S, Gambacorta M, Pacini R, Grignani Fr, Casini T, Ferrucci PF, Martelli MF, Pelicci PG, Falini B (1995). Characterization of a new monoclonal antibody (PG-M3) directed against the aminoterminal portion of the PML gene product: immunocytochemical evidence for high expression of PML proteins on activated macrophages, endothelial cells, and epithelia. Blood, in press.
45. Gaub MP, Lutz Y, Ruberte E, Petkovich M, Brand N, Chambon P (1989). Antibodies specific to the retinoic acid human nuclear receptors α and β. Proc Natl Acad Sci USA 86:3089-3094.
46. Reitsma, PH, Rothberg PG, Astrin SM, Trial J, Bar-Shavit Z, Hall A, Teitelbaun SL (1983). Regulation of myc gene expression in HL-60 leukemia cells by a vitamin D metabolite. Nature 306:492-494
47. Nilsson B (1984). Probable in vivo induction of differentiation by retinoic acid of promyelocytes in acute promyelocytic leukaemia. Br J Haematol 57:365-371
49. Chomienne C, Ballerini P, Balitrand N, Daniel MT, Fenaux P, Castaigne S, Degos L (1990). All-trans retinoic acid in acute promyelocytic leukemias. II. In vitro studies: structure-function relationship. Blood 76:1710-1717
50. Miller WH Jr, Kakizuka A, Frankel SR, Warrel RP Jr, DeBlasio A, Levine K, Evans RM, Dmitrovsky E (1992). Reverse transcription polymerase chain reaction for the rearranged retinoic acid receptor α clarifies diagnosis and detects minimal residual disease in acute promyelocytic leukemia. Proc Natl Acad Sci USA89:2694-2698
51. Dermime S, Grignani Fr, Clerici M, Nervi C, Sozzi G, Talamo GP, Marchesi E, Formelli F, Parmiani G, Pelicci PG, Gambacorti-Passerini C (1993). Occurrence of resistance to retinoic acid in the acute promyelocytic leukemia cell line NB306 is associated with altered expression of the PML/RARα protein. Blood 82: 1573-1577.
52. Grignani Fr, Fagioli M, Alcalay M, Longo L, Pandolfi PP, Donti E, Biondi A, Lo Coco F, Grignani F, Pelicci PG (1994): Acute promyelocytic leukemia: from genetics to treatment. Blood 83:10-25

Involvement of the TEL Gene in Hematologic Malignancy by Diverse Molecular Genetic Mechanisms

T.R. Golub, G.F. Barker, K. Stegmaier and D. G. Gilliland
Division of Hematology/Oncology, Department of Medicine, Brigham and Women's Hospital and Children's Hospital, Harvard Medical School, Boston, MA 02115

I. Multistep Pathogenesis of Acute Myeloid Leukemia

Cloning of chromosome translocation breakpoints have identified abnormal gene products which contribute to the pathogenesis of hematologic malignancy. Examples include the PML-RARα fusion associated with t(15;17) acute promyelocytic leukemia (APML)(1-3), the AML1-ETO fusion associated with t(8;21) acute myeloid leukemia (4), the CBFß-MYHII fusion associated with inv(16) acute myelomonocytic leukemia and eosinophilia (5), and MLL fusions at 11q23 breakpoints with various fusion proteins associated with acute myeloid leukemias (6).

Evidence which strongly supports a role for these fusion proteins in pathogenesis of acute myeloid leukemia includes the invariant association of chromosomal translocations with specific disease phenotypes and the striking conservation of translocation breakpoints. For example, all patients with APML have t(15;17), and t(15;17) is invariably associated with the phenotype of APML but not with any other subtypes of AML. The consequence of t(15;17) is fusion of the PML and RARα genes; the breakpoint occurs within the same intron of RARα in all patients, and occurs in one of two introns within the PML gene. All of the resultant fusions preserve functional domains of PML-RARα . Similar conservation of functional domains and association with specific phenotypes is seen in chromosomal translocations associated with the other subtypes of AML noted above.

Although these data provide convincing evidence that chromosomal translocations are necessary for manifestation of the leukemic phenotype, there is equally convincing evidence that leukemia, like other human cancers, requires more than one molecular genetic abnormality. Multistep pathogenesis of leukemia is supported first by rare but informative families with inherited propensity to develop leukemia (7) As for pedigrees with inherited forms of breast cancer and colon cancer, it is likely that affected members of pedigrees with inherited leukemia carry a germ line mutation which predisposes to the acquisition of a second mutation in somatic tissue, which in turn gives rise to the malignant phenotype. One example is the inheritance of a mutant DNA repair gene, MSH2, which predisposes affected members of hereditary non-polyposis coli pedigrees to the development of colon cancer (8). A second observation which supports multistep pathogenesis of AML is the phenomenon of clonal remission (9). Clonal remission occurs infrequently after induction of remission from AML using chemotherapeutic agents. Patients with clonal remission meet all diagnostic criteria for remission including normal blood counts and differential, normal morphology and normal cytogenetics, but can still be documented to have clonal hematopoiesis by X-chromosome inactivation based assays. These findings are consistent with persistent somatic mutation which gives rise to a clonal proliferative advantage, but is not sufficient to give rise to the clinical phenotype of AML.

A third argument in support of multistep pathogenesis of AML are the myelodysplastic syndromes (MDS). MDS are a heterogeneous group of hematologic disorders characterized by pancytopenia and dysplastic growth of hematopoietic progenitors, and can be considered a transition state between normal hematopoiesis and leukemia. Patients with MDS have clonal hematopoiesis as evidenced by X-chromosome inactivation based assays, and by clonal karyotype abnormalities including del 5q, del 7q, del 12p, and del 20q among others. MDS may be indolent for months to years, but frequently progresses to AML associated with acquisition of additional cytogenetic abnormalities. Thus, at least two molecular genetic abnormalities appear to be necessary for the development of the malignant phenotype in these cases.

Cloning and characterizing the genes responsible for familial leukemia and myelodysplastic syndromes should provide insight into the early molecular genetic events that govern the transition from normal hematopoiesis to acute leukemia. However, there are few pedigrees large enough for generalized linkage analysis of familial leukemia, and myelodysplastic syndromes have proven difficult to analyze due to the large size of the deletions commonly associated with MDS.

II. Chronic myelomonocytic leukemia (CMML) associated with t(5;12)

CMML, one of the FAB subtypes of MDS, is characterized by dysplastic proliferation of monocytes, and progression to AML. A recurring translocation in CMML occurs between chromosome bands 5q33 and 12p13. t(5;12)(q33;p13) is of particular interest because it occurs in regions of chromosome 5q and 12p13 which are abnormal in a significant number of patients with hematologic malignancy. For example, approximately 10% of cases of acute lymphoblastic leukemia (ALL) of childhood are associated with 12p13 deletions (10, 11). Since ALL is the most common cancer of children (del) (12p) is perhaps the most frequent cytogenetic abnormality in pediatric malignancy. The t(5;12)(q33;12p13) may therefore serve to localize genes involved in CMML, as well as in other hematologic malignancies.

A. Cloning the t(5;12) breakpoint associated with CMML

We identified a patient with t(5;12)(q33;p13) and CMML, who subsequently developed AML associated with acquisition of t(8;21)(q22:q22) in addition to t(5;12). The t(8;21) is identical at the cytogenetic level to t(8;21) seen in de novo AML. t(5;12) thus appears to satisfy criteria for an early molecular genetic abnormality which gives rise to AML by virtue of its recurring association with CMML. In this specific example, t(5;12) appears to be an early mutation antedating development of t(8;21) associated AML.

The t(5;12) breakpoint was cloned without benefit of a cell line or patient cDNA library (12). Limitations in available clinical material led us to consider strategies which would maximize the blood and bone marrow specimens available. Fluorescence-in-situ-hybridization (FISH) was performed with ordered chromosome 5q cosmid probes to localize the breakpoint between the genes encoding c-fms and ribosomal protein S14. PCR primers specific for c-fms and RPS14 were used simultaneously to screen the CEPH megaYAC Library, and identified a 600Kb YAC 745d10 containing both c-fms and RPS14. 745d10 spanned the translocation by FISH, confirming the localization of the breakpoint and delineating a 600Kb genomic region within which the breakpoint must lie. Long range genomic maps were prepared by pulsed-field gel electrophoresis, and localized the breakpoint within the platelet-derived growth factor ß (PDGFRß) gene. Ribonuclease protection assays (RPA) were then used to localize the breakpoint with PDGFRß RNA using PDGFRß specific probes on patient bone marrow RNA. RNA analysis localized a partial transcript for the 3' end of the PDGFRß gene, beginning near the transmembrane domain of PDGFRß and extending through the tyrosine kinase domains. Northern blot analysis of patient bone marrow using PDGFRß probes showed a single 5 Kb transcript, which was larger than would be predicted by a partial PDGFRß transcript, consistent with a PDGFRß fusion partner derived from chromosome 12p13.

B. The PDGFRß fusion partner in t(5;12) CMML is a novel *ETS*-like gene TEL.

Anchored PCR on patient marrow RNA was performed using nested PDGFRß primers to obtain a partial cDNA for the chromosome 12 fusion partner. Involvement of the partial cDNA in the t(5;12) translocation was confirmed by RPA of patient bone marrow, and the partial cDNA was used to screen a human K562 cDNA library to obtain a full length cDNA.

The PDGFRß fusion partner is a novel ETS-like gene, *TEL*. TEL is aa 452 a protein which contains two functional domains: (i) a 3' DNA binding domain which defines ETS family members, and (ii) a 5' predicted helix-loop-helix (HLH) domain which is shared by approximately one third of ETS family members (Fig 1).

Figure 1. TEL is a member of the ets family of transcription factors, and contains a 3' helix-loop-helix (HLH) domain and a 5' DNA binding domain

The TEL HLH domain is conserved among other ETS family members, including ets-1 and fli-1, and the Drosophila gene yan/pok, a transcriptional repressor.

The consequence of the t(5;12) translocation is a fusion transcript whose expression is driven by the *TEL* promoter, and results in fusion of the TEL HLH domain in frame to the PDGFRß transmembrane and tyrosine kinase domains (Fig. 2).

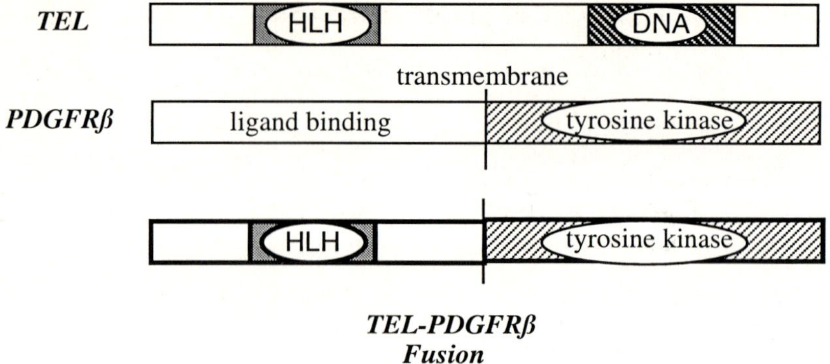

Figure 2. The consequence of the t(5;12) associated with CMML is fusion of the TEL HLH domain to the transmembrane and tyrosine kinase domains of the PDGFRß

C. Mechanisms of transformation of TEL-PDGFRß

The structure of the TEL-PDGFRß fusion suggests several possible mechanisms of transformation. Wild-type PDGFRß is known to signal a variety of cellular responses, including mitogenesis, on binding to its dimeric ligand, PDGF. PDGF mediates dimerization of PDGFRß, which activates the tyrosine kinase leading to autophosphorylation of the receptor on tyrosine residues. Phosphorylated tyrosines on the PDGFRß serve as docking sites for a number of proteins which initiate signal transduction cascades, including SRC, SYP/Grb2, PI3 kinase, and PLCγ. It is plausible based on the known function of wildtype PDGFRß that the HLH domain of TEL-PDGFRß mediates dimerization and constitutive activation of the PDGFRß tyrosine kinase domain.

The TEL-PDGFRß fusion was first tested for transforming activity in cultured mammalian cell lines. TEL-PDGFRß confers factor independent growth to the IL-3 dependent hematopoietic cell line, Ba/F3. Consistent with a model of PDGFRß tyrosine kinase activation, TEL-PDGFRß is constitutively phosphorylated in factor independent Ba/F3 cells transfected with TEL-PDGFRß.

Another possible model for TEL-PDGFRß transforming activity is that TEL has tumor suppressor activity, and that TEL-PDGFRß interferes with wildtype function. In this model, the TEL HLH domain would mediate heterodimerization between TEL-PDGFRß and wild type TEL, leading to TEL loss of function. Indirect evidence which supports TEL loss of function in pathogenesis of malignancy is provided below, and includes translocations involving TEL in which the other TEL allele is deleted, such as the TEL-AML1 fusion. In these cases, there is no functional TEL in the leukemic cells: one TEL allele is deleted and the other is disrupted by translocation (13). Other data supporting a role for TEL loss of function in hematologic malignancy is frequent loss of heterozygosity at the TEL gene locus in ALL (14).

III. TEL is frequently involved in translocations at the 12p13 locus

One rationale for cloning the t(5;12) translocation breakpoint was to determine whether the translocation would identify genes in other translocations involving chromosome 12p13. To test this possibility, additional patients with cytogenetic evidence of 12p13 rearrangements were analyzed for evidence of involvement of the TEL gene locus.

Yeast artificial chromosomes (YACs) containing the TEL gene were used to analyze patients with cytogenetic evidence of 12p13 abnormalities (15). The majority of patients (26/34) were shown to have abnormalities at the TEL gene locus. RPA was used to map to translocation breakpoints within TEL, and disclosed an unusual distribution of breakpoints within the TEL gene (Fig 3). As noted earlier, translocation breakpoints within a given gene are usually highly conserved, even when different fusion partners have been identified. For example, the MLL gene at chromosome band 11q23 is associated with AML and has numerous fusion partners. However, the breakpoint with MLL is highly conserved, regardless of the fusion partner. In contrast, there are at least three different breakpoints within the TEL gene which give rise to fusion products which express different functional domains of TEL.

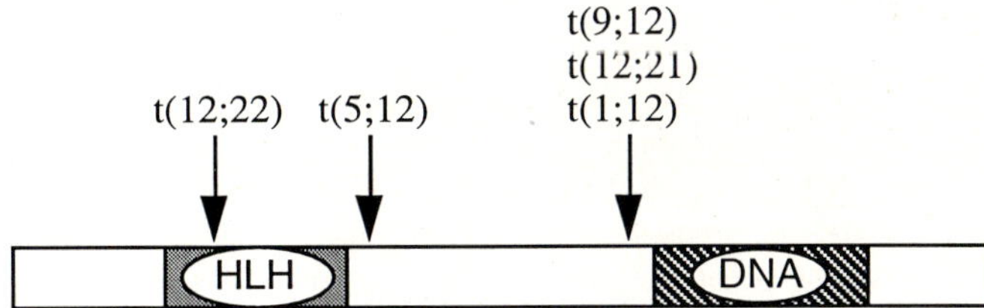

Figure 3. Diverse translocation breakpoints in TEL mapped by ribonuclease protection assays.

For example, in contrast with the structure of the TEL-PDGFRß which involves the TEL HLH domain, patients evaluated in our laboratory with the t(12;22) showed evidence of abnormal expression of the TEL DNA binding domain driven by the promoter of a chromosome 22 gene. The t(12;22) breakpoint has been cloned by Grosveld et al and gives rise to a fusion transcript containing the MNI gene fused in frame to the TEL DNA binding domain (16). The MN1-TEL fusion is analogous in structure to the EWS-fli1 fusion associated with t(11;22) Ewings sarcoma (17), and the TLS-ERG fusion associated with t(16;21) leukemia (18), in which an ETS-family DNA binding domain is abnormally expressed.

IV. TEL is fused to the protooncogene ABL in t(9;12;14) acute undifferentiated leukemia

RPA was used as described above to delineate a breakpoint within the TEL gene in a patient with a complex t(9;12;14) translocation and acute undifferentiated leukemia with myeloid markers (AMoL). Anchored PCR with nested TEL primers was used to clone the TEL fusion partner, the ABL protooncogene on chromosome 9q34. The consequence of the translocation is fusion of the TEL HLH domain inframe to exon 2 of ABL (Fig 4).

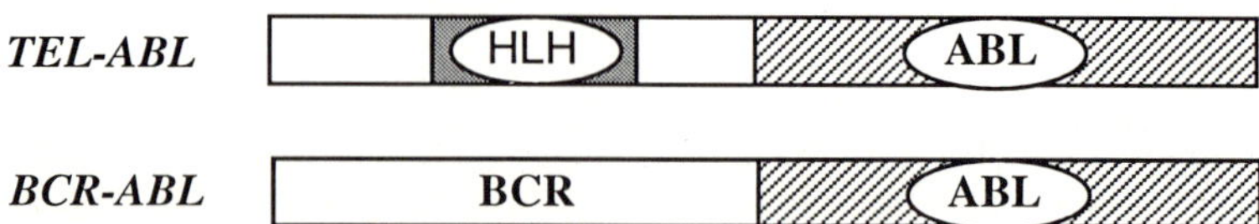

Figure 4. TEL-ABLand BCR-ABl fusions associated with t(9;12) and t(9;22) translocations, respectively In each case, fusion occurs in frame at exon 2 of ABL

The TEL-ABL fusion has several interesting features. TEL-ABL is similar in structure to the well characterized BCR-ABL fusion associated with chronic myelogenous leukemia (CML) and t(9;22). TEL is the only other fusion partner that has been identified for ABL, and has important similarities and differences from BCR. For example, BCR contains a 5' predicted coiled-coil interaction motif which is necessary for tyrosine kinase and transforming activity of BCR-ABL. The coiled-coil motif BCR and the putative HLH domain of TEL may both therefore serve dimerization or multimerization functions as a mechanism for constitutive activation of tyrosine kinase activity. The theme of dimerization leading to constitutive tyrosine kinase activity and transformation might then be shared by TEL-PDGFRß, TEL-ABL and BCR-ABL. Consistent with this hypothesis, TEL-ABL transforms Rat1 fibroblasts and is constitutively phosphorylated when stably expressed in these cells. In addition, like TEL-PDGFRß, TEL-ABL is capable of conferring IL-3 independent growth to Ba/F3 cells.

As another example of the usefulness of new fusion genes to elucidate functional domains which are relevant to transforming activity, TEL-ABL lacks a Grb2 binding site on the TEL moiety. BCR-ABL contains a Y177 Grb2 binding site on the BCR portion of the fusion, whose role in transformation has been debated (19). Since TEL-ABL lacks a Grb2 binding site, at a minimum it can be stated that transformation of cultured mammalian cells mediated by ABL fusions does not require a functional Grb2 binding site.

V. TEL is fused to the transcription factor AML1 in t(12;21) acute lymphoblastic leukemia (ALL)

As noted above, another TEL breakpoint involving the TEL HLH domain was identified in patients with ALL and t(12;21). The translocation breakpoint was cloned using anchored PCR with TEL specific primers. Based on our previous experience with cloning of TEL-PDGFRß and TEL-ABL, one might have predicted a tyrosine kinase fusion partner for TEL. However, in the case of t(12;21), TEL is fused inframe to the transcription factor AML1 (13), (Fig 5). The AML1 gene on chromosome 21q22 was first cloned by virtue of its involvement with t(8;21) and t(3;21) associated with de novo AML and therapy-related AML, respectively (20). AML1 contains two functional domains; (i) a DNA binding domain with homology to the Drosophila pair-rule gene *runt*, and (ii) a 3' transcriptional activation domain. The TEL-AML1 fusion consists of the TEL HLH domain fused in frame to AML1 at intron 2, with expression of both the AML1 DNA binding domain and the AML1 transcriptional activation domain.

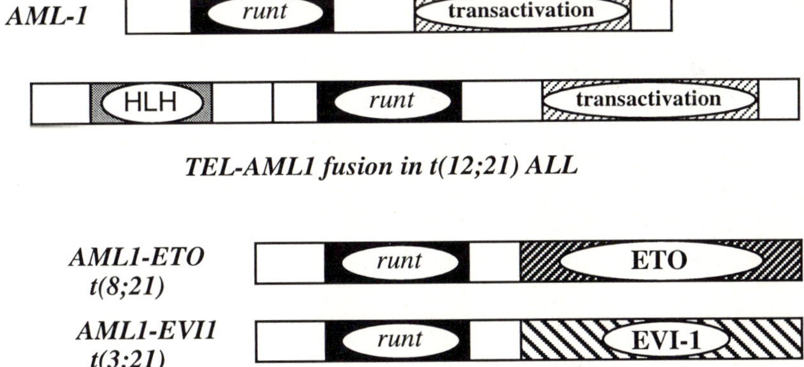

Figure 5. The TEL-AML1 fusion associated with t(12;21) differs from other AML1 fusions: TEL-AML1 is associated with ALL rather than AML, and involves expression of both the *runt* and transactivation domains of AML1

TEL-AML1 is fascinating from several perspectives. First, it suggests that the TEL HLH domain can contribute to pathogenesis of leukemia when fused either to a tyrosine kinase or to a transcription factor. Second, the structure of the TEL-AML1 fusion differs significantly from the AML1 fusions involved in t(8;21) and t(3;21) translocations. In these translocations, the 5' end of the AML1 gene including the DNA binding domain, is fused to one of several partners just 3' of the *runt* domain (Fig 5). Fusion partners include ETO in t(8;21), and various partners in t(3;21) including EVI-1, EAP and MDS 1. In each of these fusions, the AML1 transactivation domain is lost. In contrast, in the TEL-AML1 fusion, the full length AML1 gene is expressed, including the *runt* and transactivation domains. AML1 had previously only been associated with myeloid leukemias (hence the name of the gene). In two cases of TEL-AML1 reported from our laboratory, and two cases subsequently reported by Romana et al, have been associated with lymphoid leukemias (21). In part, the difference in lineage specificity of TEL-AML1 versus other AML1 fusions can be explained by the t(8;21) and t(3:21) AML1 fusions being driven by the AML1 promoter, whereas the TEL-AML1 fusion is driven by the TEL promoter. However, at a minimum it is clear that AML1 can contribute to the pathogenesis of both myeloid and lymphoid malignancies. Fourth, in each case of TEL-AML1 fusions characterized thus far (13), the other TEL allele is deleted. Thus, in these leukemic cells, there is no functional TEL: one TEL allele is deleted and the other is disrupted by translocation. Based in part on this observation, the possibility that TEL loss of function might contribute to pathogenesis of leukemia was evaluated in ALL patients, as described in the next section.

VI. Loss of heterozygosity at the TEL gene locus in pediatric ALL

To evaluate the possibility that TEL loss of function might contribute to pathogenesis, a patient population that has frequent deletions in the 12p13 region was chosen for analysis. Approximately 10% of pediatric ALL cases have 12p13 deletions. Since the most common childhood malignancy is ALL, del(12p13) is among the most common molecular genetic abnormality of childhood cancer.

To determine whether the loss of TEL function could be implicated in pathogenesis of ALL, we first determined the frequency of loss of heterozygosity (LOH) at the TEL gene locus. Genomic DNA was prepared from 81 pediatric patients at the time of diagnosis of ALL. Polymorphic microsatellite markers, D12S89 and D12S98 which flanked the TEL gene, were then tested for LOH. LOH of two microsatellite markers which flank the TEL gene would provide convincing evidence for LOH at the TEL locus. As controls to confirm that ALL patients with a single microsatellite band had loss of heterozygosity, rather than simply being homozygous for that marker, paired leukemia and remission samples were analyzed. Patients were considered informative only when remission samples documented that the patient was heterozygous at that locus.

As seen in Table 1, approximately 15% of ALL patients had LOH at the TEL gene locus (14).

Table 1 Loss of heterozygosity (LOH) at the *TEL* locus in pediatric ALL

Marker	No. Patients	No Informative	Patients with LOH
D12S89	81	63/81 (78%)	9/63 (14%)
D12S98	81	53/81 (65%)	9/53 (17%)

Of note, only one of the 9 patients with TEL LOH had cytogenetic evidence of 12p13 loss. This is in consonance with most studies of LOH in malignancy in which cytogenetic analysis underestimates LOH at most loci. Taken together, these findings suggest that TEL LOH may occur in as many as 20-25% of ALL patients.

To further delineate the region of LOH on 12p13, additional microsatellite markers telomeric and centromeric to TEL were evaluated. The region of LOH includes TEL, but extends to the centromere and also invariably includes the gene KIP1, encoding for the protein p27. p27 is a cyclin-dependent kinase (CDK) inhibitor which regulates the G1/s transition in the cell cycle. Other CDK inhibitors, such as p15 and p16, have been strongly implicated in pathogenesis of cancer through loss of function. p27 is thus a superb candidate for loss of function in ALL. Sequence analysis of the residual alleles for TEL and KIP1 are underway in these patients, as well as efforts to identify other transcription units in this region.

VII. TEL contributes to pathogenesis of leukemia by diverse molecular genetic mechanisms

In summary, we have presented evidence that the TEL gene, which we first cloned in association with t(5;12) CMML (12), can contribute to pathogenesis of leukemia by remarkably diverse mechanisms. The TEL HLH domain may be fused to the tyrosine kinase domains of PDGFRß and ABL in myeloid leukemias, or may be fused to the transcription factor AML1 in lymphoid malignancies.

In contrast, the TEL DNA binding domain may be abnormally expressed in t(12;22) leukemias in a manner analogous to other ETS DNA binding domain fusions, such as EWS-fli1 and TLS-ERG. Finally, indirect evidence has been presented that TEL loss of function may contribute to pathogenesis of leukemia, including complete loss of functional TEL in t(12;21) ALL, and LOH at the TEL locus on 12p13 in pediatric ALL.

The diversity of molecular genetic mechanisms by which TEL can be transforming suggests an important role for TEL in cell growth and differentiation. Further analysis of TEL, and its related oncogenic fusion genes, may provide further insight into the role of TEL in normal physiology of mammalian cells.

References

1. de The H, Chomienne C, Lanotte M, Degos L, Dejean A. 1990. The t(15;17) translocation of acute promyelocytic leukaemia fuses the retinoic acid receptor α gene to a novel transcribed locus. **Nature 347**:558-561.
2. de The H, Lavau C, Marcio A, Chomienne C, Degos L, Dejean A. 1991. The PML-RAR-alpha fusion mRNA generated by the t(15;17) translocation in acute promyelocytic leukemia encodes a functionally altered RAR. **Cell 66**:675.
3. Warrell RP, de The H, Wang Z-Y, Degos L. 1993. Acute promyelocytic leukemia. **New England Journal of Medicine 329**:177-189.
4. Erickson P, Gao J, Chang KS, Look T, Whisenant E, Raimondi S, Lasher J, Rowley J, Drabkin H. 1992. Identification of breakpoints in t(8;21) acute myelogenous leukemia and isolation of a fusion transcript, AML1/ETO, with similarity to Drosophila segmentation gene, runt. **Blood 80**:1825-1831.

5. Liu P, Tarle SA, Hajra A, Claxton DF, Marlton P, Freedman M, Siciliano MJ, Collins FS. 1993. Fusion between transcription factor /PEBP2B and a myosin heavy chain in acute myeloid leukemia. **Science 261**:1041-1044.
6. Thirman MJ, Gill HJ, Burnett RC, Mbangkollo D, McCabe NR, Kobayashi H, Ziemin-van der Poel S, Kaneko Y, Morgan R, Sandberg AA, Chaganti RSK, Larson RA, Le Beau MM, Diaz MO, Rowley JD. 1993. Rearrangement of the MLL gene in acute lymphoblastic and acute myeloid leukemias with 11q23 chromosomal translocations. **New England Journal of Medicine 329**:909-914.
7. Dowton SB, Beardsley D, Jamison D, Blattner S, Li FP. 1985. Studies of a familial platelet disorder. **Blood 65**:557-563.
8. Bronner CE, Baker SM, Morrison PT, Warren G, Smith LG, Lescoe MK, Kane M, Earabino C, Lipford J, Lindblom A, Tannergard P, Bollag RJ, Godwin AR, Ward DC, Nordenskjold M, Fishel R, Kolodner R, Liskay RM. 1994. Mutation in the DNA mismatch repair gene homologue *hMLH1* is associated with hereditary non-polyposis colon cancer. **Nature 368**:258-261.
9. Busque L, Gilliland DG. 1993. Clonal evolution in acute myeloid leukemia. **Blood 82**:337-342.
10. Raimondi SC. 1993. Current status of cytogenetic research in childhood acute lymphoblastic leukemia. **Blood 81**:2237-51.
11. Raimondi SC, Williams DL, Callihan T, Peiper S, Rivera GK, Murphy SB. 1986. Nonrandom involvement of the 12p12 breakpoint in chromosome abnormalities of childhood acute lymphoblastic leukemia. **Blood 68**:69-75.
12. Golub TR, Barker GF, Lovett M, Gilliland DG. 1994. Fusion of PDGF receptor beta to a novel ets-like gene, tel, in chronic myelomonocytic leukemia with t(5;12) chromosomal translocation. **Cell 77**:307-316.
13. Golub TR, Barker GF, Bohlander SK, Hiebert SW, Ward DC, Bray-Ward P, Morgan E, Raimondi SC, Rowley JD, D.G. G. 1995. Fusion of the TEL gene on 12p13 to the AML1 gene on 21q22 in acute lymphoblastic leukemia. **Proc Natl Acad Sci in press.**
14. Stegmaier K, Pendse S, Barker GF, Bray-Ward P, Ward DC, Montgomery KT, Krauter KS, Reynolds C, Sklar J, Donnelly M, Bohlander SK, Rowley JD, Sallan SE, Gilliland DG, Golub TR. 1995. Frequent loss of heterozygosity at the *TEL* gene locus in acute lymphoblastic leukemia of childhood. **Blood 86**:38-44.
15. Sato Y, Suto Y, Pietenpol J, Golub T, Gilliland DG. in press. *TEL* and *KIP1* define the smallest region of deletions on 12p13 in hematopoietic malignancies. **Blood** .
16. Buijs A, Sherr S, van Baal S, van Bezouw S, van der Plas D, van Kessel AG, Riegman P, Deprez RL, Zwarthoff E, Hagemeijer A, Grosveld G. 1995. Translocation (12;22)(p13;q11) in myeloproliferative disorders results in fusion of the ETS-like *TEL* gene on 12p13 to the *MN1* gene on 22q11. **Oncogene 10**:1511-1519.
17. Sorensen PHB, Lessnick AL, Lopez-Terrada D, Liu XF, Triche TJ, Denny CT. 1994. A second Ewing's sarcoma translocation, t(21;22), fuses the EWS gene to another ETS-family transcription factor, ERG. **Nature Genetics 6**:146-151.
18. Shimizu K, Ichikawa H, Tojo A, Kaneko Y, Maseki N, Hayashi Y, Ohira M, Asano S, Ohki M. 1993. An *ets*-related gene, *ERG*, is rearranged in human myeloid leukemia with t(16;21) chromosomal translocation. **Proceedings of the National Academy of Science USA 90**:10280-10284.
19. Pendergast AM, Quilliam LA, Cripe LD, Bessing CH, Dai Z, Li N, Der CJ, Sclessinger J, Gishizky ML. 1993. BCR-ABL-induced oncogenesis is mediated by direct interaction with the SH2 domain of the GRB-2 adapter protein. **Cell 75**:175-185.
20. Nucifora G, Begy CR, Erickson P, Drabkin HA, Rowley JD. 1993. The 3;21 translocation in myelodysplasia results in a fusion transcript between the AML1 gene and the gene for ETO, a highly conserved protein associated with the Epstein-Barr virus small RNA EBER 1. **Proc Natl Acad Sci USA 90**:7784-7788.
21. Romana SP, Mauchauffe M, Leconiat M, Chumakov I, Le Paslier D, Berger R, Bernard OA. 1995. The t(12;21) of acute lymphoblastic leukemia results in a tel-AML1 gene fusion. **Blood 85**: 3662-3670.

Transforming Properties of the Leukemic Inv(16) Fusion Gene *CBFB-MYH11*.

A. Hajra, P. P. Liu, and F. S. Collins
Laboratory of Gene Transfer, National Center for Human Genome Research, National Institutes of Health, Bethesda, MD 20892, USA

Introduction

Inv(16) and Acute Myeloid Leukemia

A characteristic pericentric inversion of chromosome 16 [inv(16)(p13q22)] is consistently associated with acute myelomonocytic leukemia with eosinophilia (AML M4Eo). This type of leukemia comprises 8% of AML cases and affects approximately 2,000 people each year in the United States. In addition to inv(16), a number of other unique features are associated with AML M4Eo, including the presence of morphologically and histochemically abnormal eosinophils, a relatively good prognosis, and possibly frequent involvement of the central nervous system (for review see Liu et al. 1995)

The consistent association of inv(16) with M4Eo AML strongly suggests that it is important in the pathogenesis of M4Eo AML. By analogy with studies of other chromosome translocations, cloning of the genes disrupted by the inv(16) rearrangement would be predicted to identify oncogenes important in the development of M4Eo AML. Functional analysis of such oncogenes might yield new insights into the pathogenesis of M4Eo AML, explain the molecular basis for the unique characteristics of this malignancy, or suggest new therapeutic methods for this disease.

Inv(16) generates a chimeric CBFβ-SMMHC Protein

Using positional cloning techniques, we have shown (Liu et al. 1993) that inv(16) fuses the *MYH11* gene (16p13) encoding smooth muscle myosin heavy chain (SMMHC) (Matsuoka et al. 1993) to the 3' end of the*CBFB* gene (16q22) encoding the β subunit of core binding factor (CBFβ) (Ogawa et al. 1993a; Wang et al. 1993). This rearrangement results in the production of a 5'-*CBFB-MYH11*-3' fusion mRNA transcript encoding a CBFβ-SMMHC chimeric protein (referred to hereafter as the inv16 protein). *CBFB-MYH11* transcripts are consistently associated with inv(16) leukemia, suggesting that they play a critical role in inv(16) leukemia (for review see Liu et al. 1995).

CBF (also known as PEBP2) was first identified as a murine viral enhancer binding protein and subsequently shown to regulate the transcription of T-cell- and myeloid-specific genes. This transcription factor binds DNA at the consensus sequence YGYGGT (where Y indicates pyrimidine). CBF has recently been shown to cooperate with the transcription factors Ets-1 and c-Myb in DNA binding and transcriptional activation (Hernandez-Munain and Krangel 1995; Sun et al. in press; Wotton et al. 1994)

Cloning of CBF cDNAs (Ogawa et al 1993a, b; Wang et al 1993) has revealed that this transcription factor consists of an α and a β subunit. Three different CBFα subunits have been identified to date. Although encoded by separate genes with distinct patterns of expression (Levanon et al 1994), they all share a region mediating DNA binding and association with the CBFβ subunit that is homologous to the *Drosophila* segmentation protein Runt (Kagoshima et al 1993). One of the CBFα genes, *CBFA2* (previously known as *AML1*), is disrupted by the leukemia-associated t(8;21) and t(3;21) translocations (Mitani et al 1994; Nucifora et al 1994; Nucifora and Rowley 1994). The involvement of both CBF subunits in myeloid leukemia suggests that this transcription factor is critical for normal myelopoiesis.

CBFβ increases the DNA binding affinity of CBFα without contacting DNA directly (Ogawa et al. 1993a; Wang et al. 1993). CBFβ is encoded by a single, ubiquitously expressed gene (Bae et al. 1994; Hajra and Collins 1995; Ogawa et al. 1993a) that is conserved through evolution (Hajra and Collins 1995; J. P. Gergen, personal communication), although it does not contain sequences found in other known genes. The longest known *CBFB* transcript encodes a protein of 187 amino acids, and all but the last 22 amino acids are fused to the SMMHC tail region by the inv(16) rearrangement (Liu et al. 1993).

SMMHC (Matsuoka et al. 1993) is a member of the myosin II class of proteins. It forms two of the subunits of the hexameric smooth muscle myosin. SMMHC consists of an amino-terminal globular domain which binds actin and has ATPase activity, and a tail region which dimerizes to form a long rod of α-helical coiled-coil structure and drives the assembly of myosin into bipolar filaments (Morano 1992). The tail region becomes fused to CBFβ as a result of inv(16). Several different breakpoints have been reported in the tail-encoding region of *MYH11*, but all resulting transcripts maintain the correct reading frame of the SMMHC tail region (Liu et al. 1995), strongly suggesting that the coiled-coil motif of this domain has a critical function in the fusion protein.

Since the inv16 protein is a chimera between a transcription factor subunit that does not bind DNA and a structural domain of myosin, it is not immediately obvious how it might cause leukemia. Therefore, we were interested in studying the function of this chimeric protein. We report here the results of these studies, which suggest that the inv16 protein transforms cells by interfering with the normal function of CBF, thereby disrupting the expression pattern of CBF-regulated genes.

Characteristics of the Inv16 Protein

cDNAs encoding full-length inv16 or normal CBFβ proteins were constructed and used as templates to generate the corresponding proteins by in vitro transcription and translation (Liu et al. 1994). The proteins are illustrated schematically in Fig. 1. The availability of antibodies recognizing normal CBFβ and SMMHC allowed us to use immunoblots to confirm that the in vitro generated inv16 and CBFβ proteins were of the correct size (67 and 22 kD, respectively). Electrophoretic mobility shifts assays (EMSAs) provided a convenient and useful method to examine the properties of the CBFβ domain of the inv16 protein. Like CBFβ, the inv16 protein cannot bind CBF binding sites alone, but can associate with CBFα to increase its DNA-binding affinity in a manner indistinguishable from that of CBFβ (Liu et al. 1994). The DNA-binding ability of inv16:CBFα heterodimers to bind CBF sites is identical to that of normal CBF. However, inv16:CBFα/DNA complexes are larger in size than normal CBF/DNA complexes. In fact, the majority of inv16:CBFα/DNA complexes are apparently so large that they fail to migrate out of the gel wells (Liu et al. 1994). This may be due in part to the larger size of the inv16 protein compared to CBFβ, but this is insufficient to account for the dramatic difference in gel migration between the two protein/DNA complexes.

The apparent very high molecular mass of inv16:CBFα/DNA complexes led us to speculate that they may represent inv16 protein multimers formed through the SMMHC domain. Glutaraldehyde cross-linking studies demonstrated that the inv16 protein could indeed self-associate into higher order oligomers, while CBFβ was exclusively monomeric (Hajra et al. manuscript in preparation). The low solubility of the inv16 protein is also consistent with it being able to form multimers (Hajra et al. 1995), and raises the possibility that the protein/DNA complexes in the gel wells may arise from protein precipitation during the EMSA binding reactions (Hajra et al. in press). The results from these in vitro studies suggest that the inv16 protein is a truly chimeric protein, retaining the ability of CBFβ to associate with CBFα , as well as the ability of SMMHC to form multimers.

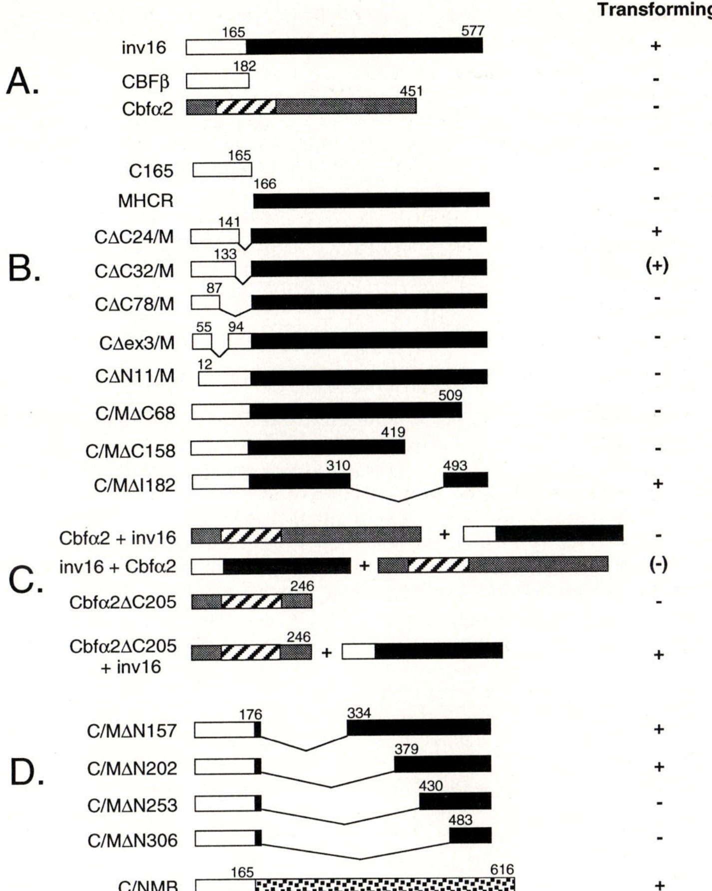

Fig. 1: Full-length and mutant cDNA constructs described in this article. White boxes indicate CBFβ sequences, black boxes indicate SMMHC sequences, grey boxes indicate Cbfα2 sequences (with the hatched region indicating the Runt domain), and the speckled boxes indicate NMMHCB sequence. Numbers above the boxes list the amino acid location of the indicated boundary. Lines below the boxes indicates regions which become fused together in the constructs. When two constructs are shown on the same line, the one shown first is the one that was first transfected into the cells. On the right is indicated whether the constructs can transform NIH 3T3 cells: +, cells expressing the construct acquire a transformed phenotype; -, cells expressing the construct retain a nontransformed phenotype; (+), cells expressing the construct form smaller foci and soft agarose colonies than other transformed cells; (-), cells expressing the construct are able to form small numbers of foci but cannot grow in soft agarose nor form tumors in nude mice. (A) constructs encoding full-length Cbfα2, CBFβ, and the inv16 protein. (B) Inv16 protein deletion mutants. (C) Coexpression of the inv16 protein and full-length or truncated Cbfα2. (D) Inv16 mutants with deletions of the SMMHC domain and a CBFβ-NMMHCB fusion construct.

Transformation Properties of the Inv16 Protein

We next investigated how the inv16 protein might functionally interact with endogenous CBF in cells. We stably expressed the inv16 protein, and overexpressed CBFα (murine Cbfα2) and CBFβ in NIH 3T3 cells. NIH 3T3 cells express normal CBF, including CBFβ and low levels of all three different CBFα subunits (Bae et al. 1994; Ogawa et al. 1993a, b). Immunoblots using the appropriate anti-CBFα, -CBFβ, or -SMMHC antibodies were used to confirm stable overexpression of the transfected constructs (Hajra et al. 1995; Hajra et al. in press). NIH 3T3 cells stably expressing the inv16 protein acquired a transformed phenotype, as indicated by their ability to form foci, grow in 0.6% soft agarose, and form tumors when injected into immunodeficient nude mice (Hajra et al. 1995). Cells stably overexpressing CBFα, CBFβ or the SMMHC rod region alone retained a normal, nontransformed phenotype (Hajra et al. 1995; Hajra et al. in press).

In EMSAs, extracts from cells expressing empty vector formed a normal CBF/DNA complex. This complex was unaltered in cells overexpressing CBFβ (Hajra et al. 1995), and was much more intense in cells overexpressing CBFα (Hajra et al. in press). In cells expressing the inv16 protein, however, this complex was almost completely replaced by a very large protein/DNA complex that remained in the gel wells, as described in the previous section (Hajra et al. 1995).

To investigate the transactivation properties of CBF in the cell lines, we cotransfected reporter constructs with the chloramphenicol acetyl transferase gene driven by viral enhancers containing CBF binding sites. Two different reporter constructs were studied: the Moloney murine leukemia virus (MoMLV) enhancer and the Rous sarcoma virus (RSV) enhancer. Relative to the levels in cells expressing empty vector, transactivation of both these reporter constructs was unaltered in cells overexpressing CBFβ, but increased by varying degrees in cells overexpressing CBFα (Liu et al. 1994; Hajra et al. in press; Hajra et al. manuscript in preparation). Expression of the inv16 protein had varying effects: transactivation of the MoMLV enhancer was decreased ten-fold (Liu et al. 1994; Hajra et al. in press), but transactivation of the RSV enhancer was increased five-fold (Hajra et al. manuscript in preparation). We confirmed that the observed transactivation effects were indeed mediated through CBF sites in the MoMLV reporter construct by using a construct containing mutated CBF sites as control (Liu et al. 1994; Hajra et al. in press).

Regions of Inv16 Protein Necessary for Cellular Transformation

To define functionally important regions of the inv16 protein, we analyzed the transforming properties of a series of deletion mutants (Fig. 1B). Some of these deletion mutants had deletions within the CBFβ domain predicted to affect heterodimerization with CBFα to varying degrees. Others had truncations of the carboxyl-terminus of the SMMHC domain, which is known to be important for the formation of myosin bipolar filaments. The predicted functional properties of these mutants were confirmed by EMSAs (to assess ability to associate with CBFα) and solubility assays (to assess multimerizing ability) (Hajra et al. 1995). Cell extracts expressing nonmultimerizing mutants did not form the characteristic EMSA complex in the gel wells. These extracts instead formed complexes that were larger than the normal CBF/DNA complex, but which were able to migrate into the gel matrix (Hajra et al. 1995).

Phenotypic analysis of cells expressing the deletion mutants indicated that only those mutants which retained the ability to both associate with CBFα and form multimers could transform NIH 3T3 cells; loss of either function abolished all transforming properties (Hajra et al. 1995). Cells expressing a mutant (CΔC32/M) with a very weak ability to associate with CBFα formed smaller foci and soft agarose colonies than cells expressing other transforming mutants. This mutant is identical to a variant inv(16) *CBFB-MYH11* fusion recently identified in three patients by Shurtleff et al. (in press). Transforming mutants had the same effect on CBF transactivation as the full-length inv16 protein, while nontransforming mutants had no effect on transactivation of the reporter constructs (Hajra et al. in press). These results indicate that the SMMHC domain plays an active role in the function of the inv16 protein, and emphasize the chimeric nature of the protein's functional properties.

Effect of CBFα Overexpression on the Transforming Properties of the Inv16 Protein

We further studied the role of CBFα in cellular transformation by inv16 protein. We investigated the effect overexpression of CBFα (Cbfα2) would have on subsequent expression of the inv16 protein. We hypothesized that if the inv16 protein transforms cells by a dominant positive effect on the CBF function overexpression of Cbfα2 would be expected to result in an increased number of transformed cells. However, if the inv16 protein transforms cells through a dominant negative mechanism overexpression of Cbfα2 would be expected to reduce or prevent cellular transformation. Finally, if the inv16 protein interacts with another protein besides CBFα during cellular transformation, transformation efficiency should be unaffected by the presence of overexpressed Cbfα2, or may be decreased if the other protein competes with Cbfα2 for binding to the inv16 protein.

The Cbfα2-overexpressing cells described in the previous section were stably transfected with an expression vector encoding the inv16 protein (Hajra et al. in press). Since Cbfα2 was encoded by a plasmid containing the hyg^r marker, while the inv16 protein expression vector had a neo^1 marker, we were able to select for stable expression of both constructs by growing the cells in hygromycin B and G418. Cells expressing both Cbfα2 and the inv16 protein retained a normal, nontransformed phenotype, and were unable to form foci, grow in soft agarose, or form tumors in nude mice (Figure 1C; Hajra et al. in press). Control cells, expressing the inv16 protein and an empty hyg^r vector, did acquire a transformed phenotype. In addition, Cbfα2 overexpression in cells already transformed by the inv16 protein could revert the cells to a less transformed phenotype. These protective effects were specific to transformation by the inv16 protein, since cells overexpressing Cbfα2 could still be transformed by an activated *Ha-ras* gene (Hajra et al. in press).

Extracts from cells protected from transformation formed both the normal CBF/DNA complex and the large inv16:CBFα complex in EMSAs. Transactivation of the previously described reporter constructs returned to normal levels in these cells. It therefore appeared that Cbfα2 overexpression suppressed the transforming properties of inv16 protein by restoring normal CBF gene regulation. To confirm this, we examined the protective effects of a nonfunctional Cbfα2 mutant (Cbfα2ΔC205, see Figure 1C). This mutant lacked the last 205 amino acids of Cbfα2, deleting the transactivation domain while leaving DNA-binding domain intact. Similar mutants have been shown to lack transactivation properties and act as a dominant negative repressor of CBF gene activation (Bae et al. 1994; Meyers et al. 1995). The Cbfα2ΔC205 construct was stably expressed in clonal cell lines, at comparable levels as full-length Cbfα2 (Hajra et al. in press). Cbfα2ΔC205 could associate with CBFβ and completely abolish CBF transactivation of the MoMLV and RSV enhancers (Hajra et al. in press, A. L., P. L., and F. C., unpublished data). Cells expressing Cbfα2ΔC205 had the same morphology as cells expressing full-length Cbfα2, and did not acquire a transformed phenotype. However, stable expression of the inv16 protein in Cbfα2ΔC205-expressing cells caused the cells to acquire a transformed phenotype (Hajra et al. in press).

These results suggest that suppression of inv16 transformation by CBFα is dependent on the transactivation properties of the latter protein. A dominant negative inhibition of CBFα transactivation by itself is not sufficient for cellular transformation. This may be due to the inv16 protein and Cbfα2ΔC205 having different mechanisms of interfering with CBFα, or may be due to cellular effects of inv16 protein that is unrelated to CBF function, perhaps involving the SMMHC domain.

Properties of the Myosin Heavy Chain Domain Necessary for Inv16 Protein Transformation

We next investigated how much of the SMMHC domain is required for the transformation properties of the inv16 protein. We constructed a series of mutants which had deletions of varying size within the SMMHC domain (Figure 1D; Hajra et al. manuscript in preparation). We confirmed stable expression of the mutants by immunoblots, and examined the phenotype of NIH 3T3 cells expressing these mutants as described above. The C/MΔN202 mutant, containing 210 amino acids of the SMMHC domain, was able to transform NIH 3T3 cells as efficiently as the full-length inv16 protein. On the other hand, a mutant (C/MΔN253) containing 159 amino acids of the SMMHC domain lost all transforming properties. The nontransforming C/MΔN253 mutant was poorly soluble in cells, indicating that it was still capable of forming multimers. However, glutaraldehyde cross-linking studies indicated that the C/MΔN253 mutant did not form large oligomers in vitro as efficiently as the C/MΔN202 mutant, which had the same oligomerization ability as the full-length inv16 protein (Hajra et al. manuscript in preparation). Interestingly, these cross-linking assays also showed that another deletion mutant (C/MΔN306), incapable of forming multimers, could still form homodimers. This most likely occurs because the remaining SMMHC domain in the mutant (106 amino acids) is sufficient to mediate coiled-coil interactions.

Despite their similar multimerizing abilities, the transforming C/MΔN202 and the nontransforming C/MΔN253 mutants had very different effects on cellular CBF function. In EMSAs, extracts from cells expressing C/MΔN202 formed only the characteristic large protein/DNA complex in the gel wells, as described above. Cell extracts expressing C/MΔN253 formed this same complex, but these extracts also formed small amounts of the normal CBF/DNA complex. C/MΔN202 had the same effects on transactivation of CBF-responsive reporter constructs as the full-length inv16 protein, while C/MΔN253 had no effect on CBF transactivation at all (Hajra et al. manuscript in preparation).

Since the SMMHC domain appeared to be critical to the transforming properties of the inv16 protein, we investigated whether other myosin heavy chains could functionally substitute for this domain. We therefore replaced the SMMHC domain of the inv16 protein with the analogous rod region from the nonmuscle myosin heavy chain 1 (NMMHCB) protein. Cells stably expressing CBFβ-NMMHCB acquired a transformed phenotype, including the ability to form foci, grow in soft agarose, and form tumors in nude mice. Cells expressing CBFβ-NMMHCB had the same CBF DNA binding and transactivation properties as cells expressing the inv16 protein, including the formation of large characteristic EMSA complexes in the gel well, and inhibition of MoMLV enhancer transactivation (Hajra et al. manuscript in preparation).

These results indicate that multimerization of the inv16 protein is necessary but not sufficient for cellular transformation. Alteration of normal CBF function is the best predictor of the transforming potential of inv16 protein deletion mutants, emphasizing that interference with CBF is the critical component of cellular transformation by the inv16 protein. The function of the SMMHC domain reflects general properties shared by all myosin heavy chains.

Conclusions

The results described here support the interference model we have previously proposed to explain the action of the inv16 protein (Liu et al. 1995). This model is summarized in Fig. 2. According to this model, the inv16 protein homodimerizes and self-assembles into multimeric complexes through a process mediated by the SMMHC domain. The inv16 protein multimers then associate with CBFα subunits, competing with endogenous CBFβ. The α subunit is present in limiting amounts in NIH 3T3 cells, since CBF function is correlated with CBFα expression levels but not with that of CBFβ (see above). The cellular overexpression of the inv16 protein ensures its association with the majority of endogenous CBFα. The different subcellular localizations of the inv16 protein (nuclear) and CBFβ (cytoplasmic) may also facilitate the preferential interaction of the oncoprotein with CBFα (Lu et al. 1995; C. Wijmenga, P. L. and F. C., unpublished data). The inv16:CBFα multimers bind to CBF sites in the transfected cells, and these large, DNA-bound multimers may be able to interfere with other transcription factors that bind to sites near the CBF site, by either disrupting protein-DNA or protein-protein interactions. If these adjacent proteins are transcriptional coactivators , the result is decreased expression of the CBF-responsive promoter (as seen with the MoMLV enhancer). However, if the adjacent proteins are transcriptional repressors, the result may actually be an increase in gene expression (as seen with the RSV enhancer). CBF-responsive promoters may be particularly susceptible to this type of interference, since CBF appears to frequently require cooperation with other transcription factors for full gene activation (Hernandez-Munain and Krangel 1995; Sun et al in press; Wotton et al. 1994).

The effect of the inv16 protein on CBF function is apparently promoter-specific, and is dependent on the specific arrangement of binding sites for CBF and other proteins within each CBF-regulated promoter. Cellular transformation is probably the consequence of a unique combination of over- and under-expression of certain CBF-regulated genes. Identification of genes regulated differently by CBF and the inv16 protein will provide insights into which cellular pathways are altered in transformed cells.

The model described here explains all the present data on the functional properties of the inv16 protein. The nontransforming C/MΔN253 mutant may form multimers that are too small or have the wrong conformation to interfere with adjacent transcription factors. In cells overexpressing Cbfα2, the amount of CBFα is no longer limiting, and instead is actually in great excess. The inv16 protein becomes saturated with CBFα, and the remainder associates with CBFβ. This results in increased amounts of normal CBF sufficient to counteract the disruptive effects of the inv16 protein. The effects of the inv16 protein appear to be partially due to a mass action effect arising from its high expression levels. Further studies, therefore, are needed to determine how this oncoprotein exerts a dominant effect in leukemic AML M4Eo cells in which both inv16 and CBFβ proteins are produced from genes with the same promoter and should be present at similar levels unless other mechanisms reduce the level of CBFβ (e.g., RNA and/or protein stability; subcellular localization).

The model described here provides a theoretical basis for further investigations of the function of the inv16 protein. Future studies on the inv16 protein will follow at least two directions. The first direction will lead to more detailed investigations of some of the observations described here. This will include the identification of target genes, determination of how DNA-binding differs between inv16:CBFα and normal CBF using EMSAs and DNase footprinting, and studies to elucidate how DNA-bound inv16 interacts with adjacently binding transcription factors. The other direction of future studies on the inv16 protein will explore the function of this protein specifically in myeloid cells. Information about the inv16 protein function gained from studies on NIH 3T3 cells suggest a number of specific hypotheses which can be tested in myeloid

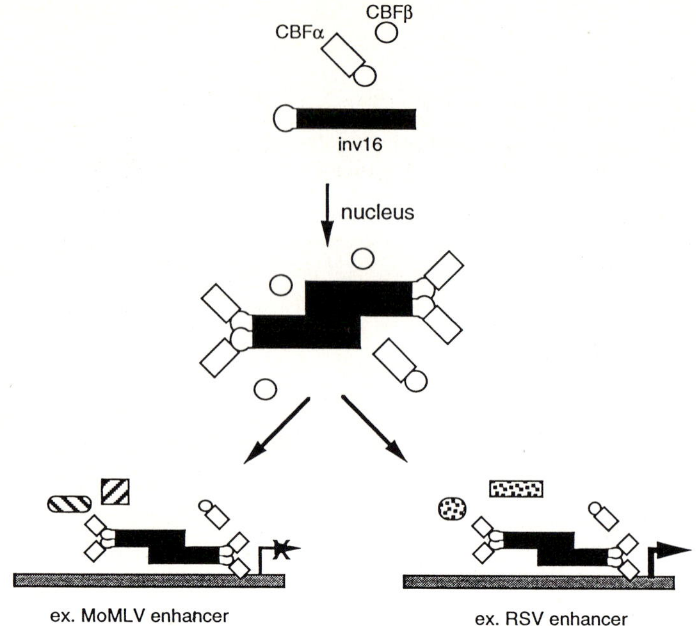

Fig. 2: Model of how the inv16 protein alters normal CBF function. White circles indicate CBFβ, tilted white rectangles indicate CBFα, and horizontal black rectangles indicate the SMMHC rod region. The inv16 protein can self-associate into large multimers. It is not known whether multimerization occurs in the cytoplasm or nucleus, or in both locations. In the nucleus, the inv16 protein associates with CBFα. Inv16:CBFα multimers can bind DNA (grey boxes) at CBF sites, preventing DNA binding by normal CBF and interfering with the binding of other transcription factors to adjacent DNA sites. If these adjacent transcription factors are gene activators (hatched shapes), this disruption inhibits expression of the target gene. This is seen with the MoMLV cnhancer. If, on the other hand, the adjacent transcription factor is a gene repressor (speckled shapes), the result is a superactivation of the target gene. This is apparently occurring with the RSV enhancer. Cellular transformation is the cumulative effect of increased and decreased expression of specific CBF-regulated genes.

cells. These studies could involve the overexpression of the inv16 protein in myeloid cell lines and the creation of animal models, such as *CBFB-MYH11* transgenic mice. These studies could also investigate the relationship of the inv16 protein to the abnormal eosinophils present in AML M4Eo.

Acknowledgements

We would like to thank our colleagues Cisca Wijmenga, Paula Gregory, and Trevor Blake for many helpful discussions, suggestions, technical help, and the use of C.W.'s unpublished data. We would also like to thank Nancy Speck and Bob Adelstein for providing various reagents and very valuable advice.

References

Bae S-C, Ogawa E, Maruyama M, Oka H, Satake M, Shigesada K, Jenkins NA, Gilbert DJ, Copeland NG, Ito Y (1994) PEBP2αB/mouse AML1 consists of multiple isoforms that possess differential transactivation potentials. Mol Cell Biol 14:3242-3252

Hajra A, Collins F S (1995) Structure of the leukemia-associated human CBFB gene. Genomics 26:571-579

Hajra A, Liu PP, Wang Q, Kelley CA, Stacy T, Adelstein RS, Speck NA, Collins FS (1995) The leukemic core binding factor β-smooth muscle myosin heavy chain (CBFβ-SMMHC) chimeric protein requires both CBFβ and myosin heavy chain domains for transformation of NIH 3T3 cells. Proc Natl Acad Sci USA 92:1926-1930

Hajra A, Liu PP, Speck NA, Collins FS Overexpression of CBFα reverses cellular transformation by the CBFβ-SMMHC chimeric oncoprotein. Mol Cell Biol in press

Hernandez-Munain C and Krangel MS (1995) c-Myb and core-binding factor/PEBP2 display functional synergy but bind independently to adjacent sites in the T-cell receptor δ enhancer. Mol Cell Biol 15:3090-3099

Kagoshima H, Shigesada K, Satake M, Ito Y, Miyoshi H, Ohki M, Pepling M, Gergen P (1993) The Runt domain identifies a new family of heteromeric transcriptional regulators. Trends Genet 9:338-341

Levanon D, Negreanu V, Bernstein Y, Bar-Am I, Avivi L, Groner Y (1994) AML1, AML2, and AML3, the human members of the runt domain gene-family: cDNA structure, expression, and chromosomal localization. Genomics 23:425-432

Liu P, Tarlé SA, Hajra A, Claxton DF, Marlton P, Freedman M, Siciliano MJ, Collins FS (1993) Fusion between transcription factor CBFβ/PEBP2β and a myosin heavy chain in acute myeloid leukemia. Science 261:1041-1044

Liu P, Seidel N, Bodine D, Speck N, Tarlé S, Collins FS (1994) Acute myeloid leukemia with inv(16) produces a chimeric transcription factor with a myosin heavy chain tail. Cold Spring Harbor Symp.Quant.Biol 59:547-553

Liu PP, Hajra A, Wijmenga C, Collins FS (1995) Molecular pathogenesis of the chromosome 16 inversion in the M4Eo subtype of AML. Blood 85:2289-2302

Lu J, Maruyama M, Satake M, Bae S-C, Ogawa E, Kagoshima H, Shigesada K, Ito Y (1995) Subcellular localization of the α and β subunits of acute myeloid leuekmia-linked transcription factor PEBP2/CBF. Mol Cell Biol 15:1651-1661

Matsuoka R, Yoshida MC, Furutani Y, Imamura S, Kanda N, Yanagisawa M, Masaki T, Takao A (1993) Human smooth muscle heavy chain gene mapped to chromosomal region 16q12. Am J Med Genet 46:61-67

Meyers S, Lenny N, Hiebert SW (1995) The t(8;21) fusion protein interferes with AML-1B-dependent transcriptional activation. Mol Cell Biol 15:1974-1982

Mitani K, Ogawa S, Tanaka T, Miyoshi H, Kurokawa M, Mano H, Yazaki Y, Ohki M, Hirai H (1994) Generation of the AML1-EVI-1 fusion gene in the t(3;21)(q26;q22) causes blastic crisis in chronic myelocytic leukemia. EMBO J 13:504-510

Morano IL (1992) Molecular biology of smooth muscle. J Hypertens 10:411-416

Nucifora G, Begy CR, Kobyashi H, Roulston D, Claxton D, Pedersen-Bjergaard J, Parganas E, Ihle JN, Rowley JD (1994) Consistent intergenic splicing and production of multiple transcripts between AML1 at 21q22 and unrelated genes at 3q26 in (3;21)(q26;q22) transloactions. Proc Natl Acad Sci USA 91:4004-4008

Nucifora G, Rowley JD (1994) The AML1 and ETO genes in acute myeloid leukemia with a t(8;21) Leuk Lymphoma 14:353-362

Ogawa E, Inuzuka M, Maruyama M, Statke M, Naito-Fujimoto M, Ito Y, Shigesada K (1993a) Molecular cloning and characterization of PEBP2β, the heterodimeric partner of a novel Drosophila runt-related DNA binding protein PEBP2α. Virology 194:314-331

Ogawa E, Maruyama M, Kagoshima H, Inuzuka M, Lu J, Satake M, Shigesada K, Ito Y (1993b) PEBP2/PEA2 represents a family of transcription factors homologous to the products of the Drosophila runt gene and the human AML1 gene. Proc Natl Acad Sci USA 90:6859-6863

Shurtleff SA, Meyers S, Hiebert SW, Raimondi SC, Head DR, Willman CL, Wolman S, Carroll AJ, Behm F, Hulshof MG, Motroni TA, Liu P, Collins FS, Downing JR Heterogeneity in CBFβ/MYH11 fusion messages encoded by the inv(16)(p13q22) and the t(16;16)(p13;q22) translocation in acute myelogenous leukemia. Blood in press

Sun W, Graves BJ, Speck NA Transactivation of the Moloney murine leukemia virus and T cell receptor β chain enhancers by CBF and Ets requires intact binding sites for both proteins. J Virol in press

Wang S, Wang Q, Crute BE, Melnikova IN, Keller SR, Speck NA (1993) Cloning and characterization of subunits of the T-cell receptor and murine leukemia virus enhancer core-binding factor. Mol Cell Biol 13:3324-3339

Wotton D, Ghysdael J, Wang S, Speck NA, Owen, MJ (1994) Cooperative binding of Ets-1 and Core Binding Factor to DNA. Mol Cell Biol 14:840-850

Current Topics in Microbiology and Immunology

Volumes published since 1989 (and still available)

Vol. 169: **Kolakofsky, Daniel (Ed.):** Bunyaviridae. 1991. 34 figs. X, 256 pp.
ISBN 3-540-53061-4

Vol. 170: **Compans, Richard W. (Ed.):** Protein Traffic in Eukaryotic Cells. Selected Reviews. 1991. 14 figs. X, 186 pp.
ISBN 3-540-53631-0

Vol. 171: **Kung, Hsing-Jien; Vogt, Peter K. (Eds.):** Retroviral Insertion and Oncogene Activation. 1991. 18 figs. X, 179 pp.
ISBN 3-540-53857-7

Vol. 172: **Chesebro, Bruce W. (Ed.):** Transmissible Spongiform Encephalopathies. 1991. 48 figs. X, 288 pp.
ISBN 3-540-53883-6

Vol. 173: **Pfeffer, Klaus; Heeg, Klaus; Wagner, Hermann; Riethmüller, Gert (Eds.):** Function and Specificity of γ/δ TCells. 1991. 41 figs. XII, 296 pp.
ISBN 3-540-53781-3

Vol. 174: **Fleischer, Bernhard; Sjögren, Hans Olov (Eds.):** Superantigens. 1991. 13 figs. IX, 137 pp. ISBN 3-540-54205-1

Vol. 175: **Aktories, Klaus (Ed.):** ADP-Ribosylating Toxins. 1992. 23 figs. IX, 148 pp.
ISBN 3-540-54598-0

Vol. 176: **Holland, John J. (Ed.):** Genetic Diversity of RNA Viruses. 1992. 34 figs. IX, 226 pp. ISBN 3-540-54652-9

Vol. 177: **Müller-Sieburg, Christa; Torok-Storb, Beverly; Visser, Jan; Storb, Rainer (Eds.):** Hematopoietic Stem Cells. 1992. 18 figs. XIII, 143 pp. ISBN 3-540-54531-X

Vol. 178: **Parker, Charles J. (Ed.):** Membrane Defenses Against Attack by Complement and Perforins. 1992. 26 figs. VIII, 188 pp. ISBN 3-540-54653-7

Vol. 179: **Rouse, Barry T. (Ed.):** Herpes Simplex Virus. 1992. 9 figs. X, 180 pp.
ISBN 3-540-55066-6

Vol. 180: **Sansonetti, P. J. (Ed.):** Pathogenesis of Shigellosis. 1992. 15 figs. X, 143 pp.
ISBN 3-540-55058-5

Vol. 181: **Russell, Stephen W.; Gordon, Siamon (Eds.):** Macrophage Biology and Activation. 1992. 42 figs. IX, 299 pp.
ISBN 3-540-55293-6

Vol. 182: **Potter, Michael; Melchers, Fritz (Eds.):** Mechanisms in B-Cell Neoplasia. 1992. 188 figs. XX, 499 pp.
ISBN 3-540-55658-3

Vol. 183: **Dimmock, Nigel J.:** Neutralization of Animal Viruses. 1993. 10 figs. VII, 149 pp.
ISBN 3-540-56030-0

Vol. 184: **Dunon, Dominique; Mackay, Charles R.; Imhof, Beat A. (Eds.):** Adhesion in Leukocyte Homing and Differentiation. 1993. 37 figs. IX, 260 pp.
ISBN 3-540-56756-9

Vol. 185: **Ramig, Robert F. (Ed.):** Rotaviruses. 1994. 37 figs. X, 380 pp.
ISBN 3-540-56761-5

Vol. 186: **zur Hausen, Harald (Ed.):** Human Pathogenic Papillomaviruses. 1994. 37 figs. XIII, 274 pp. ISBN 3-540-57193-0

Vol. 187: **Rupprecht, Charles E.; Dietzschold, Bernhard; Koprowski, Hilary (Eds.):** Lyssaviruses. 1994. 50 figs. IX, 352 pp. ISBN 3-540-57194-9

Vol. 188: **Letvin, Norman L.; Desrosiers, Ronald C. (Eds.):** Simian Immunodeficiency Virus. 1994. 37 figs. X, 240 pp.
ISBN 3-540-57274-0

Vol. 189: **Oldstone, Michael B. A. (Ed.):** Cytotoxic T-Lymphocytes in Human Viral and Malaria Infections. 1994. 37 figs. IX, 210 pp. ISBN 3-540-57259-7

Vol. 190: **Koprowski, Hilary; Lipkin, W. Ian (Eds.):** Borna Disease. 1995. 33 figs. IX, 134 pp. ISBN 3-540-57388-7

Vol. 191: **ter Meulen, Volker; Billeter, Martin A. (Eds.):** Measles Virus. 1995. 23 figs. IX, 196 pp. ISBN 3-540-57389-5

Vol. 192: **Dangl, Jeffrey L. (Ed.):** Bacterial Pathogenesis of Plants and Animals. 1994. 41 figs. IX, 343 pp. ISBN 3-540-57391-7

Vol. 193: **Chen, Irvin S. Y.; Koprowski, Hilary; Srinivasan, Alagarsamy; Vogt, Peter K. (Eds.):** Transacting Functions of Human Retroviruses. 1995. 49 figs. IX, 240 pp. ISBN 3-540-57901-X

Vol. 194: **Potter, Michael; Melchers, Fritz (Eds.):** Mechanisms in B-cell Neoplasia. 1995. 152 figs. XXV, 458 pp. ISBN 3-540-58447-1

Vol. 195: **Montecucco, Cesare (Ed.):** Clostridial Neurotoxins. 1995. 28 figs. XI., 278 pp. ISBN 3-540-58452-8

Vol. 196: **Koprowski, Hilary; Maeda, Hiroshi (Eds.):** The Role of Nitric Oxide in Physiology and Pathophysiology. 1995. 21 figs. IX, 90 pp. ISBN 3-540-58214-2

Vol. 197: **Meyer, Peter (Ed.):** Gene Silencing in Higher Plants and Related Phenomena in Other Eukaryotes. 1995. 17 figs. IX, 232 pp. ISBN 3-540-58236-3

Vol. 198: **Griffiths, Gillian M.; Tschopp, Jürg (Eds.):** Pathways for Cytolysis. 1995. 45 figs. IX, 224 pp. ISBN 3-540-58725-X

Vol. 199/I: **Doerfler, Walter; Böhm, Petra (Eds.):** The Molecular Repertoire of Adenoviruses I. 1995. 51 figs. XIII, 280 pp. ISBN 3-540-58828-0

Vol. 199/II: **Doerfler, Walter; Böhm, Petra (Eds.):** The Molecular Repertoire of Adenoviruses II. 1995. 36 figs. XIII, 278 pp. ISBN 3-540-58829-9

Vol. 199/III: **Doerfler, Walter; Böhm, Petra (Eds.):** The Molecular Repertoire of Adenoviruses III. 1995. 51 figs. XIII, 310 pp. ISBN 3-540-58987-2

Vol. 200: **Kroemer, Guido; Martinez-A., Carlos (Eds.):** Apoptosis in Immunology. 1995. 14 figs. XI, 242 pp. ISBN 3-540-58756-X

Vol. 201: **Kosco-Vilbois, Marie H. (Ed.):** An Antigen Depository of the Immune System: Follicular Dendritic Cells. 1995. 39 figs. IX, 209 pp. ISBN 3-540-59013-7

Vol. 202: **Oldstone, Michael B. A.; Vitković, Ljubiša (Eds.):** HIV and Dementia. 1995. 40 figs. XIII, 279 pp. ISBN 3-540-59117-6

Vol. 203: **Sarnow, Peter (Ed.):** Cap-Independent Translation. 1995. 31 figs. XI, 183 pp. ISBN 3-540-59121-4

Vol. 204: **Saedler, Heinz; Gierl, Alfons (Eds.):** Transposable Elements. 1995. 42 figs. IX, 234 pp. ISBN 3-540-59342-X

Vol. 205: **Littman, Dan. R. (Ed.):** The CD4 Molecule. 1995. 29 figs. XIV, 184 pp. ISBN 3-540-59344-6

Vol. 206: **Chisari, Francis V.; Oldstone, Michael B. A. (Eds.):** Transgenic Models of Human Viral and Immunological Disease. 1995. 53 figs. XII, 350 pp. ISBN 3-540-59341-1

Vol. 207: **Prusiner, Stanley B. (Ed.):** Prions Prions Prions. 1995. 42 figs. VI, 180 pp. ISBN 3-540-59343-8

Vol. 208: **Farnham, Peggy (Ed.):** Transcriptional Control of Cell Growth. 1995. 17 figs. VIII, 250 pp. ISBN 3-540-60113-9

Vol. 209: **Miller, Virginia L. (Ed.):** Bacterial Invasiveness. 1995. 16 figs. IX, 180 pp. ISBN 3-540-60065-5

Vol. 210: **Potter, M.; Rose, N. R. (Eds.):** Immunology of Silicones. 1996. 136 figs. XX, 430 pp. ISBN 3-540-60272-0

Printing: Saladruck, Berlin
Binding: Buchbinderei Lüderitz & Bauer, Berlin